GEOMETRIC MECHANICS

Part II: Rotating, Translating and Rolling

2nd Edition

GEOMETRIC MECHANICS

Part II: Rotating, Translating and Rolling

2nd Edition

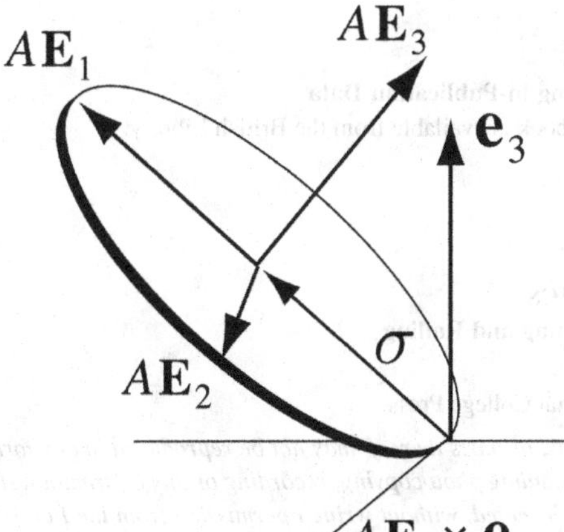

DARRYL D HOLM

Imperial College London, UK

Imperial College Press

ICP

Published by

Imperial College Press
57 Shelton Street
Covent Garden
London WC2H 9HE

Distributed by

World Scientific Publishing Co. Pte. Ltd.
5 Toh Tuck Link, Singapore 596224
USA office: 27 Warren Street, Suite 401-402, Hackensack, NJ 07601
UK office: 57 Shelton Street, Covent Garden, London WC2H 9HE

British Library Cataloguing-in-Publication Data
A catalogue record for this book is available from the British Library.

GEOMETRY MECHANICS
Part II: Rotating, Translating and Rolling
(2nd Edition)

ISBN-13 978-1-84816-777-3
ISBN-10 1-84816-777-6
ISBN-13 978-1-84816-778-0 (pbk)
ISBN-10 1-84816-778-4 (pbk)

Printed in Singapore.

To Justine, for her love, kindness and patience.
Thanks for letting me think this was important.

Contents

Preface xv

1 Galileo 1

1.1 Principle of Galilean relativity 2

1.2 Galilean transformations 3

1.2.1 Admissible force laws for an N-particle system 6

1.3 Subgroups of the Galilean transformations 8

1.3.1 Matrix representation of $SE(3)$ 9

1.4 Lie group actions of $SE(3)$ 11

1.5 Lie group actions of $G(3)$ 12

1.5.1 Matrix representation of $G(3)$ 14

1.6 Lie algebra of $SE(3)$ 15

1.7 Lie algebra of $G(3)$ 17

2 Newton, Lagrange, Hamilton and the rigid body 19

2.1 Newton 21

2.1.1 Newtonian form of free rigid rotation 21

2.1.2 Newtonian form of rigid-body motion 30

2.2 Lagrange 36

2.2.1 The principle of stationary action 36

2.3 Noether's theorem 39

2.3.1 Lie symmetries and conservation laws 39

2.3.2 Infinitesimal transformations of a Lie group 40

2.4 Lagrangian form of rigid-body motion 46

2.4.1 Hamilton–Pontryagin constrained variations 50

2.4.2 Manakov's formulation of the $SO(n)$ rigid body 54

2.4.3 Matrix Euler–Poincaré equations 55

2.4.4 An isospectral eigenvalue problem for the $SO(n)$ rigid body 56

2.4.5 Manakov's integration of the $SO(n)$ rigid body 58

2.5 Hamilton 60

2.5.1 Hamiltonian form of rigid-body motion 62

2.5.2 Lie–Poisson Hamiltonian rigid-body dynamics 63

2.5.3 Lie–Poisson bracket 64

2.5.4 Nambu's $\mathbb{R}^3$ Poisson bracket 65

2.5.5 Clebsch variational principle for the rigid body 69

2.5.6 Rotating motion with potential energy 72

3 Quaternions 77

3.1 Operating with quaternions 78

3.1.1 Multiplying quaternions using Pauli matrices 79

3.1.2 Quaternionic conjugate 82

3.1.3 Decomposition of three-vectors 85

3.1.4 Alignment dynamics for Newton's second law 86

3.1.5 Quaternionic dynamics of Kepler's problem 90

3.2 Quaternionic conjugation 93

3.2.1 Cayley–Klein parameters 93

3.2.2 Pure quaternions, Pauli matrices and $SU(2)$ 99

3.2.3 Tilde map: $\mathbb{R}^3 \simeq su(2) \simeq so(3)$ 102

3.2.4 Dual of the tilde map: $\mathbb{R}^{3^*} \simeq su(2)^* \simeq so(3)^*$ 103

3.2.5 Pauli matrices and Poincaré's sphere $\mathbb{C}^2 \to S^2$ 103

3.2.6 Poincaré's sphere and Hopf's fibration 105

3.2.7 Coquaternions 108

4 Adjoint and coadjoint actions 111

4.1 Cayley–Klein dynamics for the rigid body 112
4.1.1 Cayley–Klein parameters, rigid-body dynamics 112
4.1.2 Body angular frequency 113
4.1.3 Cayley–Klein parameters 115
4.2 Actions of quaternions, Lie groups and Lie algebras 116
4.2.1 AD, Ad, ad, Ad^* and ad^* actions of quaternions 117
4.2.2 AD, Ad, and ad for Lie algebras and groups 118
4.3 Example: The Heisenberg Lie group 124
4.3.1 Definitions for the Heisenberg group 124
4.3.2 Adjoint actions: AD, Ad and ad 126
4.3.3 Coadjoint actions: Ad^* and ad^* 127
4.3.4 Coadjoint motion and harmonic oscillations 129

5 The special orthogonal group $SO(3)$ 131

5.1 Adjoint and coadjoint actions of $SO(3)$ 132
5.1.1 Ad and ad operations for the hat map 132
5.1.2 AD, Ad and ad actions of $SO(3)$ 133
5.1.3 Dual Lie algebra isomorphism 135

6 Adjoint and coadjoint semidirect-product group actions 141

6.1 Special Euclidean group $SE(3)$ 142
6.2 Adjoint operations for $SE(3)$ 144
6.3 Adjoint actions of $SE(3)$'s Lie algebra 148
6.3.1 The ad action of $se(3)$ on itself 148
6.3.2 The ad^* action of $se(3)$ on its dual $se(3)^*$ 149

6.3.3 Left versus right 151

6.4 Special Euclidean group $SE(2)$ 153

6.5 Semidirect-product group $SL(2,\mathbb{R})Ⓢ\mathbb{R}^2$ 156

6.5.1 Definitions for $SL(2,\mathbb{R})Ⓢ\mathbb{R}^2$ 156

6.5.2 AD, Ad, and ad actions 158

6.5.3 Ad^* and ad^* actions 160

6.5.4 Coadjoint motion relation 162

6.6 Galilean group 164

6.6.1 Definitions for $G(3)$ 164

6.6.2 AD, Ad, and ad actions of $G(3)$ 165

6.7 Iterated semidirect products 167

7 Euler–Poincaré and Lie–Poisson equations on $SE(3)$ 169

7.1 Euler–Poincaré equations for left-invariant Lagrangians under $SE(3)$ 170

7.1.1 Legendre transform from $se(3)$ to $se(3)^*$ 172

7.1.2 Lie–Poisson bracket on $se(3)^*$ 172

7.1.3 Coadjoint motion on $se(3)^*$ 173

7.2 Kirchhoff equations on $se(3)^*$ 176

7.2.1 Looks can be deceiving: The heavy top 178

8 Heavy-top equations 181

8.1 Introduction and definitions 182

8.2 Heavy-top action principle 183

8.3 Lie–Poisson brackets 184

8.3.1 Lie–Poisson brackets and momentum maps 185

8.3.2 Lie–Poisson brackets for the heavy top 186

8.4 Clebsch action principle 187

8.5 Kaluza–Klein construction 188

9 The Euler–Poincaré theorem 193

9.1 Action principles on Lie algebras 194
9.2 Hamilton–Pontryagin principle 198
9.3 Clebsch approach to Euler–Poincaré 199
9.3.1 Defining the Lie derivative 201
9.3.2 Clebsch Euler–Poincaré principle 202
9.4 Lie–Poisson Hamiltonian formulation 206
9.4.1 Cotangent-lift momentum maps 207

10 Lie–Poisson Hamiltonian form of a continuum spin chain 209

10.1 Formulating continuum spin chain equations 210
10.2 Euler–Poincaré equations 212
10.3 Hamiltonian formulation 213

11 Momentum maps 221

11.1 The momentum map 222
11.2 Cotangent lift 224
11.3 Examples of momentum maps 226
11.3.1 The Poincaré sphere $S^2 \in S^3$ 237
11.3.2 Overview 242

12 Round, rolling rigid bodies 245

12.1 Introduction 246
12.1.1 Holonomic versus nonholonomic 246
12.1.2 The Chaplygin ball 248
12.2 Nonholonomic Hamilton–Pontryagin variational principle 252
12.2.1 HP principle for the Chaplygin ball 256

12.2.2 Circular disk rocking in a vertical plane 265
12.2.3 Euler's rolling and spinning disk 268
12.3 Nonholonomic Euler–Poincaré reduction 275
12.3.1 Semidirect-product structure 276
12.3.2 Euler–Poincaré theorem 278
12.3.3 Constrained reduced Lagrangian 282

A Geometrical structure of classical mechanics 287

A.1 Manifolds 288
A.2 Motion: Tangent vectors and flows 296
A.2.1 Vector fields, integral curves and flows 297
A.2.2 Differentials of functions: The cotangent bundle 299
A.3 Tangent and cotangent lifts 300
A.3.1 Summary of derivatives on manifolds 301

B Lie groups and Lie algebras 305

B.1 Matrix Lie groups 306
B.2 Defining matrix Lie algebras 310
B.3 Examples of matrix Lie groups 312
B.4 Lie group actions 314
B.4.1 Left and right translations on a Lie group 316
B.5 Tangent and cotangent lift actions 317
B.6 Jacobi–Lie bracket 320
B.7 Lie derivative and Jacobi–Lie bracket 323
B.7.1 Lie derivative of a vector field 323
B.7.2 Vector fields in ideal fluid dynamics 325

C Enhanced coursework 327

C.1 Variations on rigid-body dynamics 328

C.1.1 Two times 328
C.1.2 Rotations in complex space 334
C.1.3 Rotations in four dimensions: $SO(4)$ 338
C.2 $\mathbb{C}^3$ oscillators 343
C.3 Momentum maps for $GL(n,\mathbb{R})$ 348
C.4 Motion on the symplectic Lie group $Sp(2)$ 354
C.5 Two coupled rigid bodies 359

D Poincaré's 1901 paper 363

Bibliography 367

Index 385

Preface

Introduction

This is the 2nd edition of the text for a course of 33 lectures in geometric mechanics, taught annually by the author to fourth-year undergraduates in their last term in applied mathematics at Imperial College London. The text mimics the lectures, which attempt to provide an air of immediacy and flexibility in which students may achieve insight and proficiency in using one of the fundamental approaches for solving a variety of problems in geometric mechanics. This is the Euler–Poincaré approach, which uses the Lie group invariance of Hamilton's principle to produce symmetry-reduced motion equations and reveal their geometrical meaning. It has been taught to students with various academic backgrounds from mathematics, physics and engineering.

Each chapter of the text is presented as a line of inquiry, often by asking sequences of related questions such as "What is angular velocity?", "What is kinetic energy?", "What is angular momentum?", and so forth. In adopting such an inquiry-based approach, one focuses on a sequence of exemplary problems, each of whose solution facilitates taking the next step. The present text takes those steps, forgoing any attempt at mathematical rigour. Readers interested in a more rigorous approach are invited to consult some of the many works cited in the bibliography which treat the subject in that style. This book is meant to be an intermediate introduction to geometric mechanics that bridges standard textbooks and more advanced study.

Prerequisites

The prerequisites are standard for an advanced undergraduate student. Namely, the student should be familiar with the linear algebra of vectors and matrices, ordinary differential equations, multivariable calculus and have some familiarity with variational principles and canonical Poisson brackets in classical mechanics at the level of a second- or third-year undergraduate in mathematics, physics or engineering. An undergraduate background in physics is particularly helpful, because all the examples of rotating, spinning and rolling rigid bodies treated here from a geometric viewpoint would be familiar from undergraduate physics classes.

How to read this book

Most of the book is meant to be read in sequential order from front to back. The 120 Exercises are an important aspect of the text. These are shaded, indented and marked with ★.

Exercise. ★

Their 55 Worked Answers are indented and marked with ▲.

Key theorems, results and remarks are placed in frames.

The three appendices provide supplementary material, such as condensed summaries of the essentials of manifolds (Appendix A) and Lie groups (Appendix B) for students who may wish to acquire a bit more mathematical background. In addition, the appendices provide material for supplementary lectures that extend the course material. Examples include variants of rotating motion that depend on more than one time variable, as well as rotations in complex space and in higher dimensions in Appendix C. The appendices also contain ideas for additional homework and exam problems that go beyond the many exercises and examples sprinkled throughout the text.

Description of contents

Galilean relativity and the idea of a uniformly moving reference frame are explained in Chapter 1. Freely rotating motion is then treated in Chapters 2, 3 and 4, first by reviewing Newton's and Lagrange's approaches, then by following Hamilton's approach via quaternions and Cayley–Klein parameters, not Euler angles.

Hamilton's rules for multiplication of quaternions introduced the adjoint and coadjoint actions that lie at the heart of geometric mechanics. For the rotations and translations in $\mathbb{R}^3$ studied in Chapters 5 and 6, the adjoint and coadjoint actions are both equivalent to the vector cross product. Poincaré [Po1901] opened the field of geometric mechanics by noticing that these actions define the motion generated by any Lie group.

When applied to Hamilton's principle defined on the tangent space of an arbitrary Lie group, the adjoint and coadjoint actions studied in Chapter 6 result in the Euler–Poincaré equations derived in Chapter 7. Legendre-transforming the Lagrangian in Hamilton's principle summons the Lie–Poisson Hamiltonian formulation of dynamics on a Lie group. The Euler–Poincaré equations provide the framework for all of the applications treated in this text. These applications include finite-dimensional dynamics of three-dimensional rotations and translations in the special Euclidean group $SE(3)$. The Euler–Poincaré problem on $SE(3)$ recovers Kirchhoff's classic treatment in modern form of the dynamics of an ellipsoidal body moving in an incompressible fluid flow without vorticity.

The Euler–Poincaré formulation of Kirchhoff's problem on $SE(3)$ in Chapter 7 couples rotations and translations, but it does not yet introduce potential energy. The semidirect-product structure of $SE(3)$, however, introduces the key idea for incorporating potential energy. Namely, the same semidirect-product structure is also invoked in passing from rotations of a free rigid body to rotations of a heavy top with a *fixed* point of support under gravity. Thereby, semidirect-product reduction becomes a central focus of the text.

The heavy top treated in Chapter 8 is a key example, because it introduces the dual representation of the action of a Lie algebra on a vector space. This is the diamond operation $(\diamond)$, by which the forces and torques produced by potential energy gradients are represented in the Euler–Poincaré framework in Chapters 9 and 10. The diamond operation is then found in Chapter 11 to lie at the heart of the standard (cotangent-lift) momentum map.

This observation reveals the relation between the results of reduction by Lie symmetry on the Lagrangian and Hamiltonian sides. Namely,

> Lie symmetry reduction on the Lagrangian side produces the Euler–Poincaré equation, whose formulation on the Hamiltonian side as a Lie–Poisson equation governs the dynamics of the momentum map associated with the cotangent lift of the Lie algebra action of that Lie symmetry on the configuration manifold.

The chief purpose of this book is to explain that statement, so that it may be understood by undergraduate students in mathematics, physics and engineering.

In the Euler–Poincaré framework, the adjoint and coadjoint actions combine with the diamond operation to provide a powerful tool for investigating other applications of geometric mechanics, including nonholonomic constraints discussed in Chapter 12. In the same chapter, nonholonomic mechanics is discussed in the context of two classic problems, known as Chaplygin's ball (an unbalanced rolling ball) and Euler's disk (a spinning, falling, rolling flat coin). In these classic examples, the semidirect-product structure couples rotations, translations and potential energy together with the rolling constraint.

What is new in the 2nd edition?

The organisation of the 1st edition has been preserved in the 2nd edition. However, the substance of the text has been rewritten throughout to improve the flow and enrich the development of the material. Some examples of the new improvements include the following:

- The Galilean group and the implications of Galilean relativity for Noether's theorem have been developed further.
- The role of Noether's theorem has been given added emphasis throughout, with various applications for Euler–Poincaré systems, Lie–Poisson systems and nonholonomically constrained systems, such as the rolling bodies treated in Chapter 12.
- Additional examples of adjoint and coadjoint actions of Lie groups have been worked out. These include the Heisenberg group, and the semidirect-product Lie group $SL(2,R)\circledS\mathbb{R}^2$, which is the group of motions of ellipses that translate, rotate and dilate in the plane, while preserving their area.
- Manakov's approach of regarding the rigid body as an isospectral eigenvalue problem has been developed further and additional examples of its application have been given, particularly for the Euler top and the Lagrange top.
- A section has been added about coquaternions. The coquaternions comprise a representation of $SP(2)$, while the more well-known quaternions form a representation of $SU(2)$.
- Momentum maps are discussed in more depth in the 2nd edition. For example, dual pairs of momentum maps are discussed in Chapter 11 in the context of the Hopf fibration, under the reduction of $\mathbb{C}^2$ by S^1 from the right and by $SU(2)$ from the left.
- Additional enhanced coursework has been provided, including treatments of Euler–Poincaré equations for (i) the articulated motion of two coupled rigid bodies; and (ii) geodesic

motion on the symplectic group $SP(2)$, which leads to a ***Bloch–Iserles equation*** [BlIs2006].

- A new English translation is provided in Appendix D of Poincaré's famous two-page paper [Po1901]. There, Poincaré derived the Euler–Poincaré equations by using Lie's theory of infinitesimal transformations for variations in Hamilton's principle. Thus began modern geometric mechanics.

Acknowledgements

I am enormously grateful to my students, who have been partners in this endeavour, for their comments, questions and interest. I am also grateful to my friends and collaborators for their camaraderie in first forging many of these tools. We are in this together. Thanks for all the fun we've had. I appreciate the many helpful remarks and recommendations regarding the text from Poul Hjorth (who also drew most of the figures), David Ellis, Rossen Ivanov, Byungsoo Kim, Vakhtang Putkaradze, Peter Olver, Tudor Ratiu, Cristina Stoica, Tanya Schmah and Cesare Tronci.

In addition, I am grateful to the reviewers of the 1st edition, for their uniformly encouraging comments and critiques. Of course, the errors that still remain are all my own. I am also grateful to Tasha D'Cruz, Amy Hendrickson, and Kwong Lai Fun for their diligent cooperation with me in the process of editing and type-setting this 2nd edition.

Finally, I am grateful for three decades of camaraderie with my late friend Jerry Marsden, who inspired me in many ways.

1 GALILEO

Contents

1.1 Principle of Galilean relativity 2
1.2 Galilean transformations 3
1.2.1 Admissible force laws for an N-particle system 6
1.3 Subgroups of the Galilean transformations 8
1.3.1 Matrix representation of $SE(3)$ 9
1.4 Lie group actions of $SE(3)$ 11
1.5 Lie group actions of $G(3)$ 12
1.5.1 Matrix representation of $G(3)$ 14
1.6 Lie algebra of $SE(3)$ 15
1.7 Lie algebra of $G(3)$ 17

1.1 Principle of Galilean relativity

Galileo Galilei

Principles of relativity address the problem of how events that occur in one place or state of motion are observed from another. And if events occurring in one place or state of motion look different from those in another, how should one determine the laws of motion?

Galileo approached this problem via a thought experiment which imagined observations of motion made inside a ship by people who could not see outside. He showed that the people isolated inside a uniformly moving ship would be *unable to determine by measurements made inside it whether they were moving!*

> ...have the ship proceed with any speed you like, so long as the motion is uniform and not fluctuating this way and that. You will discover not the least change in all the effects named, nor could you tell from any of them whether the ship was moving or standing still.
>
> – Galileo, *Dialogue Concerning the Two Chief World Systems* [Ga1632]

Galileo's thought experiment showed that a man who is below decks on a ship cannot tell whether the ship is docked or is moving uniformly through the water at constant velocity. He may observe water dripping from a bottle, fish swimming in a tank, butterflies flying, etc. Their behaviour will be just the same, whether the ship is moving or not.

Definition 1.1.1 (Galilean transformations) *Transformations of reference location, time, orientation or state of uniform translation at constant velocity are called **Galilean transformations**.*

Definition 1.1.2 (Uniform rectilinear motion) *Coordinate systems related by Galilean transformations are said to be in* ***uniform rectilinear motion*** *relative to each other.*

Galileo's thought experiment led him to the following principle.

Definition 1.1.3 (Principle of Galilean relativity) *The laws of motion are independent of reference location, time, orientation or state of uniform translation at constant velocity. Hence, these laws are invariant (i.e., they do not change their forms) under Galilean transformations.*

Remark 1.1.1 (Two tenets of Galilean relativity) Galilean relativity sets out two important tenets:

- It is impossible to determine who is actually at rest.
- Objects continue in uniform motion unless acted upon.

The second tenet is known as ***Galileo's law of inertia***. It is also the basis for ***Newton's first law of motion***. □

1.2 Galilean transformations

Definition 1.2.1 (Galilean transformations) *Galilean transformations of a coordinate frame consist of space-time translations, rotations and reflections of spatial coordinates, as well as Galilean "boosts" into uniform rectilinear motion.*

In three dimensions, the Galilean transformations depend smoothly on ten real parameters, as follows:

- *Space-time translations,*

$$g_1(\mathbf{r}, t) = (\mathbf{r} + \mathbf{r}_0, t + t_0) \,.$$

These possess four real parameters: $(\mathbf{r}_0, t_0) \in \mathbb{R}^3 \times \mathbb{R}$, for the three dimensions of space, plus time.

- *Spatial rotations and reflections,*

$$g_2(\mathbf{r}, t) = (O\mathbf{r}, t)\,,$$

for any linear **orthogonal transformation** $O : \mathbb{R}^3 \mapsto \mathbb{R}^3$ *with* $O^T = O^{-1}$. *These have three real parameters, for the three axes of rotation and reflection. Because the inverse of an orthogonal transformation is its transpose* ($O^{-1} = O^T$) *they preserve both the lengths and relative orientations of vectors. It has two connected components corresponding to the positive and negative values of the determinant,* $\det O = \pm 1$, *which changes sign under reflections.*

- *Galilean boosts into uniform rectilinear motion,*

$$g_3(\mathbf{r}, t) = (\mathbf{r} + \mathbf{v}_0 t, t)\,.$$

These have three real parameters: $\mathbf{v}_0 \in \mathbb{R}^3$, *for the three components of the velocity boost vector.*

Definition 1.2.2 (Group) *A* ***group*** *G is a set of elements that possesses a binary product (multiplication),* $G \times G \to G$, *such that the following properties hold:*

- *The product gh of g and h is associative, that is,* $(gh)k = g(hk)$.
- *An identity element exists,* $e : eg = g$ *and* $ge = g$, *for all* $g \in G$.
- *The inverse operation exists,* $G \to G$, *so that* $gg^{-1} = g^{-1}g = e$.

Definition 1.2.3 (Lie group) *A* ***Lie group*** *is a group that depends smoothly on a set of parameters. That is, a Lie group is both a group and a smooth manifold, for which the group operation is by composition of smooth invertible functions.*

Proposition 1.2.1 (Lie group property) *Galilean transformations form a Lie group, modulo reflections.*

Proof. Any Galilean transformation

$$g \in G(3) : \mathbb{R}^3 \times \mathbb{R} \mapsto \mathbb{R}^3 \times \mathbb{R}$$

may be expressed uniquely as a composition of the three basic transformations $\{g_1, g_2, g_3\} \in G(3)$. Consequently, the set of elements comprising the transformations $\{g_1, g_2, g_3\} \in G(3)$ closes under the binary operation of composition. The Galilean transformations also possess an identity element $e : eg_i = g_i = g_ie$, $i = 1, 2, 3$, and each element g possesses a unique inverse g^{-1}, so that $gg^{-1} = e = g^{-1}g$.

These properties, plus associativity, define a group. The smooth dependence of the group of Galilean transformations on its ten parameters means that the ***Galilean group*** $G(3)$ is a Lie group (except for the reflections, which are discrete, not smooth). ■

Remark 1.2.1 Compositions of Galilean boosts and translations commute. That is,

$$g_1 g_3 = g_3 g_1 \,.$$

However, the order of composition does matter in Galilean transformations when rotations and reflections are involved. For example, the action of the Galilean group composition $g_1g_3g_2$ on $(\mathbf{r}, t)$ from the left is given by

$$g(\mathbf{r}, t) \quad = \quad (O\mathbf{r} + t\mathbf{v}_0 + \mathbf{r}_0, t + t_0)\,,$$

for

$$g = g_1(\mathbf{r}_0, t_0) g_3(\mathbf{v}_0) g_2(O) =: g_1 g_3 g_2 \,.$$

However, the result for another composition, say $g_1g_2g_3$, would in general be different. □

Exercise. Write the corresponding transformations for $g_1g_2g_3$, $g_1g_3g_2$, $g_2g_1g_3$ and $g_3g_2g_1$, showing how they depend on the order in which the rotations, boosts and translations are composed. Write the inverse transformation for each of these compositions of left actions. ★

Answer. The various compositions of translations $g_1(\mathbf{r}_0, t_0)$, rotations $g_2(O)$ and boosts $g_3(\mathbf{v}_0)$ in general yield different results, as

$$\begin{aligned}
g_1g_2g_3(\mathbf{r},t) &= \Big(O(\mathbf{r}+t\mathbf{v}_0)+\mathbf{r}_0,\, t+t_0\Big)\,,\\
g_1g_3g_2(\mathbf{r},t) &= \Big(O\mathbf{r}+t\mathbf{v}_0+\mathbf{r}_0, t+t_0\Big)\,,\\
g_2g_1g_3(\mathbf{r},t) &= \Big(O(\mathbf{r}+t\mathbf{v}_0+\mathbf{r}_0),\, t+t_0\Big)\,,\\
g_3g_2g_1(\mathbf{r},t) &= \Big(O(\mathbf{r}+\mathbf{r}_0)+t\mathbf{v}_0,\, t+t_0\Big)\,.
\end{aligned}$$

The inverses are $(g_1g_2g_3)^{-1} = g_3^{-1}g_2^{-1}g_1^{-1}$, etc. ▲

Remark 1.2.2 (Decomposition of the Galilean group) Because the rotations take vectors into vectors, any element of the transformations $g_1g_2g_3$, $g_2g_1g_3$ and $g_3g_2g_1$ in the Galilean group may be written uniquely in the simplest form, as $g_1g_3g_2$.

Thus, any element of the Galilean group may be written uniquely as a rotation, followed by a space translation, a Galilean boost and a time translation. The latter three may be composed in any order, because they commute with each other. □

Exercise. What properties are preserved by the Galilean group? ★

Answer. The Galilean group $G(3)$ preserves the results of measuring length and time intervals, and relative orientation in different frames of motion related to each other by Galilean transformations. ▲

1.2.1 Admissible force laws for an N-particle system

For a system of N interacting particles, Newton's second law of motion (the law of acceleration) determines the motion resulting from the force $\mathbf{F}_j$ exerted on the jth particle by the other $N-1$ particles as

$$m_j\ddot{\mathbf{r}}_j = \mathbf{F}_j(\mathbf{r}_k - \mathbf{r}_l, \dot{\mathbf{r}}_k - \dot{\mathbf{r}}_l)\,, \quad \text{with } j,k,l = 1,2,\ldots,N \text{ (no sum)}\,.$$

This force law is independent of reference location, time or state of uniform translation at constant velocity. It will also be independent of reference orientation and thus will be ***Galilean-invariant***, provided the forces $\mathbf{F}_j$ transform under rotations and parity reflections as ***vectors***

$$m_jO\ddot{\mathbf{r}}_j = O\mathbf{F}_j = \mathbf{F}_j\Big(O(\mathbf{r}_k - \mathbf{r}_l), O(\dot{\mathbf{r}}_k - \dot{\mathbf{r}}_l)\Big)\,, \tag{1.2.1}$$

for any orthogonal transformation O.

This requirement for Galilean invariance that the force in Newton's law of acceleration transforms as a vector is the reason that vectors are so important in classical mechanics.

For example, Newton's law of gravitational motion is given by

$$m_j\ddot{\mathbf{r}}_j = \sum_{k\neq j}\mathbf{F}_{jk}\,, \tag{1.2.2}$$

in which the gravitational forces $\mathbf{F}_{jk}$ between $(j,\,k)$ particle pairs are given by

$$\mathbf{F}_{jk} = \frac{\gamma\, m_j m_k}{|\mathbf{r}_{jk}|^3}\mathbf{r}_{jk}\,, \quad \text{with} \quad \mathbf{r}_{jk} = \mathbf{r}_j - \mathbf{r}_k\,, \tag{1.2.3}$$

and γ is the gravitational constant.

Exercise. Prove that Newton's law (1.2.2) for gravitational forces (1.2.3) is Galilean-invariant. That is, prove that Newton's law of gravitational motion takes the same form in any Galilean reference frame. ★

1.3 Subgroups of the Galilean transformations

Definition 1.3.1 (Subgroup) *A* ***subgroup*** *is a subset of a group whose elements also satisfy the defining properties of a group.*

Exercise. List the subgroups of the Galilean group that do not involve time. ★

Answer. The subgroups of the Galilean group that are independent of time consist of

- Spatial translations $g_1(\mathbf{r}_0)$ acting on $\mathbf{r}$ as $g_1(\mathbf{r}_0)\mathbf{r} = \mathbf{r} + \mathbf{r}_0$.
- Proper rotations $g_2(O)$ with $g_2(O)\mathbf{r} = O\mathbf{r}$ where $O^T = O^{-1}$ and $\det O = +1$. This subgroup is called $SO(3)$, the ***special orthogonal group*** in three dimensions.
- Rotations and reflections $g_2(O)$ with $O^T = O^{-1}$ and $\det O = \pm 1$. This subgroup is called $O(3)$, the ***orthogonal group*** in three dimensions.
- Spatial translations $g_1(\mathbf{r}_0)$ with $\mathbf{r}_0 \in \mathbb{R}^3$ compose with proper rotations $g_2(O) \in SO(3)$ acting on a vector $\mathbf{r} \in \mathbb{R}^3$ as

$$E(O, \mathbf{r}_0)\mathbf{r} = g_1(\mathbf{r}_0)g_2(O)\mathbf{r} = O\mathbf{r} + \mathbf{r}_0\,,$$

  where $O^T = O^{-1}$ and $\det O = +1$. This subgroup is called $SE(3)$, the ***special Euclidean group*** in three dimensions. Its ***action*** on $\mathbb{R}^3$ is written abstractly as $SE(3) \times \mathbb{R}^3 \to \mathbb{R}^3$.
- Spatial translations $g_1(\mathbf{r}_0)$ compose with proper rotations and reflections $g_2(O)$, as $g_1(\mathbf{r}_0)g_2(O)$ acting on $\mathbf{r}$. This subgroup is called $E(3)$, the ***Euclidean group*** in three dimensions. ▲

Remark 1.3.1 Spatial translations and rotations do not commute in general. That is, $g_1 g_2 \neq g_2 g_1$, unless the direction of translation and axis of rotation are collinear. □

1.3.1 Matrix representation of $SE(3)$

As we have seen, the special Euclidean group in three dimensions $SE(3)$ acts on a position vector $\mathbf{r} \in \mathbb{R}^3$ by

$$E(O, \mathbf{r}_0)\mathbf{r} = O\mathbf{r} + \mathbf{r}_0 \,.$$

A 4×4 ***matrix representation*** of this action may be found by noticing that its right-hand side arises in multiplying the matrix times the *extended* vector $(\mathbf{r}, 1)^T$ as

$$\begin{pmatrix} O & \mathbf{r}_0 \\ 0 & 1 \end{pmatrix} \begin{pmatrix} \mathbf{r} \\ 1 \end{pmatrix} = \begin{pmatrix} O\mathbf{r} + \mathbf{r}_0 \\ 1 \end{pmatrix}.$$

Therefore we may identify a group element of $SE(3)$ with a 4×4 matrix,

$$E(O, \mathbf{r}_0) = \begin{pmatrix} O & \mathbf{r}_0 \\ 0 & 1 \end{pmatrix}.$$

The group $SE(3)$ has six parameters. These are the angles of rotation about each of the three spatial axes by the orthogonal matrix $O \in SO(3)$ with $O^T = O^{-1}$ and the three components of the vector of translations $\mathbf{r}_0 \in \mathbb{R}^3$.

The ***group composition law*** for $SE(3)$ is expressed as

$$\begin{aligned} E(\tilde{O}, \tilde{\mathbf{r}}_0)E(O, \mathbf{r}_0)\mathbf{r} &= E(\tilde{O}, \tilde{\mathbf{r}}_0)(O\mathbf{r} + \mathbf{r}_0) \\ &= \tilde{O}(O\mathbf{r} + \mathbf{r}_0) + \tilde{\mathbf{r}}_0 \,, \end{aligned}$$

with $(O\,, \tilde{O}) \in SO(3)$ and $(\mathbf{r}\,, \tilde{\mathbf{r}}_0) \in \mathbb{R}^3$. This formula for group composition may be represented by matrix multiplication from the left as

$$E(\tilde{O},\tilde{\mathbf{r}}_0)E(O,\mathbf{r}_0) = \begin{pmatrix} \tilde{O} & \tilde{\mathbf{r}}_0 \\ 0 & 1 \end{pmatrix}\begin{pmatrix} O & \mathbf{r}_0 \\ 0 & 1 \end{pmatrix} = \begin{pmatrix} \tilde{O}O & \tilde{O}\mathbf{r}_0+\tilde{\mathbf{r}}_0 \\ 0 & 1 \end{pmatrix},$$

which may also be expressed by simply writing the top row,

$$(\tilde{O}\,,\,\tilde{\mathbf{r}}_0)(O\,,\,\mathbf{r}_0) = (\tilde{O}O\,,\,\tilde{O}\mathbf{r}_0+\tilde{\mathbf{r}}_0)\,.$$

The identity element (e) of $SE(3)$ is represented by

$$e = E(I,\mathbf{0}) = \begin{pmatrix} I & \mathbf{0} \\ 0 & 1 \end{pmatrix},$$

or simply $e = (I,\mathbf{0})$. The inverse element is represented by the matrix inverse

$$E(O,\mathbf{r}_0)^{-1} = \begin{pmatrix} O^{-1} & -O^{-1}\mathbf{r}_0 \\ 0 & 1 \end{pmatrix}.$$

In this matrix representation of $SE(3)$, one checks directly that

$$E(O,\mathbf{r}_0)^{-1}E(O,\mathbf{r}_0) = \begin{pmatrix} O^{-1} & -O^{-1}\mathbf{r}_0 \\ 0 & 1 \end{pmatrix}\begin{pmatrix} O & \mathbf{r}_0 \\ 0 & 1 \end{pmatrix} = \begin{pmatrix} I & \mathbf{0} \\ 0 & 1 \end{pmatrix} = (I,\mathbf{0}) = e\,.$$

In the shorter notation, the inverse may be written as

$$(O\,,\,\mathbf{r}_0)^{-1} = (O^{-1}\,,\,-O^{-1}\mathbf{r}_0)$$

and $O^{-1} = O^T$ since the 3×3 matrix $O\in SO(3)$ is orthogonal.

Remark 1.3.2 The inverse operation of $SE(3)$ involves composition of the inverse for rotations with the inverse for translations. This entwining means that the group structure of $SE(3)$ is not simply a direct product of its two subgroups $\mathbb{R}^3$ and $SO(3)$. □

1.4 Lie group actions of $SE(3)$

Group multiplication in $SE(3)$ is denoted as

$$(\tilde{O}\,,\,\tilde{\mathbf{r}}_0)(O\,,\,\mathbf{r}_0) = (\tilde{O}O\,,\,\tilde{O}\mathbf{r}_0 + \tilde{\mathbf{r}}_0)\,. \tag{1.4.1}$$

This notation demonstrates the following group properties of $SE(3)$:

- Translations in the subgroup $\mathbb{R}^3 \subset SE(3)$ act on each other by vector addition,

$$\mathbb{R}^3 \times \mathbb{R}^3 \mapsto \mathbb{R}^3 : \ (I\,,\,\tilde{\mathbf{r}}_0)(I\,,\,\mathbf{r}_0) = (I\,,\,\mathbf{r}_0 + \tilde{\mathbf{r}}_0)\,.$$

- Rotations in the subgroup $SO(3) \subset SE(3)$ act on each other by composition,

$$SO(3) \times SO(3) \mapsto SO(3) : \ (\tilde{O}\,,\,\mathbf{0})(O\,,\,\mathbf{0}) = (\tilde{O}O\,,\,\mathbf{0})\,.$$

- Rotations in the subgroup $SO(3) \subset SE(3)$ act ***homogeneously*** on the vector space of translations in the subgroup $\mathbb{R}^3 \subset SE(3)$,

$$SO(3) \times \mathbb{R}^3 \mapsto \mathbb{R}^3 : \ (\tilde{O}\,,\,\mathbf{0})(I\,,\,\mathbf{r}_0) = (\tilde{O}\,,\,\tilde{O}\mathbf{r}_0)\,.$$

  That is, the action of the subgroup $SO(3) \subset SE(3)$ on the subgroup $\mathbb{R}^3 \subset SE(3)$ maps $\mathbb{R}^3$ into itself. The translations $\mathbb{R}^3 \subset SE(3)$ are thus said to form a ***normal,*** or ***invariant subgroup*** of the group $SE(3)$.

- Every element of $(O\,,\,\mathbf{r}_0)$ of $SE(3)$ may be represented uniquely by composing a translation acting from the left on a rotation. That is, each element may be decomposed into

$$(O\,,\,\mathbf{r}_0) = (I\,,\,\mathbf{r}_0)(O\,,\,\mathbf{0})\,,$$

  for a *unique* $\mathbf{r}_0 \in \mathbb{R}^3$ and $O \in SO(3)$. Conversely, one may uniquely represent

$$(O\,,\,\mathbf{r}_0) = (O\,,\,\mathbf{0})(I\,,\,O^{-1}\mathbf{r}_0)\,,$$

by composing a rotation acting from the left on a translation. This equivalence endows the Lie group $SE(3)$ with a semidirect-product structure,

$$SE(3) = SO(3) \circledS \mathbb{R}^3 . \tag{1.4.2}$$

Definition 1.4.1 (Semidirect-product Lie group) *A Lie group G that may be decomposed uniquely into a normal subgroup N and a subgroup H such that every group element may be written as*

$$g = nh \quad \text{or} \quad g = hn \quad \text{(in either order)}, \tag{1.4.3}$$

for unique choices of $n \in N$ and $h \in H$, is called a **semidirect product** *of H and N, denoted here by* Ⓢ, *as in*

$$G = H \circledS N .$$

When the normal subgroup N is a vector space, the action of a semidirect-product group on itself is given as in formula (1.4.1) for $SE(3)$. If the normal subgroup N is not a vector space, then the operation of addition in formula (1.4.1) is replaced by the composition law for N.

1.5 Lie group actions of $G(3)$

The Galiliean group in three dimensions $G(3)$ has ten parameters $(O \in SO(3)\,,\, \mathbf{r}_0 \in \mathbb{R}^3\,,\, \mathbf{v}_0 \in \mathbb{R}^3\,,\, t_0 \in \mathbb{R}\,)$. The Galilean group is *also* a semidirect-product Lie group, which may be written as

$$G(3) = SE(3) \circledS \mathbb{R}^4 = \Big(SO(3) \circledS \mathbb{R}^3 \Big) \circledS \mathbb{R}^4 . \tag{1.5.1}$$

That is, the subgroup of Euclidean motions consisting of rotations and Galilean velocity boosts $(O, \mathbf{v}_0) \in SE(3)$ acts homogeneously on the subgroups of space and time translations $(\mathbf{r}_0, t_0) \in \mathbb{R}^4$ which commute with each other.

Exercise. Compute explicitly the *inverse* of the Galilean group element $g = g_1 g_3 g_2$ obtained by representing the action of the Galilean group as matrix multiplication $G(3) \times \mathbb{R}^4 \to \mathbb{R}^4$ on the extended vector $(\mathbf{r}, t, 1)^T \in \mathbb{R}^4$,

$$g_1 g_3 g_2 \begin{pmatrix} \mathbf{r} \\ t \\ 1 \end{pmatrix} = \begin{pmatrix} O & \mathbf{v}_0 & \mathbf{r}_0 \\ \mathbf{0} & 1 & t_0 \\ \mathbf{0} & 0 & 1 \end{pmatrix} \begin{pmatrix} \mathbf{r} \\ t \\ 1 \end{pmatrix} \tag{1.5.2}$$
$$= \begin{pmatrix} O\mathbf{r} + t\mathbf{v}_0 + \mathbf{r}_0 \\ t + t_0 \\ 1 \end{pmatrix}.$$

★

Answer. Write the product $g = g_1 g_3 g_2$ as

$$g = g_1 g_3 g_2 = \begin{pmatrix} I & \mathbf{0} & \mathbf{r}_0 \\ \mathbf{0} & 1 & t_0 \\ \mathbf{0} & 0 & 1 \end{pmatrix} \begin{pmatrix} I & \mathbf{v}_0 & \mathbf{0} \\ \mathbf{0} & 1 & 0 \\ \mathbf{0} & 0 & 1 \end{pmatrix} \begin{pmatrix} O & \mathbf{0} & \mathbf{0} \\ \mathbf{0} & 1 & 0 \\ \mathbf{0} & 0 & 1 \end{pmatrix}.$$

Then, the product $g^{-1} = (g_1 g_3 g_2)^{-1} = g_2^{-1} g_3^{-1} g_1^{-1}$ appears in matrix form as

$$g^{-1} = \begin{pmatrix} O^{-1} & \mathbf{0} & \mathbf{0} \\ \mathbf{0} & 1 & 0 \\ \mathbf{0} & 0 & 1 \end{pmatrix} \begin{pmatrix} I & -\mathbf{v}_0 & \mathbf{0} \\ \mathbf{0} & 1 & 0 \\ \mathbf{0} & 0 & 1 \end{pmatrix} \begin{pmatrix} I & \mathbf{0} & -\mathbf{r}_0 \\ \mathbf{0} & 1 & -t_0 \\ \mathbf{0} & 0 & 1 \end{pmatrix}$$
$$= \begin{pmatrix} O^{-1} & -O^{-1}\mathbf{v}_0 & -O^{-1}(\mathbf{r}_0 - t\mathbf{v}_0) \\ \mathbf{0} & 1 & -t_0 \\ \mathbf{0} & 0 & 1 \end{pmatrix}.$$

▲

Exercise. Write the corresponding matrices for the Galilean transformations for $g_1g_2g_3$, $g_2g_1g_3$ and $g_3g_2g_1$. ★

1.5.1 Matrix representation of $G(3)$

The formula for group composition $G(3) \times G(3) \to G(3)$ may be represented by matrix multiplication from the left as

$$\begin{pmatrix} \tilde{O} & \tilde{\mathbf{v}}_0 & \tilde{\mathbf{r}}_0 \\ \mathbf{0} & 1 & \tilde{t}_0 \\ \mathbf{0} & 0 & 1 \end{pmatrix} \begin{pmatrix} O & \mathbf{v}_0 & \mathbf{r}_0 \\ \mathbf{0} & 1 & t_0 \\ \mathbf{0} & 0 & 1 \end{pmatrix} \tag{1.5.3}$$
$$= \begin{pmatrix} \tilde{O}O & \tilde{O}\mathbf{v}_0 + \tilde{\mathbf{v}}_0 & \tilde{O}\mathbf{r}_0 + \tilde{\mathbf{v}}_0 t_0 + \tilde{\mathbf{r}}_0 \\ \mathbf{0} & 1 & \tilde{t}_0 + t_0 \\ \mathbf{0} & 0 & 1 \end{pmatrix},$$

which may be expressed more succinctly as

$$(\tilde{O}, \tilde{\mathbf{v}}_0, \tilde{\mathbf{r}}_0, \tilde{t}_0)(O, \mathbf{v}_0, \mathbf{r}_0, t_0) \tag{1.5.4}$$
$$= (\tilde{O}O,\ \tilde{O}\mathbf{v}_0 + \tilde{\mathbf{v}}_0,\ \tilde{O}\mathbf{r}_0 + \tilde{\mathbf{v}}_0 t_0 + \tilde{\mathbf{r}}_0,\ \tilde{t}_0 + t_0).$$

Exercise. Check the semidirect-product structure (1.5.1) for the Lie group $G(3) = SE(3) \circledS \mathbb{R}^4$, by writing explicit matrix expressions for $g = nh$ and $g = hn$ with $h = SE(3)$ and $n = \mathbb{R}^4$. ★

Answer. In verifying the semidirect-product structure condition (1.4.3) that $g = nh$ or $g = hn$ in either order, we write explicitly

$$\begin{aligned} (O, \mathbf{v}_0, \mathbf{r}_0, t_0) &= (I, 0, \mathbf{r}_0, t_0)\,(O, \mathbf{v}_0, 0, 0) \\ &= (O, \mathbf{v}_0, 0, 0)\,(I, 0, O^{-1}(\mathbf{r}_0 - \mathbf{v}_0 t_0), t_0)\,. \end{aligned} \tag{1.5.5}$$

▲

1.6 Lie algebra of $SE(3)$

A 4×4 matrix representation of tangent vectors for $SE(3)$ at the identity may be found by first computing the derivative of a general group element $(O(s), \mathbf{r}_0(s))$ along the group path with parameter s and bringing the result back to the identity at $s = 0$,

$$\left[\begin{pmatrix} O(s) & \mathbf{r}_0(s) \\ 0 & 1 \end{pmatrix}^{-1} \begin{pmatrix} O'(s) & \mathbf{r}_0'(s) \\ 0 & 0 \end{pmatrix}\right]_{s=0}$$

$$= \begin{pmatrix} O^{-1}(0)O'(0) & O^{-1}(0)\mathbf{r}_0'(0) \\ 0 & 0 \end{pmatrix} =: \begin{pmatrix} \widehat{\Xi} & \mathbf{r}_0 \\ 0 & 0 \end{pmatrix},$$

where in the last step we have dropped the unnecessary superscript prime ($'$). The quantity $\widehat{\Xi} = O^{-1}(s)O'(s)|_{s=0}$ is a 3×3 skew-symmetric matrix, since O is a 3×3 orthogonal matrix. Thus, $\widehat{\Xi}$ may be written using the ***hat map***, defined by

$$\widehat{\Xi} = \begin{pmatrix} 0 & -\Xi_3 & \Xi_2 \\ \Xi_3 & 0 & -\Xi_1 \\ -\Xi_2 & \Xi_1 & 0 \end{pmatrix}, \tag{1.6.1}$$

in terms of a vector $\boldsymbol{\Xi} \in \mathbb{R}^3$ with components Ξ_i, with $i = 1, 2, 3$. Infinitesimal rotations are expressed by the vector cross product,

$$\widehat{\Xi}\mathbf{r} = \boldsymbol{\Xi} \times \mathbf{r}\,. \tag{1.6.2}$$

The matrix components of $\widehat{\Xi}$ may also be written in terms of the components of the vector $\boldsymbol{\Xi}$ as

$$\widehat{\Xi}_{jk} = \left(O^{-1}\frac{dO}{ds}\right)_{jk}\Bigg|_{s=0} = -\,\Xi_i \epsilon_{ijk}\,,$$

where ϵ_{ijk} with $i, j, k = 1, 2, 3$ is the totally antisymmetric tensor with $\epsilon_{123} = 1$, $\epsilon_{213} = -1$, etc. One may compute directly, for a fixed vector $\mathbf{r}$,

$$\frac{d}{ds}e^{s\widehat{\Xi}}\mathbf{r} = \widehat{\Xi}e^{s\widehat{\Xi}}\mathbf{r} = \boldsymbol{\Xi} \times e^{s\widehat{\Xi}}\mathbf{r}\,.$$

Consequently, one may evaluate, at $s=0$,

$$\frac{d}{ds}e^{s\widehat{\Xi}}\mathbf{r}\bigg|_{s=0} = \widehat{\Xi}\mathbf{r} = \boldsymbol{\Xi}\times\mathbf{r}\,.$$

This expression recovers the expected result in (1.6.2) in terms of the exponential notation. It means the quantity $\mathbf{r}(s) = \exp(s\widehat{\Xi})\mathbf{r}$ describes a finite, right-handed rotation of the initial vector $\mathbf{r} = \mathbf{r}(0)$ by the angle $s|\boldsymbol{\Xi}|$ around the axis pointing in the direction of $\boldsymbol{\Xi}$.

Remark 1.6.1 (Properties of the hat map) The hat map arises in the infinitesimal rotations

$$\widehat{\Xi}_{jk} = (O^{-1}dO/ds)_{jk}|_{s=0} = -\,\Xi_i\epsilon_{ijk}\,.$$

The hat map is an isomorphism:

$$(\mathbb{R}^3, \times) \mapsto (\mathfrak{so}(3),\, [\,\cdot\,,\,\cdot\,])\,.$$

That is, the hat map identifies the composition of two vectors in $\mathbb{R}^3$ using the cross product with the commutator of two skew-symmetric 3×3 matrices. Specifically, we write for any two vectors $\mathbf{Q}, \boldsymbol{\Xi} \in \mathbb{R}^3$,

$$-\,(\mathbf{Q}\times\boldsymbol{\Xi})_k = \epsilon_{klm}\Xi^{\,l}Q^m = \widehat{\Xi}_{km}\,Q^m\,.$$

That is,

$$\boldsymbol{\Xi}\times\mathbf{Q} = \widehat{\Xi}\mathbf{Q} \quad \text{for all} \quad \boldsymbol{\Xi},\,\mathbf{Q}\in\mathbb{R}^3\,.$$

The following formulas may be easily verified for $\mathbf{P}, \mathbf{Q}, \boldsymbol{\Xi} \in \mathbb{R}^3$:

$$\begin{aligned}
(\mathbf{P}\times\mathbf{Q})\widehat{\ } &= \left[\widehat{P},\widehat{Q}\right],\\
\left[\widehat{P},\widehat{Q}\right]\boldsymbol{\Xi} &= (\mathbf{P}\times\mathbf{Q})\times\boldsymbol{\Xi}\,,\\
\mathbf{P}\cdot\mathbf{Q} &= -\frac{1}{2}\,\mathrm{trace}\left(\widehat{P}\widehat{Q}\right).
\end{aligned}$$

□

Remark 1.6.2 The commutator of infinitesimal transformation matrices given by the formula

$$\left[\begin{pmatrix}\widehat{\Xi}_1 & \mathbf{r}_1\\ 0 & 0\end{pmatrix},\begin{pmatrix}\widehat{\Xi}_2 & \mathbf{r}_2\\ 0 & 0\end{pmatrix}\right] = \begin{pmatrix}\widehat{\Xi}_1\widehat{\Xi}_2 - \widehat{\Xi}_2\widehat{\Xi}_1 & \widehat{\Xi}_1\mathbf{r}_2 - \widehat{\Xi}_2\mathbf{r}_1\\ 0 & 0\end{pmatrix}$$

provides a matrix representation of $se(3)$, the Lie algebra of the Lie group $SE(3)$. In vector notation, this becomes

$$\left[\begin{pmatrix}\Xi_1\times & \mathbf{r}_1\\ 0 & 0\end{pmatrix},\begin{pmatrix}\Xi_2\times & \mathbf{r}_2\\ 0 & 0\end{pmatrix}\right]=\begin{pmatrix}(\Xi_1\times\Xi_2)\times & \Xi_1\times\mathbf{r}_2-\Xi_2\times\mathbf{r}_1\\ 0 & 0\end{pmatrix}.$$

□

Remark 1.6.3 The $se(3)$ matrix commutator yields

$$\begin{aligned}\left[(\widehat{\Xi}_1\,,\,\mathbf{r}_1)\,,\,(\widehat{\Xi}_2\,,\,\mathbf{r}_2)\right] &= \left(\widehat{\Xi}_1\widehat{\Xi}_2-\widehat{\Xi}_2\widehat{\Xi}_1,\,\widehat{\Xi}_1\mathbf{r}_2-\widehat{\Xi}_2\mathbf{r}_1\right)\\ &= \left(\left[\widehat{\Xi}_1,\,\widehat{\Xi}_2\right],\,\widehat{\Xi}_1\mathbf{r}_2-\widehat{\Xi}_2\mathbf{r}_1\right),\end{aligned}$$

which is the classic expression for the Lie algebra of a semidirect-product Lie group. □

1.7 Lie algebra of $G(3)$

A 5×5 matrix representation of tangent vectors for $G(3)$ at the identity may be found by computing the derivative of a general group element $(O(s),\mathbf{v}_0(s),\mathbf{r}_0(s),t_0(s))$ along the group path with parameter s and bringing the result back to the identity at $s=0$,

$$\begin{aligned}&\left[\begin{pmatrix}O(s) & \mathbf{v}_0(s) & \mathbf{r}_0(s)\\ 0 & 1 & t_0(s)\\ 0 & 0 & 1\end{pmatrix}^{-1}\begin{pmatrix}O'(s) & \mathbf{v}_0'(s) & \mathbf{r}_0'(s)\\ 0 & 0 & t_0'(s)\\ 0 & 0 & 0\end{pmatrix}\right]_{s=0}\\ &=\left.\begin{pmatrix}O^{-1}(s)O'(s) & O^{-1}(s)\mathbf{v}_0'(s) & O^{-1}(s)(\mathbf{r}_0'(s)-\mathbf{v}_0'(s)t_0'(s))\\ 0 & 0 & t_0'(s)\\ 0 & 0 & 0\end{pmatrix}\right|_{s=0}\\ &=\begin{pmatrix}\widehat{\Xi} & \mathbf{v}_0 & \mathbf{r}_0-\mathbf{v}_0t_0\\ 0 & 0 & t_0\\ 0 & 0 & 0\end{pmatrix}=:\left(\widehat{\Xi},\,\mathbf{v}_0,\,\mathbf{r}_0,\,t_0\right),\end{aligned}$$

in terms of the 3×3 skew-symmetric matrix $\widehat{\Xi}=O^{-1}(s)O'(s)|_{s=0}$. For notational convenience, the superscript primes that would have appeared on the tangents of the Galilean shift parameters $\mathbf{r}_0'(0)$, $\mathbf{v}_0'(0)$ and $t_0'(0)$ at the identity $s = 0$ have been dropped in the last line and replaced by the simpler forms $\mathbf{r}_0$, $\mathbf{v}_0$, t_0, respectively.

Exercise. (Galilean Lie algebra commutator) Verify the commutation relation

$$\Big[(\widehat{\Xi}_1,\, \mathbf{v}_1,\, \mathbf{r}_1,\, t_1),\, (\widehat{\Xi}_2,\, \mathbf{v}_2,\, \mathbf{r}_2,\, t_2)\Big]$$

$$= \Big(\big[\widehat{\Xi}_1,\, \widehat{\Xi}_2\big],\, \widehat{\Xi}_1\mathbf{v}_2 - \widehat{\Xi}_2\mathbf{v}_1,\, \widehat{\Xi}_1(\mathbf{r}_2, \mathbf{v}_2, t_2) - \widehat{\Xi}_2(\mathbf{r}_1, \mathbf{v}_1, t_1),\, 0\Big),$$

where

$$\widehat{\Xi}_1(\mathbf{r}_2, \mathbf{v}_2, t_2) - \widehat{\Xi}_2(\mathbf{r}_1, \mathbf{v}_1, t_1)$$
$$= \Big(\widehat{\Xi}_1(\mathbf{r}_2 - \mathbf{v}_2 t_2) + \mathbf{v}_1 t_2\Big) - \Big(\widehat{\Xi}_2(\mathbf{r}_1 - \mathbf{v}_1 t_1) + \mathbf{v}_2 t_1\Big)\,.$$

★

According to the principle of Galilean relativity, the laws of mechanics must take the same form in any uniformly moving reference frame. That is, the expressions of these laws must be invariant in form under Galilean transformations. In this chapter, we have introduced the Galilean transformations, shown that they comprise a Lie group, found its subgroups, endowed them with a matrix representation, and identified their group structure mathematically as a nested semidirect product.

Rigid motion in $\mathbb{R}^3$ corresponds to a smoothly varying sequence of changes of reference frame along a time-dependent path in the special Euclidean Lie group, $SE(3)$. This is the main subject of the text.

2 NEWTON, LAGRANGE, HAMILTON AND THE RIGID BODY

Contents

2.1	Newton	21
	2.1.1 Newtonian form of free rigid rotation	21
	2.1.2 Newtonian form of rigid-body motion	30
2.2	Lagrange	36
	2.2.1 The principle of stationary action	36
2.3	Noether's theorem	39
	2.3.1 Lie symmetries and conservation laws	39
	2.3.2 Infinitesimal transformations of a Lie group	40
2.4	Lagrangian form of rigid-body motion	46
	2.4.1 Hamilton–Pontryagin constrained variations	50
	2.4.2 Manakov's formulation of the $SO(n)$ rigid body	54
	2.4.3 Matrix Euler–Poincaré equations	55
	2.4.4 An isospectral eigenvalue problem for the $SO(n)$ rigid body	56
	2.4.5 Manakov's integration of the $SO(n)$ rigid body	58

2.5 Hamilton 60
2.5.1 Hamiltonian form of rigid-body motion 62
2.5.2 Lie–Poisson Hamiltonian rigid-body dynamics 63
2.5.3 Lie–Poisson bracket 64
2.5.4 Nambu's $\mathbb{R}^3$ Poisson bracket 65
2.5.5 Clebsch variational principle for the rigid body 69
2.5.6 Rotating motion with potential energy 72

2.1 Newton

2.1.1 Newtonian form of free rigid rotation

Isaac Newton

Definition 2.1.1 *In* ***free rigid rotation*** *a body rotates about its centre of mass and the pairwise distances between all points in the body remain fixed.*

Definition 2.1.2 *A system of coordinates fixed in a body undergoing free rigid rotation is stationary in the rotating orthonormal basis called the* **body frame**, *introduced by Euler [Eu1758].*

The orientation of the orthonormal frame $(\mathbf{E}_1, \mathbf{E}_2, \mathbf{E}_3)$ fixed in the rotating body relative to a basis $(\mathbf{e}_1, \mathbf{e}_2, \mathbf{e}_3)$ fixed in space depends smoothly on time $t \in \mathbb{R}$. In the fixed spatial coordinate system, the body frame is seen as the moving frame

$$(O(t)\mathbf{E}_1,\, O(t)\mathbf{E}_2,\, O(t)\mathbf{E}_3)\,,$$

where $O(t) \in SO(3)$ defines the attitude of the body relative to its reference configuration according to the following matrix multiplication on its three unit vectors:

$$\mathbf{e}_a(t) = O(t)\mathbf{E}_a\,, \quad a = 1, 2, 3. \tag{2.1.1}$$

Here the unit vectors $\mathbf{e}_a(0) = \mathbf{E}_a$ with $a = 1, 2, 3$ comprise at initial time $t = 0$ an orthonormal basis of coordinates and $O(t)$ is a special ($\det O(t) = 1$) orthogonal ($O^T(t)O(t) = Id$) 3×3 matrix. That is, $O(t)$ is a continuous function defined along a curve parameterised by time t in the special orthogonal matrix group $SO(3)$. At the initial time $t = 0$, we may take $O(0) = Id$, without any loss.

As the orientation of the body is evolving according to (2.1.1), each basis vector in the set,

$$\mathbf{e}(t) \in \{\mathbf{e}_1(t), \mathbf{e}_2(t), \mathbf{e}_3(t)\},$$

preserves its (unit) length. We prove this and fix notation by writing

$$\begin{aligned} 1 &= |\mathbf{e}(t)|^2 := \mathbf{e}(t)\cdot\mathbf{e}(t) := \mathbf{e}(t)^T\mathbf{e}(t) = (O(t)\mathbf{E})^T O(t)\mathbf{E} \\ &= \mathbf{E}^T O^T(t) O(t)\mathbf{E} = \mathbf{E}^T (Id)\mathbf{E} = |\mathbf{E}|^2, \end{aligned} \tag{2.1.2}$$

which follows because $O(t)$ is orthogonal; that is, $O^T(t)O(t) = Id$.

The basis vectors in the orthonormal frame $\mathbf{e}_a(0) = \mathbf{E}_a$ define the initial orientation of the set of rotating points with respect to some choice of fixed spatial coordinates at time $t = 0$. Each point $\mathbf{r}(t)$ in the subsequent rigid motion may be represented in either fixed or rotating coordinates as

$$\mathbf{r}(t) = r_0^A(t)\mathbf{e}_A(0) \quad \text{in the fixed basis,} \tag{2.1.3}$$
$$= r^a \mathbf{e}_a(t) \quad \text{in the rotating basis.} \tag{2.1.4}$$

The fixed basis is called the ***spatial frame*** and the rotating basis is the ***body frame***.

The constant components r^a of a position vector relative to the rotating basis are related to its *initial* spatial position as

$$r^a = O^a_A(0) r_0^A(0).$$

This is simply $r^a = \delta^a_A r_0^A(0)$ for the choice $O(0) = Id$ in which the two coordinate bases are initially aligned. The components of any vector $\mathbf{J}$ in the fixed (spatial) frame are related to those in the moving (body) frame by the mutual rotation of their axes in (2.1.1) at any time. That is,

$$\mathbf{J} = \mathbf{e}_a(0) J^a_{space}(t) = \mathbf{e}_a(t) J^a_{body} = O(t)\mathbf{e}_a(0) J^a_{body}, \tag{2.1.5}$$

or equivalently, as in Equation (2.1.1),

$$\mathbf{J}_{space}(t) = O(t)\mathbf{J}_{body}. \tag{2.1.6}$$

Lemma 2.1.1 *The velocity $\dot{\mathbf{r}}(t)$ of a point $\mathbf{r}(t)$ in free rigid rotation depends linearly on its position relative to the centre of mass.*

Proof. In particular, $\mathbf{r}(t) = r^a O(t)\mathbf{e}_a(0)$ implies

$$\dot{\mathbf{r}}(t) = r^a \dot{\mathbf{e}}_a(t) = r^a \dot{O}(t)\mathbf{e}_a(0) =: r^a \dot{O}O^{-1}(t)\mathbf{e}_a(t) =: \widehat{\omega}(t)\mathbf{r}\,, \quad (2.1.7)$$

which is linear. ■

Remark 2.1.1 (Wide-hat notation) The wide-hat notation $(\widehat{\cdot})$ denotes a skew-symmetric 3×3 matrix. There is no danger of confusing wide-hat notation $(\widehat{\cdot})$ with narrow-hat notation $(\hat{\cdot})$, which denotes a unit vector (or, later, a unit quaternion). □

Lemma 2.1.2 (Skew-symmetry) *The* ***spatial angular velocity matrix*** *$\widehat{\omega}(t) = \dot{O}O^{-1}(t)$ in (2.1.7) is skew-symmetric, i.e.,*

$$\widehat{\omega}^T = -\,\widehat{\omega}\,.$$

Proof. Being orthogonal, the matrix $O(t)$ satisfies $OO^T = Id$. This implies that $\widehat{\omega}$ is skew-symmetric,

$$\begin{aligned} 0 = (OO^T)^{\cdot} &= \dot{O}O^T + O\dot{O}^T = \dot{O}O^T + (\dot{O}O^T)^T \\ &= \dot{O}O^{-1} + (\dot{O}O^{-1})^T = \widehat{\omega} + \widehat{\omega}^T\,. \end{aligned}$$

■

Remark 2.1.2 The skew 3×3 real matrices form a closed linear space under addition. □

Definition 2.1.3 (Commutator product of skew matrices) *The commutator product of two skew matrices $\widehat{\omega}$ and $\widehat{\xi}$ is defined as the skew matrix product*

$$[\widehat{\omega},\, \widehat{\xi}\,] := \widehat{\omega}\,\widehat{\xi} - \widehat{\xi}\,\widehat{\omega}\,. \quad (2.1.8)$$

Remark 2.1.3 This commutator product is again a skew 3×3 real matrix. □

Definition 2.1.4 (Basis set for skew matrices) *Any 3×3 antisymmetric matrix $\widehat{\omega}^T = -\widehat{\omega}$ may be written as a linear combination of the following three linearly independent basis elements for the 3×3 skew matrices:*

$$\widehat{J}_1 = \begin{pmatrix} 0 & 0 & 0 \\ 0 & 0 & -1 \\ 0 & 1 & 0 \end{pmatrix}, \quad \widehat{J}_2 = \begin{pmatrix} 0 & 0 & 1 \\ 0 & 0 & 0 \\ -1 & 0 & 0 \end{pmatrix}, \quad \widehat{J}_3 = \begin{pmatrix} 0 & -1 & 0 \\ 1 & 0 & 0 \\ 0 & 0 & 0 \end{pmatrix}.$$

That is, the element $\widehat{J}_a$ for this choice of basis has matrix components

$$(\widehat{J}_a)_{bc} = -\epsilon_{abc},$$

where ϵ_{abc} is the totally antisymmetric tensor with

$$\epsilon_{123} = +1, \qquad \epsilon_{213} = -1, \qquad \epsilon_{113} = 0, \qquad etc.$$

Lemma 2.1.3 (Commutation relations) *The skew matrix basis $\widehat{J}_a$ with $a = 1, 2, 3$ satisfies the* ***commutation relations****,*

$$[\,\widehat{J}_a,\, \widehat{J}_b\,] := \widehat{J}_a\widehat{J}_b - \widehat{J}_b\widehat{J}_a = \epsilon_{abc}\widehat{J}_c\,. \tag{2.1.9}$$

Proof. This may be verified by a direct calculation, $[\,\widehat{J}_1,\, \widehat{J}_2\,] = \widehat{J}_3$, etc. ■

Remark 2.1.4 The closure of the basis set of skew-symmetric matrices under the commutator product gives the linear space of skew-symmetric matrices a ***Lie algebra structure***. The constants ϵ_{abc} in the commutation relations among the skew 3×3 matrix basis elements are called the ***structure constants*** and the corresponding Lie algebra is called $so(3)$. This also means the abstract $so(3)$ Lie algebra may be represented by skew 3×3 matrices, which is a great convenience, as we shall see for example in Section 4.2. The Lie algebra $so(3)$ may also be defined as the tangent space to the Lie group $SO(3)$ at the identity, as discussed in Appendix B. □

Theorem 2.1.1 (Hat map) *The components of any 3×3 skew matrix $\widehat{\omega}$ may be identified with the corresponding components of a vector $\boldsymbol{\omega} \in \mathbb{R}^3$.*

Proof. In the basis (2.1.4), one writes the linear invertible relation,

$$\widehat{\omega} = \begin{pmatrix} 0 & -\omega^3 & \omega^2 \\ \omega^3 & 0 & -\omega^1 \\ -\omega^2 & \omega^1 & 0 \end{pmatrix} = \omega^a \widehat{J}_a =: \boldsymbol{\omega} \cdot \widehat{\boldsymbol{J}}, \tag{2.1.10}$$

for $a = 1, 2, 3$. This is a one-to-one invertible map, i.e., it is an ***isomorphism***, between 3×3 skew-symmetric matrices and vectors in $\mathbb{R}^3$. ■

Remark 2.1.5 The superscript hat $(\widehat{\ })$ applied to a vector identifies that vector in $\mathbb{R}^3$ with a 3×3 skew-symmetric matrix. For example, the unit vectors in the Cartesian basis set, $\{\mathbf{e}_1, \mathbf{e}_2, \mathbf{e}_3\}$, are associated with the basis elements $\widehat{J}_a$, for $a = 1, 2, 3$, in Equation (2.1.4) by $\widehat{J}_a = \widehat{\mathbf{e}}_a$, or in matrix components,

$$(\widehat{\mathbf{e}}_a)_{bc} = -\delta^d_a \epsilon_{dbc} = -\epsilon_{abc} = (\mathbf{e}_a \times)_{bc} \,.$$

□

Remark 2.1.6 The last equality in the definition of the hat map in Equation (2.1.10) introduces the convenient notation $\widehat{\boldsymbol{J}}$ that denotes the basis for the 3×3 skew-symmetric matrices $\widehat{J}_a$, with $a = 1, 2, 3$ as a vector of matrices. □

Definition 2.1.5 (Hat map for angular velocity vector) *The relation $\widehat{\omega} = \boldsymbol{\omega} \cdot \widehat{\boldsymbol{J}}$ in Equation (2.1.10) identifies the skew-symmetric 3×3 matrix $\hat{\omega}(t)$ with the **angular velocity vector** $\boldsymbol{\omega}(t) \in \mathbb{R}^3$ whose components $\omega^c(t)$, with $c = 1, 2, 3$, are given by*

$$(\dot{O}O^{-1})_{ab}(t) = \widehat{\omega}_{ab}(t) = -\,\epsilon_{abc}\,\omega^c(t)\,. \tag{2.1.11}$$

*Equation (2.1.11) defines the matrix components of the **hat map for angular velocity**.*

Remark 2.1.7 Equivalently, the hat map in Equation (2.1.10) is defined by the identity

$$\widehat{\omega}\boldsymbol{\lambda} = \boldsymbol{\omega} \times \boldsymbol{\lambda} \quad \text{for all} \quad \boldsymbol{\omega}, \boldsymbol{\lambda} \in \mathbb{R}^3.$$

Thus, we may write $\widehat{\omega} = \widehat{\omega} = \boldsymbol{\omega}\times$ to identify the vector $\boldsymbol{\omega} \in \mathbb{R}^3$ with the skew 3×3 matrix $\widehat{\omega} \in so(3)$. □

Proposition 2.1.1 *The* 3×3 *skew matrices*

$$\widehat{\omega} = \boldsymbol{\omega} \cdot \widehat{\boldsymbol{J}} \quad \textit{and} \quad \widehat{\lambda} = \boldsymbol{\lambda} \cdot \widehat{\boldsymbol{J}}$$

associated with the vectors $\boldsymbol{\omega}$ *and* $\boldsymbol{\lambda}$ *in* $\mathbb{R}^3$ *satisfy the commutation relation*

$$[\widehat{\omega}, \widehat{\lambda}] = \boldsymbol{\omega} \times \boldsymbol{\lambda} \cdot \widehat{\boldsymbol{J}} =: (\boldsymbol{\omega} \times \boldsymbol{\lambda})\widehat{\ }, \tag{2.1.12}$$

where $\boldsymbol{\omega} \times \boldsymbol{\lambda}$ *is the vector product in* $\mathbb{R}^3$.

Proof. Formula (2.1.9) implies the result, by

$$\begin{aligned} [\widehat{\omega}, \widehat{\lambda}] &= [\boldsymbol{\omega} \cdot \widehat{\boldsymbol{J}}, \boldsymbol{\lambda} \cdot \widehat{\boldsymbol{J}}] = [\omega^a \widehat{J}_a, \lambda^b \widehat{J}_b] \\ &= \omega^a \lambda^b [\widehat{J}_a, \widehat{J}_b] = \omega^a \lambda^b \epsilon_{abc} \widehat{J}_c = \boldsymbol{\omega} \times \boldsymbol{\lambda} \cdot \widehat{\boldsymbol{J}}. \end{aligned}$$

■

Remark 2.1.8 According to Proposition 2.1.1, the hat map $\widehat{\ }$: $(\mathbb{R}^3, \times) \mapsto (so(3), [\cdot, \cdot])$ allows the velocity in space (2.1.7) of a point at $\mathbf{r}$ undergoing rigid-body motion to be expressed equivalently either by a skew-matrix multiplication, or as a vector product. That is,

$$\dot{\mathbf{r}}(t) =: \widehat{\omega}(t)\mathbf{r} =: \boldsymbol{\omega}(t) \times \mathbf{r}. \tag{2.1.13}$$

Hence, free rigid motion of a point displaced by $\mathbf{r}$ from the centre of mass is a rotation in space of $\mathbf{r}$ about the time-dependent angular velocity vector $\boldsymbol{\omega}(t)$. Accordingly, $d|\mathbf{r}|^2/dt = 2\mathbf{r} \cdot \dot{\mathbf{r}} = 0$, and the displacement distance is preserved, $|\mathbf{r}|(t) = |\mathbf{r}|(0)$. □

Kinetic energy of free rigid rotation

The kinetic energy for N particles of masses m_j $j = 1, 2, \ldots, N$, mutually undergoing free rigid rotation is computed in terms of the angular velocity as

$$\begin{aligned} K &= \frac{1}{2}\sum_{j=1}^{N} m_j \dot{\mathbf{r}}_j \cdot \dot{\mathbf{r}}_j \\ &= \frac{1}{2}\sum_{j=1}^{N} m_j (\boldsymbol{\omega} \times \mathbf{r}_j) \cdot (\boldsymbol{\omega} \times \mathbf{r}_j) \\ &=: \frac{1}{2}\langle\!\langle \boldsymbol{\omega}, \boldsymbol{\omega} \rangle\!\rangle . \end{aligned}$$

Definition 2.1.6 (Symmetric mass-weighted pairing) *The kinetic energy induces a symmetric mass-weighted pairing*

$$\langle\!\langle \cdot\,, \cdot \rangle\!\rangle : \mathbb{R}^3 \times \mathbb{R}^3 \mapsto \mathbb{R} \,,$$

defined for any two vectors $\mathbf{a}, \mathbf{b} \in \mathbb{R}^3$ *as*

$$\langle\!\langle \mathbf{a}, \mathbf{b} \rangle\!\rangle := \sum_j m_j (\mathbf{a} \times \mathbf{r}_j) \cdot (\mathbf{b} \times \mathbf{r}_j) =: \mathbb{I}\,\mathbf{a} \cdot \mathbf{b} \,. \tag{2.1.14}$$

Definition 2.1.7 (Moment of inertia tensor) *The mass-weighted pairing, or inner product in (2.1.14)*

$$\langle\!\langle \mathbf{a}, \mathbf{b} \rangle\!\rangle = \mathbb{I}\,\mathbf{a} \cdot \mathbf{b} \,,$$

defines the symmetric ***moment of inertia tensor*** $\mathbb{I}$ *for the particle system.*

Definition 2.1.8 (Angular momentum of rigid motion) *The* ***angular momentum*** *is defined as the derivative of the kinetic energy with respect to angular velocity. In the present case with (2.1.14), this produces the linear relation*

$$\begin{aligned} \mathbf{J} = \frac{\partial K}{\partial \boldsymbol{\omega}} &= -\frac{1}{2}\sum_{j=1}^{N} m_j \mathbf{r}_j \times \Big(\mathbf{r}_j \times \boldsymbol{\omega} \Big) \\ &= \frac{1}{2}\sum_{j=1}^{N} m_j \Big(\big(|\mathbf{r}_j|^2 Id - \mathbf{r}_j \otimes \mathbf{r}_j \big) \Big) \boldsymbol{\omega} \\ &=: \mathbb{I}\,\boldsymbol{\omega} \,, \end{aligned} \tag{2.1.15}$$

where $\mathbb{I}$ *is the moment of inertia tensor defined by the symmetric pairing in (2.1.14).*

Conservation of angular momentum in free rigid rotation

In free rigid rotation no external torques are applied, so the angular momentum $\mathbf{J}$ is conserved. In the fixed basis $\mathbf{J} = J_0^A(t)\mathbf{e}_A(0)$ and this conservation law is expressed as

$$0 = \frac{d\mathbf{J}}{dt} = \frac{dJ_0^A}{dt}\mathbf{e}_A(0)\,, \tag{2.1.16}$$

so each component J_0^A, for $A = 1, 2, 3$ of angular momentum in the spatial frame is separately conserved. In the rotating basis $\mathbf{J} = J^a(t)\mathbf{e}_a(t)$ and angular momentum conservation becomes

$$\begin{aligned} 0 = \frac{d\mathbf{J}}{dt} &= \frac{dJ^a}{dt}\mathbf{e}_a(t) + J^a\frac{d\mathbf{e}_a(t)}{dt} \\ &= \frac{dJ^a}{dt}\mathbf{e}_a(t) + \boldsymbol{\omega} \times J^a\mathbf{e}_a(t) \\ &= \left(\frac{dJ^a}{dt} + (\boldsymbol{\omega} \times \mathbf{J})^a\right)\mathbf{e}_a(t)\,. \end{aligned} \tag{2.1.17}$$

Consequently, the components J^a, for $a = 1, 2, 3$ of angular momentum in the body frame satisfy the quadratically nonlinear *system* of Equations (2.1.17) with $\mathbf{J} = \mathbb{I}\,\boldsymbol{\omega}$.

Lemma 2.1.4 (Space vs body dynamics) *Upon denoting* $\mathbf{J}_{space} = (J_0^1,\, J_0^2,\, J_0^3)$ *in the fixed basis* $\mathbf{e}_A(0)$ *and* $\mathbf{J}_{body} = (J^1,\, J^2,\, J^3)$ *in the time-dependent basis* $\mathbf{e}_a(t)$*, with*

$$\mathbf{J}_{space} = O(t)\mathbf{J}_{body}\,, \tag{2.1.18}$$

one may summarise the two equivalent sets of Equations (2.1.16) and (2.1.17) as

$$\frac{d\mathbf{J}_{space}}{dt} = 0 \quad \text{and} \quad \frac{d\mathbf{J}_{body}}{dt} + \left(\mathbb{I}^{-1}\mathbf{J}_{body}\right) \times \mathbf{J}_{body} = 0\,. \tag{2.1.19}$$

Proof. The time derivative of relation (2.1.18) gives

$$\begin{aligned}\frac{d\mathbf{J}_{body}}{dt} &= \frac{d}{dt}\Big(O^{-1}(t)\mathbf{J}_{space}(t)\Big)\\ &= -O^{-1}\dot{O}\Big(O^{-1}(t)\mathbf{J}_{space}(t)\Big) + O^{-1}\underbrace{\frac{d\mathbf{J}_{space}}{dt}}_{\text{vanishes}}\\ &= -\widehat{\omega}_{body}\mathbf{J}_{body} = -\boldsymbol{\omega}_{body}\times\mathbf{J}_{body}\,,\end{aligned}\tag{2.1.20}$$

with $\widehat{\omega}_{body} := O^{-1}\dot{O} = \boldsymbol{\omega}_{body}\times$ and $\boldsymbol{\omega}_{body} = \mathbb{I}^{-1}\mathbf{J}_{body}$, which defines the ***body angular velocity***. This is the usual heuristic derivation of the dynamics for body angular momentum. ■

Remark 2.1.9 (Darwin, Coriolis and centrifugal forces) Many elementary mechanics texts make the following points about the various ***noninertial forces*** that arise in a rotating frame. For any vector $\mathbf{r}(t) = r^a(t)\mathbf{e}_a(t)$ the body and space time derivatives satisfy the first time-derivative relation, as in (2.1.17),

$$\begin{aligned}\dot{\mathbf{r}}(t) &= \dot{r}^a\mathbf{e}_a(t) + r^a\dot{\mathbf{e}}_a(t)\\ &= \dot{r}^a\mathbf{e}_a(t) + \boldsymbol{\omega}\times r^a\mathbf{e}_a(t)\\ &= \big(\dot{r}^a + \epsilon^a_{bc}\omega^b r^c\big)\mathbf{e}_a(t)\\ &=: \big(\dot{r}^a + (\boldsymbol{\omega}\times\mathbf{r})^a\big)\mathbf{e}_a(t)\,.\end{aligned}\tag{2.1.21}$$

Taking a second time derivative in this notation yields

$$\begin{aligned}\ddot{\mathbf{r}}(t) &= \big(\ddot{r}^{\,a} + (\dot{\boldsymbol{\omega}}\times\mathbf{r})^a + (\boldsymbol{\omega}\times\dot{\mathbf{r}})^a\big)\mathbf{e}_a(t) + \big(\dot{r}^a + (\boldsymbol{\omega}\times\mathbf{r})^a\big)\dot{\mathbf{e}}_a(t)\\ &= \big(\ddot{r}^{\,a} + (\dot{\boldsymbol{\omega}}\times\mathbf{r})^a + (\boldsymbol{\omega}\times\dot{\mathbf{r}})^a\big)\mathbf{e}_a(t) + \Big(\boldsymbol{\omega}\times\big(\dot{\mathbf{r}} + \boldsymbol{\omega}\times\mathbf{r}\big)\Big)^a\mathbf{e}_a(t)\\ &= \big(\ddot{r}^{\,a} + (\dot{\boldsymbol{\omega}}\times\mathbf{r})^a + 2(\boldsymbol{\omega}\times\dot{\mathbf{r}})^a + (\boldsymbol{\omega}\times\boldsymbol{\omega}\times\mathbf{r})\big)^a\mathbf{e}_a(t)\,.\end{aligned}$$

Newton's second law for the evolution of the position vector $\mathbf{r}(t)$ of a particle of mass m in a frame rotating with time-dependent angular velocity $\boldsymbol{\omega}(t)$ becomes

$$\mathbf{F}(\mathbf{r}) = m\big(\ddot{\mathbf{r}} + \underbrace{\dot{\boldsymbol{\omega}}\times\mathbf{r}}_{\text{Darwin}} + \underbrace{2(\boldsymbol{\omega}\times\dot{\mathbf{r}})}_{\text{Coriolis}} + \underbrace{\boldsymbol{\omega}\times(\boldsymbol{\omega}\times\mathbf{r})}_{\text{centrifugal}}\big)\,.\tag{2.1.22}$$

- The Darwin force is usually small; so it is often neglected.
- Only the Coriolis force depends on the velocity in the moving frame. The Coriolis force is very important in large-scale motions on Earth. For example, pressure balance with the Coriolis force dominates the (geostrophic) motion of weather systems that comprise the climate.
- The centrifugal force is important, for example, in obtaining orbital equilibria in gravitationally attracting systems.

□

Remark 2.1.10 The space and body angular velocities differ by

$$\widehat{\omega}_{body} := O^{-1}\dot{O} \quad \text{versus} \quad \widehat{\omega}_{space} := \dot{O}O^{-1} = O\widehat{\omega}_{body}O^{-1} .$$

Namely, $\widehat{\omega}_{body}$ is left-invariant under $O \to RO$ and $\widehat{\omega}_{space}$ is right-invariant under $O \to OR$, for any choice of matrix $R \in SO(3)$. This means that neither angular velocity depends on the initial orientation. □

Remark 2.1.11 The angular velocities $\widehat{\omega}_{body} = O^{-1}\dot{O}$ and $\widehat{\omega}_{space} = \dot{O}O^{-1}$ are respectively the left and right translations to the identity of the tangent matrix $\dot{O}(t)$ at $O(t)$. These are called the left and right tangent spaces of $SO(3)$ at its identity. □

Remark 2.1.12 Equations (2.1.19) for free rigid rotations of particle systems are prototypes of Euler's equations for the motion of a rigid body. □

2.1.2 Newtonian form of rigid-body motion

In describing rotations of a rigid body, for example a solid object occupying a spatial domain $\mathcal{B} \subset \mathbb{R}^3$, one replaces the mass-weighted sums over points in space in the previous definitions of dynamical quantities for free rigid rotation, with volume integrals weighted with a mass density as a function of position in the body. That is,

$$\sum_j m_j \to \int_{\mathcal{B}} d^3\mathbf{X}\,\rho(\mathbf{X}),$$

where $\rho(\mathbf{X})$ is the mass density at a point $\mathbf{X} \in \mathcal{B}$ fixed inside the body, as measured in coordinates whose origin is at the centre of mass.

Example 2.1.1 (Kinetic energy of a rotating solid body) *The kinetic energy of a solid body rotating about its centre of mass is given by*

$$K = \frac{1}{2}\int_{\mathcal{B}} \rho(\mathbf{X})\,|\dot{\mathbf{x}}(\mathbf{X},t)|^2\,d^3\mathbf{X}, \qquad (2.1.23)$$

where the **spatial path** *in* $\mathbb{R}^3$ *of a point* $\mathbf{X} \in \mathcal{B}$ *in the rotating body is given by*

$$\mathbf{x}(\mathbf{X},t) = O(t)\mathbf{X} \in \mathbb{R}^3 \quad \textit{with} \quad O(t) \in SO(3)\,.$$

The time derivative of this rotating motion yields the **spatial velocity**

$$\dot{\mathbf{x}}(\mathbf{X},t) = \dot{O}(t)\mathbf{X} = \dot{O}O^{-1}(t)\mathbf{x} =: \widehat{\omega}(t)\mathbf{x} =: \boldsymbol{\omega}(t) \times \mathbf{x}\,, \qquad (2.1.24)$$

as in Equation (2.1.7) for free rotation.

Kinetic energy and angular momentum of a rigid body

The kinetic energy (2.1.23) of a rigid body rotating about its centre of mass may be expressed in the spatial frame in analogy to Equation (2.1.14) for free rotation,

$$K = \frac{1}{2}\int_{O(t)\mathcal{B}} \rho(O^{-1}(t)\mathbf{x})\,|\boldsymbol{\omega}(t) \times \mathbf{x}|^2\,d^3\mathbf{x}\,. \qquad (2.1.25)$$

However, its additional time dependence makes this integral unwieldy. Instead, one takes advantage of the preservation of scalar products by rotations to write

$$|\dot{\mathbf{x}}|^2 = |O^{-1}\dot{\mathbf{x}}|^2\,,$$

and one computes as in Equation (2.1.24) the ***body velocity***

$$O^{-1}\dot{\mathbf{x}}(\mathbf{X},t) = O^{-1}\dot{O}(t)\mathbf{X} =: \widehat{\Omega}(t)\mathbf{X} =: \boldsymbol{\Omega}(t) \times \mathbf{X} \,. \tag{2.1.26}$$

Here, skew-symmetry $\widehat{\Omega}^T = -\widehat{\Omega}$ of the matrix

$$\widehat{\Omega} = O^{-1}\dot{O} = \widehat{\omega}_{body}$$

follows because the matrix O is orthogonal, that is, $OO^T = Id$. Skew-symmetry of $\widehat{\Omega}$ allows one to introduce the ***body angular velocity vector*** $\boldsymbol{\Omega}(t)$ whose components Ω_i, with $i = 1, 2, 3$, are given in body coordinates by

$$(O^{-1}\dot{O})_{jk} = \widehat{\Omega}_{jk} = -\,\Omega_i \epsilon_{ijk} \,. \tag{2.1.27}$$

In terms of body angular velocity vector $\boldsymbol{\Omega}(t)$ the kinetic energy of a rigid body becomes

$$K = \frac{1}{2}\int_{\mathcal{B}} \rho(\mathbf{X})\,|\boldsymbol{\Omega}(t) \times \mathbf{X}|^2\, d^3\mathbf{X} =: \frac{1}{2}\Big\langle\!\!\Big\langle \boldsymbol{\Omega}(t), \boldsymbol{\Omega}(t) \Big\rangle\!\!\Big\rangle \,, \tag{2.1.28}$$

where $\langle\!\langle \cdot\,, \cdot \rangle\!\rangle$ is a mass-weighted symmetric pairing, defined for any two vectors $\mathbf{a}, \mathbf{b} \in \mathbb{R}^3$ as the following integration over the body,

$$\Big\langle\!\!\Big\langle \mathbf{a}, \mathbf{b} \Big\rangle\!\!\Big\rangle := \int_{\mathcal{B}} \rho(\mathbf{X})(\mathbf{a} \times \mathbf{X}) \cdot (\mathbf{b} \times \mathbf{X})\, d^3\mathbf{X} \,. \tag{2.1.29}$$

Definition 2.1.9 (Moment of inertia tensor) *The mass-weighted pairing, or inner product in (2.1.29)*

$$\Big\langle\!\!\Big\langle \mathbf{a}, \mathbf{b} \Big\rangle\!\!\Big\rangle = \mathbb{I}\mathbf{a} \cdot \mathbf{b} \,,$$

*defines the symmetric **moment of inertia tensor** $\mathbb{I}$ for the rigid body.*

Exercise. By definition, $\mathbb{I}$ is constant in the body frame. What is its time dependence in the spatial frame? ★

Remark 2.1.13 (Principal axis frame) The moment of inertia tensor becomes *diagonal,*

$$\mathbb{I} = \operatorname{diag}(I_1, I_2, I_3)\,,$$

upon aligning the body reference coordinates with its ***principal axis frame***. In the principal axis coordinates of $\mathbb{I}$, the kinetic energy (2.1.28) takes the elegant form

$$K = \frac{1}{2}\langle\!\langle\, \boldsymbol{\Omega}\,,\, \boldsymbol{\Omega}\,\rangle\!\rangle = \frac{1}{2}\,\mathbb{I}\boldsymbol{\Omega}\cdot\boldsymbol{\Omega} = \frac{1}{2}\,(I_1\Omega_1^2 + I_2\Omega_2^2 + I_3\Omega_3^2)\,.$$

□

Definition 2.1.10 (Body angular momentum) *The **body angular momentum** is defined as the derivative of the kinetic energy (2.1.28) with respect to body angular velocity. This produces the linear relation*

$$\begin{aligned} \boldsymbol{\Pi} = \frac{\partial K}{\partial \boldsymbol{\Omega}} &= -\int_{\mathcal{B}} \rho(\mathbf{X})\,\mathbf{X}\times\Big(\mathbf{X}\times\boldsymbol{\Omega}(t)\Big)\,d^3\mathbf{X} \\ &= \left(\int_{\mathcal{B}} \rho(\mathbf{X})\,\Big(|\mathbf{X}|^2 Id - \mathbf{X}\otimes\mathbf{X}\Big)\,d^3\mathbf{X}\right)\boldsymbol{\Omega}(t) \\ &= \mathbb{I}\boldsymbol{\Omega}\,. \end{aligned} \tag{2.1.30}$$

This $\mathbb{I}$ is the continuum version of the moment of inertia tensor defined for particle systems in Equation (2.1.14). It is called the moment of inertia tensor of the rigid body.

Remark 2.1.14 In general, the body angular momentum vector $\boldsymbol{\Pi}$ is not parallel to the body angular velocity vector $\boldsymbol{\Omega}$. Their misalignment is measured by $\boldsymbol{\Pi}\times\boldsymbol{\Omega}\neq 0$. □

Angular momentum conservation

The rigid body rotates freely along $O(t)$ in the absence of any externally applied forces or torques, so Newton's second law implies that the motion of the rigid body conserves total angular momentum when expressed in the fixed space coordinates in $\mathbb{R}^3$. This conservation law is expressed in the *spatial* frame as

$$\frac{d\pi}{dt} = 0\,,$$

where $\boldsymbol{\pi}(t)$ is the ***angular momentum vector in space***. The angular momentum vector in space $\boldsymbol{\pi}(t)$ is related to the angular momentum vector in the body $\mathbf{\Pi}(t)$ by the mutual rotation of their axes at any time. That is, $\boldsymbol{\pi}(t) = O(t)\mathbf{\Pi}(t)$. Likewise, the ***angular velocity vector in space*** satisfies $\boldsymbol{\omega}(t) = O(t)\mathbf{\Omega}(t)$. The angular velocity vector in space is related to its corresponding angular momentum vector by

$$\begin{aligned}\boldsymbol{\pi}(t) = O(t)\mathbf{\Pi}(t) &= O(t)\mathbb{I}\mathbf{\Omega}(t)\\ &= \big(O(t)\mathbb{I}O^{-1}(t)\big)\,\boldsymbol{\omega}(t) =: \mathbb{I}_{space}(t)\boldsymbol{\omega}(t)\,.\end{aligned}$$

Thus, the ***moment of inertia tensor in space*** $\mathbb{I}_{space}(t)$ transforms as a symmetric tensor,

$$\mathbb{I}_{space}(t) = O(t)\mathbb{I}O^{-1}(t)\,, \tag{2.1.31}$$

so it is *time-dependent* and the relation of the spatial angular velocity vector $\boldsymbol{\omega}(t)$ to the motion of the rigid body may be found by using (2.1.26), as

$$\dot{\mathbf{x}}(\mathbf{X},t) = \dot{O}(t)\mathbf{X} =: \dot{O}O^{-1}(t)\mathbf{x} =: \widehat{\omega}(t)\mathbf{x} =: \boldsymbol{\omega}(t)\times\mathbf{x}\,. \tag{2.1.32}$$

As expected, the motion in space of a point at $\mathbf{x}$ within the rigid body is a rotation by the time-dependent angular velocity $\boldsymbol{\omega}(t)$. We may formally confirm the relation of the spatial angular velocity vector to the body angular velocity vector $\boldsymbol{\omega}(t) = O(t)\mathbf{\Omega}(t)$ by using (2.1.26) in the following calculation:

$$\begin{aligned}\mathbf{\Omega}(t)\times\mathbf{X} &= \widehat{\Omega}(t)\mathbf{X} = O^{-1}\dot{O}\mathbf{X}\\ &= O^{-1}\widehat{\omega}(t)\mathbf{x} = O^{-1}(\boldsymbol{\omega}(t)\times\mathbf{x}) = (O^{-1}\boldsymbol{\omega}(t)\times O^{-1}\mathbf{x})\,.\end{aligned}$$

Consequently, the conservation of spatial angular momentum in the absence of external torques implies

$$\begin{aligned}\frac{d\boldsymbol{\pi}}{dt} &= \frac{d}{dt}\Big(O(t)\mathbf{\Pi}\Big)\\ &= O(t)\Big(\frac{d\mathbf{\Pi}}{dt} + O^{-1}\dot{O}\,\mathbf{\Pi}\Big)\\ &= O(t)\Big(\frac{d\mathbf{\Pi}}{dt} + \widehat{\Omega}\,\mathbf{\Pi}\Big)\\ &= O(t)\Big(\frac{d\mathbf{\Pi}}{dt} + \mathbf{\Omega}\times\mathbf{\Pi}\Big) = 0\,.\end{aligned}$$

Hence, the body angular momentum satisfies ***Euler's equations*** for a rigid body,

$$\frac{d\boldsymbol{\Pi}}{dt} + \boldsymbol{\Omega} \times \boldsymbol{\Pi} = 0\,. \tag{2.1.33}$$

Remark 2.1.15 (Body angular momentum equation) Viewed in the moving frame, the rigid body occupies a fixed domain $\mathcal{B}$, so its moment of inertia tensor in that frame $\mathbb{I}$ is constant. Its body angular momentum vector $\boldsymbol{\Pi} = \mathbb{I}\boldsymbol{\Omega}$ evolves according to (2.1.33) by rotating around the body angular velocity vector $\boldsymbol{\Omega} = \mathbb{I}^{-1}\boldsymbol{\Pi}$. That is, conservation of spatial angular momentum $\boldsymbol{\pi}(t) = O(t)\boldsymbol{\Pi}(t)$ relative to a fixed frame implies the body angular momentum $\boldsymbol{\Pi}$ appears constant in a frame rotating with the body angular velocity $\boldsymbol{\Omega} = \mathbb{I}^{-1}\boldsymbol{\Pi}$.

□

Proposition 2.1.2 (Conservation laws) *The dynamics of Equation (2.1.33) conserves both the square of the body angular momentum* $|\boldsymbol{\Pi}|^2$ *and the kinetic energy* $K = \boldsymbol{\Omega} \cdot \boldsymbol{\Pi}/2$.

Proof. These two conservation laws may be verified by direct calculations:

$$\begin{aligned} \frac{d|\boldsymbol{\Pi}|^2}{dt} &= 2\boldsymbol{\Pi} \cdot \frac{d\boldsymbol{\Pi}}{dt} = 2\boldsymbol{\Pi} \cdot \boldsymbol{\Pi} \times \boldsymbol{\Omega} = 0\,, \\ \frac{d(\boldsymbol{\Omega} \cdot \boldsymbol{\Pi})}{dt} &= 2\boldsymbol{\Omega} \cdot \frac{d\boldsymbol{\Pi}}{dt} = 2\boldsymbol{\Omega} \cdot \boldsymbol{\Pi} \times \boldsymbol{\Omega} = 0\,, \end{aligned}$$

where one uses the symmetry of the moment of inertia tensor in the second line. ■

Remark 2.1.16 (Reconstruction formula) Having found the evolution of $\boldsymbol{\Pi}(t)$ and thus $\boldsymbol{\Omega}(t)$ by solving (2.1.33), one may compute the net angle of rotation $O(t)$ in body coordinates from the skew-symmetric angular velocity matrix $\widehat{\Omega}$ in (2.1.27) and its defining relation,

$$\dot{O}(t) = O\widehat{\Omega}(t)\,.$$

Solving this linear differential equation with time-dependent coefficients yields the paths of rotations $O(t) \in SO(3)$. Having these, one may finally construct the trajectories in space taken by points $\mathbf{X}$ in the body $\mathcal{B}$ given by $\mathbf{x}(\mathbf{X}, t) = O(t)\mathbf{X} \in \mathbb{R}^3$. □

2.2 Lagrange

2.2.1 The principle of stationary action

Joseph-Louis Lagrange

In Lagrangian mechanics, a mechanical system in a ***configuration space*** with generalised coordinates and velocities,

$$q^a, \dot{q}^a, \quad a = 1, 2, \ldots, 3N,$$

is characterised by its Lagrangian $L(q(t), \dot{q}(t))$ – a smooth, real-valued function. The motion of a Lagrangian system is determined by the principle of stationary action, formulated using the operation of variational derivative.

Definition 2.2.1 (Variational derivative) *The* ***variational derivative*** *of a functional $S[q]$ is defined as its linearisation in an arbitrary direction δq in the configuration space. That is, $S[q]$ is defined as*

$$\delta S[q]; = \lim_{s \to 0} \frac{S[q + s\delta q] - S[q]}{s} = \frac{d}{ds}\Big|_{s=0} S[q + s\delta q] =: \Big\langle \frac{\delta S}{\delta q}, \delta q \Big\rangle ,$$

where the pairing $\langle \cdot, \cdot \rangle$ is obtained in the process of linearisation.

Theorem 2.2.1 (Principle of stationary action) *The Euler–Lagrange equations,*

$$\big[L \big]_{q^a} := \frac{d}{dt}\frac{\partial L}{\partial \dot{q}^a} - \frac{\partial L}{\partial q^a} = 0, \tag{2.2.1}$$

follow from stationarity of the **action integral**, *S, defined as the integral over a time interval* $t \in (t_1\,,\,t_2)$,

$$S := \int_{t_1}^{t_2} L(q,\dot{q})\,dt\,. \tag{2.2.2}$$

Then the ***principle of stationary action,***

$$\delta S = 0\,,$$

implies $[\,L\,]_{q^a} = 0$, *for variations* δq^a *that vanish at the endpoints in time.*

Proof. Applying the variational derivative in Definition 2.2.1 to the action integral in (2.2.2) yields

$$\begin{aligned}
\delta S[q] &= \int_{t_1}^{t_2} \left(\frac{\partial L}{\partial q^a}\,\delta q^a + \frac{\partial L}{\partial \dot{q}^a}\,\delta \dot{q}^a \right) dt \\
&= \int_{t_1}^{t_2} \left(\frac{\partial L}{\partial q^a} - \frac{d}{dt}\frac{\partial L}{\partial \dot{q}^a} \right) \delta q^a\,dt + \left[\frac{\partial L}{\partial \dot{q}^a}\,\delta q^a \right]_{t_1}^{t_2} \\
&=: -\int_{t_1}^{t_2} \big[\,L\,\big]_{q^a}\,\delta q^a\,dt\,.
\end{aligned} \tag{2.2.3}$$

Here one integrates by parts and in the last step applies the condition that the variations δq^a vanish at the endpoints in time. Because the variations δq^a are otherwise arbitrary, one concludes that the Euler–Lagrange Equations (2.2.1) are satisfied. ■

Remark 2.2.1 The principle of stationary action is sometimes also called ***Hamilton's principle.*** □

Example 2.2.1 (Simple mechanical systems) *The Lagrangian for the motion of a* ***simple mechanical system*** *is given in the separated form,*

$$L(q,\dot{q}) = \frac{m}{2}|\dot{q}|^2 - V(q) \quad \textit{with Euclidean norm} \quad |\dot{q}|^2 := \dot{q}^b\delta_{bc}\dot{q}^c\,.$$

The Lagrangian in this case has partial derivatives

$$\frac{\partial L}{\partial \dot{q}^a} = m\delta_{ac}\dot{q}^c \quad \textit{and} \quad \frac{\partial L}{\partial q^a} = -\frac{\partial V(q)}{\partial q^a}\,.$$

Consequently, its Euler–Lagrange equations $[\,L\,]_{q^a} = 0$ *are*

$$
\begin{aligned}
[\,L\,]_{q^a} &:= \frac{d}{dt}\frac{\partial L}{\partial \dot{q}^a} - \frac{\partial L}{\partial q^a} \\
&= m\delta_{ac}\ddot{q}^c + \frac{\partial V(q)}{\partial q^a} = 0 .
\end{aligned}
$$

This is Newton's law of acceleration for a potential force.

Example 2.2.2 (Geodesic motion in a Riemannian space) *The Lagrangian for the motion of a free particle in a Riemannian space is its kinetic energy with respect to the Riemannian metric,*

$$
L(q,\,\dot{q}) = \frac{1}{2}\dot{q}^b g_{bc}(q)\dot{q}^c \,.
$$

The Lagrangian in this case has partial derivatives

$$
\frac{\partial L}{\partial \dot{q}^a} = g_{ac}(q)\dot{q}^c \quad \textit{and} \quad \frac{\partial L}{\partial q^a} = \frac{1}{2}\frac{\partial g_{bc}(q)}{\partial q^a}\dot{q}^b\dot{q}^c \,.
$$

Consequently, its Euler–Lagrange equations $[\,L\,]_{q^a} = 0$ *are*

$$
\begin{aligned}
[\,L\,]_{q^a} &:= \frac{d}{dt}\frac{\partial L}{\partial \dot{q}^a} - \frac{\partial L}{\partial q^a} \\
&= g_{ae}(q)\ddot{q}^{\,e} + \frac{\partial g_{ae}(q)}{\partial q^b}\dot{q}^b\dot{q}^e - \frac{1}{2}\frac{\partial g_{be}(q)}{\partial q^a}\dot{q}^b\dot{q}^e = 0 \,.
\end{aligned}
$$

Symmetrising the middle term and contracting with co-metric g^{ca} *satisfying* $g^{ca}g_{ae} = \delta^c_e$ *yields*

$$
\ddot{q}^{\,c} + \Gamma^c_{be}(q)\dot{q}^b\dot{q}^e = 0 \,, \tag{2.2.4}
$$

where Γ^c_{be} *are the* ***Christoffel symbols****, given in terms of the metric by*

$$
\Gamma^c_{be}(q) = \frac{1}{2}g^{ca}\left[\frac{\partial g_{ae}(q)}{\partial q^b} + \frac{\partial g_{ab}(q)}{\partial q^e} - \frac{\partial g_{be}(q)}{\partial q^a}\right] . \tag{2.2.5}
$$

These Euler–Lagrange equations are the ***geodesic equations*** *of a free particle moving in a Riemannian space.*

2.3 Noether's theorem

2.3.1 Lie symmetries and conservation laws

Emmy Noether

Recall from Definition 1.2.3 that a Lie group depends smoothly on its parameters. (See Appendix B for more details.)

Definition 2.3.1 (Lie symmetry) *A smooth transformation of variables* $\{t, q\}$ *depending on a single parameter* s *defined by*

$$\{t, q\} \mapsto \{\, \bar{t}(t, q, s),\, \bar{q}(t, q, s)\}\,,$$

that leaves the action $S = \int L\,dt$ *invariant is called a* **Lie symmetry** *of the action.*

Theorem 2.3.1 (Noether's theorem) *Each Lie symmetry of the action for a Lagrangian system defined on a manifold* M *with Lagrangian* L *corresponds to a constant of the motion [No1918].*

Example 2.3.1 *Suppose the variation of the action in (2.2.3) vanishes* ($\delta S = 0$) *because of a Lie symmetry which does not preserve the endpoints. Then on solutions of the Euler–Lagrange equations, the endpoint term must vanish for another reason. For example, if the Lie symmetry leaves time invariant, so that*

$$\{t, q\} \mapsto \{\, t,\, \bar{q}(t, q, s)\}\,,$$

then the endpoint term must vanish,

$$\left[\frac{\partial L}{\partial \dot{q}}\delta q\right]_{t_1}^{t_2} = 0\,.$$

Hence, the quantity

$$A(q, \dot{q}, \delta q) = \frac{\partial L}{\partial \dot{q}^a}\,\delta q^a$$

is a ***constant of motion*** *for solutions of the Euler–Lagrange equations. In particular, if* $\delta q^a = c^a$ *for constants* c^a, $a = 1, \ldots, n$, *that is, for spatial translations in* n *dimensions, then the quantities* $\partial L / \partial \dot{q}^a$ *(the corresponding momentum components) are constants of motion.*

Remark 2.3.1 This result first appeared in Noether [No1918]. See, e.g., [Ol2000, SaCa1981] for good discussions of the history, framework and applications of Noether's theorem. We shall see in a moment that Lie symmetries that reparameterise time may also yield constants of motion. □

2.3.2 Infinitesimal transformations of a Lie group

Definition 2.3.2 (Infinitesimal Lie transformations)

Sophus Lie

Consider the Lie group of transformations

$$\{t, q\} \mapsto \{\bar{t}(t, q, s), \bar{q}(t, q, s)\},$$

and suppose the identity transformation is arranged to occur for $s = 0$. *The derivatives with respect to the group parameters* s *at the identity,*

$$\begin{aligned} \tau(t, q) &= \frac{d}{ds}\bigg|_{s=0} \bar{t}(t, q, s), \\ \xi^a(t, q) &= \frac{d}{ds}\bigg|_{s=0} \bar{q}^a(t, q, s), \end{aligned}$$

are called the ***infinitesimal transformations*** *of the action of a Lie group on the time and space variables.*

Thus, at linear order in a Taylor expansion in the group parameter s one has

$$\bar{t} = t + s\tau(t, q), \quad \bar{q}^a = q^a + s\xi^a(t, q), \tag{2.3.1}$$

where τ and ξ^a are functions of coordinates and time, but do not depend on velocities. Then, to first order in s the velocities of the transformed trajectories are computed as

$$\frac{d\bar{q}^a}{d\bar{t}} = \frac{\dot{q}^a + s\dot{\xi}^a}{1 + s\dot{\tau}} = \dot{q}^a + s(\dot{\xi}^a - \dot{q}^a\dot{\tau}), \tag{2.3.2}$$

where order $O(s^2)$ terms are neglected and one defines the total time derivatives

$$\dot{\tau} \equiv \frac{\partial \tau}{\partial t}(t,q) + \dot{q}^b \frac{\partial \tau}{\partial q^b}(t,q) \quad \text{and} \quad \dot{\xi}^a \equiv \frac{\partial \xi^a}{\partial t}(t,q) + \dot{q}^b \frac{\partial \xi^a}{\partial q^b}(t,q).$$

Remark 2.3.2 The result (2.3.2) for the infinitesimal transformation of the derivative when both the independent and dependent variables are transformed is the operation introduced by Lie of ***prolongation*** of the infinitesimal Lie group actions (2.3.1). Prolongation is a generalisation of the tangent lift explained in Definition A.3.3. □

We are now in a position to prove Noether's Theorem 2.3.1.

Proof. The variation of the action corresponding to the Lie symmetry with infinitesimal transformations (2.3.1) is

$$\begin{aligned}
\delta S &= \int_{t_1}^{t_2} \left(\frac{\partial L}{\partial t}\delta t + \frac{\partial L}{\partial q^a}\delta q^a + \frac{\partial L}{\partial \dot{q}^a}\delta \dot{q}^a + L\frac{d\delta t}{dt} \right) dt \\
&= \int_{t_1}^{t_2} \left(\frac{\partial L}{\partial t}\tau + \frac{\partial L}{\partial q^a}\xi^a + \frac{\partial L}{\partial \dot{q}^a}(\dot{\xi}^a - \dot{q}^a\dot{\tau}) + L\dot{\tau} \right) dt \\
&= \int_{t_1}^{t_2} \left(\frac{\partial L}{\partial q^a} - \frac{d}{dt}\frac{\partial L}{\partial \dot{q}^a} \right)(\xi^a - \dot{q}^a\tau) + \frac{d}{dt}\left(L\tau + \frac{\partial L}{\partial \dot{q}^a}(\xi^a - \dot{q}^a\tau) \right) dt \\
&= \int_{t_1}^{t_2} [\,L\,]_{q^a}(\xi^a - \dot{q}^a\tau) + \frac{d}{dt}\left[\frac{\partial L}{\partial \dot{q}^a}\,\xi^a - \left(\frac{\partial L}{\partial \dot{q}^a}\,\dot{q}^a - L \right)\tau \right] dt\,.
\end{aligned}$$

Thus, stationarity $\delta S = 0$ and the Euler–Lagrange equations $[\,L\,]_{q^a} = 0$ imply

$$0 = \left[\frac{\partial L}{\partial \dot{q}^a}\,\xi^a - \left(\frac{\partial L}{\partial \dot{q}^a}\,\dot{q}^a - L \right)\tau \right]_{t_1}^{t_2},$$

so that the quantity

$$\begin{aligned} C(t,q,\dot{q}) &= \frac{\partial L}{\partial \dot{q}^a}\,\xi^a - \Big(\frac{\partial L}{\partial \dot{q}^a}\,\dot{q}^a - L\Big)\tau &\quad (2.3.3)\\ &\equiv \langle\, p\,,\,\delta q\,\rangle - H\,\delta t &\quad (2.3.4) \end{aligned}$$

has the same value at every time along the solution path. That is, $C(t,q,\dot{q})$ is a constant of the motion. ■

Remark 2.3.3 The abbreviated notation in Equation (2.3.4) for δq and δt is standard. If δt is absent and δq is a constant (corresponding to translations in space in a certain direction) then $\delta S = 0$ implies that the ***linear momentum*** $p = \partial L/\partial \dot{q}$ in that direction is conserved for solutions of the Euler–Lagrange equations $[\,L\,]_q = 0$. □

Exercise. Show that conservation of energy results from Noether's theorem if, in Hamilton's principle, the variations are chosen as

$$\delta q(t) = \frac{d}{ds}\Big|_{s=0} q(\,\bar{t}(t,q,s))\,,$$

corresponding to symmetry of the Lagrangian under space-time-dependent transformations of time along a given curve, so that $t \to \bar{t}(t,q,s)$ and $q(t) \to q(\,\bar{t}(t,q,s))$ with $\bar{t}(t,q,0) = t$. ★

Answer. Under reparameterisations of time along the curve

$$q(t) \to q(\,\bar{t}(t,q,s))\,,$$

the action $S = \int_{t_1}^{t_2} L(q,\dot{q})\,dt$ changes infinitesimally according to

$$\delta S = \Big[\Big(L(q,\dot{q}) - \frac{\partial L}{\partial \dot{q}}\,\dot{q}\Big)\delta t\Big]_{t_1}^{t_2}\,,$$

with variations in position and time defined by

$$\begin{aligned} \delta q(t) &= \frac{d}{ds}\bigg|_{s=0} q(\bar{t}(t,q,s)) \\ &= \dot{q}(t)\delta t \quad \text{and} \quad \delta t = \frac{d\bar{t}(t,q,s)}{ds}\bigg|_{s=0} . \end{aligned}$$

For translations in time, δt is a constant and stationarity of the action $\delta S = 0$ implies that the energy

$$E(t,q,\dot{q}) \equiv \frac{\partial L}{\partial \dot{q}^a}\,\dot{q}^a - L \tag{2.3.5}$$

is a constant of motion along solutions of the Euler–Lagrange equations. ▲

Exercise. (Infinitesimal Galilean transformations) From their finite transformations in Definition 1.2.1, compute the infinitesimal transformations of the Galilean group under composition of first rotations, then boosts, then translations in space and time, in the case when they act on a velocity-space-time point $(\dot{\mathbf{q}}, \mathbf{q}, t)$. ★

Answer. The composition of translations $g_1(\mathbf{q}_0(s), t_0(s))$, Galilean boosts $g_3(\mathbf{v}_0(s))$ and rotations $g_2(O(s))$ acting on a velocity-space-time point $(\dot{\mathbf{q}}, \mathbf{q}, t)$ is given by

$$g_1 g_3 g_2(\dot{\mathbf{q}}, \mathbf{q}, t) = \Big(O(s)\dot{\mathbf{q}} + \mathbf{v}_0(s), O(s)\mathbf{q} + t\mathbf{v}_0(s) + \mathbf{q}_0(s), t + t_0(s)\Big) .$$

One computes the infinitesimal transformations as

$$\begin{aligned} \tau &= \frac{dt}{ds}\bigg|_{s=0} = t_0 , \\ \boldsymbol{\xi} &= \frac{d\mathbf{q}}{ds}\bigg|_{s=0} = \mathbf{q}_0 + \mathbf{v}_0 t + \boldsymbol{\Xi} \times \mathbf{q} , \\ \dot{\boldsymbol{\xi}} - \dot{\mathbf{q}}\dot{\tau} &= \frac{d\dot{\mathbf{q}}}{ds}\bigg|_{s=0} = \mathbf{v}_0 + \boldsymbol{\Xi} \times \dot{\mathbf{q}} . \end{aligned}$$

The infinitesimal velocity transformation may also be computed from Equation (2.3.2).

Consequently, the infinitesimal transformation by the Galilean group of a function $F(t, \mathbf{q}, \dot{\mathbf{q}})$ is given by operation of the following vector field, obtained as the first term in a Taylor series,

$$\frac{d}{ds}\bigg|_{s=0} F(t(s), \mathbf{q}(s), \dot{\mathbf{q}}(s)) \qquad (2.3.6)$$

$$= t_0 \frac{\partial F}{\partial t} + \left(\mathbf{q}_0 + \mathbf{v}_0 t + \boldsymbol{\Xi} \times \mathbf{q}\right) \cdot \frac{\partial F}{\partial \mathbf{q}}$$

$$+\left(\mathbf{v}_0 + \boldsymbol{\Xi} \times \dot{\mathbf{q}}\right) \cdot \frac{\partial F}{\partial \dot{\mathbf{q}}} .$$

▲

Exercise. (Galilean Lie symmetries) Since the Galilean transformations form a Lie group, one may expect them to be a source of Lie symmetries of the Lagrangian in Hamilton's principle. Compute the corresponding Noether conservation laws. ★

Answer. As we have already seen, symmetries under space and time translations imply conservation of linear momentum and energy, respectively. Likewise, symmetry under rotations implies conservation of angular momentum.

Suppose the Lagrangian in Hamilton's principle (2.2.2) is invariant under S^1 rotations about a spatial direction $\boldsymbol{\Xi}$. The infinitesimal transformation of such a rotation is $\delta\mathbf{q} = \boldsymbol{\Xi} \times \mathbf{q}$. In this case, the conserved Noether quantity (2.3.4) is

$$J^{\Xi}(\mathbf{q}, \dot{\mathbf{q}}) = \frac{\partial L}{\partial \dot{\mathbf{q}}} \cdot \delta\mathbf{q} = \mathbf{q} \times \frac{\partial L}{\partial \dot{\mathbf{q}}} \cdot \boldsymbol{\Xi} , \qquad (2.3.7)$$

which is the angular momentum about the Ξ-axis.

Finally, symmetry under Galilean boosts implies vanishing of total momentum for a system of N particles, labelled by an index $j = 1, 2, \ldots, N$. The last statement may be proved explicitly from Noether's theorem and the infinitesimal Galilean boost transformations, as follows:

$$\begin{aligned}
\delta S &= \int_{t_1}^{t_2} \left(\frac{\partial L}{\partial t}\delta t + \sum_j \left(\frac{\partial L}{\partial \mathbf{q}^j} \cdot \delta \mathbf{q}^j + \frac{\partial L}{\partial \dot{\mathbf{q}}^j} \cdot \delta \dot{\mathbf{q}}^j \right) + L\frac{d\delta t}{dt} \right) dt \\
&= \int_{t_1}^{t_2} \sum_j \left(\frac{\partial L}{\partial \mathbf{q}^j} \cdot \mathbf{v}_0 t + \frac{\partial L}{\partial \dot{\mathbf{q}}^j} \cdot \mathbf{v}_0 \right) dt \\
&= \int_{t_1}^{t_2} \sum_j \left(\frac{\partial L}{\partial \mathbf{q}^j} - \frac{d}{dt}\frac{\partial L}{\partial \dot{\mathbf{q}}^j} \right) \cdot \mathbf{v}_0 t + \frac{d}{dt} \sum_j \left(\frac{\partial L}{\partial \dot{\mathbf{q}}^j} \cdot \mathbf{v}_0 t \right) dt \\
&= \int_{t_1}^{t_2} - \sum_j [\, L \,]_{\mathbf{q}^j} \cdot \mathbf{v}_0 t \, dt + \left[\left(\sum_j \frac{\partial L}{\partial \dot{\mathbf{q}}^j} \right) \cdot \mathbf{v}_0 t \right]_{t_1}^{t_2} .
\end{aligned}$$

So the Euler–Lagrange equations $[\, L \,]_{\mathbf{q}^j} = 0$ and stationarity $\delta S = 0$ for any time $t \in [t_1, t_2]$ together imply

$$\mathbf{P}t \cdot \mathbf{v}_0 = \text{constant}, \quad \text{with} \quad \mathbf{P} := \sum_j \frac{\partial L}{\partial \dot{\mathbf{q}}^j} .$$

Let's explore the meaning of this result for a system of N particles with constant total mass $M = \sum_j m_j$. The centre of mass of the system is defined as

$$\mathbf{Q}_{CM} := M^{-1} \sum_j m_j \mathbf{q}^j .$$

For a simple mechanical system with Lagrangian

$$L(\mathbf{q}, \dot{\mathbf{q}}) = \frac{1}{2}\Big(\sum_j m_j |\dot{\mathbf{q}}^j|^2 \Big) - V(\{\mathbf{q}^j\}) ,$$

one finds that

$$\begin{aligned}\mathbf{P} &= \sum_j \frac{\partial L}{\partial \dot{\mathbf{q}}^j} = \sum_j m_j \dot{\mathbf{q}}^j \\ &= \frac{d}{dt}\sum_j m_j \mathbf{q}^j(t) =: \frac{d}{dt}\left(M\mathbf{Q}_{CM}\right).\end{aligned}$$

Hence, for such a system Noether's theorem yields

$$\mathbf{P}t - M\mathbf{Q}_{CM} = 0,$$

provided the action principle is *also* invariant under spatial translations, so that $\mathbf{P}$ is *also* a constant of the motion. Consequently, the motion of a simple mechanical system of particles may always be taken as being relative to a fixed centre of mass. ▲

2.4 Lagrangian form of rigid-body motion

In the absence of external torques, Euler's equations in (2.1.33) for rigid-body motion in principal axis coordinates are

$$\begin{aligned}I_1\dot{\Omega}_1 &= (I_2 - I_3)\Omega_2\Omega_3, \\ I_2\dot{\Omega}_2 &= (I_3 - I_1)\Omega_3\Omega_1, \\ I_3\dot{\Omega}_3 &= (I_1 - I_2)\Omega_1\Omega_2,\end{aligned} \tag{2.4.1}$$

or, equivalently,

$$\mathbb{I}\dot{\mathbf{\Omega}} = \mathbb{I}\mathbf{\Omega} \times \mathbf{\Omega}\,, \tag{2.4.2}$$

where $\mathbf{\Omega} = (\Omega_1, \Omega_2, \Omega_3)$ is the body angular velocity vector and I_1, I_2, I_3 are the moments of inertia in the principal axis frame of the rigid body. We ask whether these equations may be expressed using Hamilton's principle on $\mathbb{R}^3$. For this, we will need to define the variational derivative of a functional $S[(\mathbf{\Omega}]$.

Definition 2.4.1 (Variational derivative) *The variational derivative of a functional $S[(\boldsymbol{\Omega}]$ is defined as its linearisation in an arbitrary direction $\delta\boldsymbol{\Omega}$ in the vector space of body angular velocities. That is,*

$$\delta S[\boldsymbol{\Omega}] := \lim_{s\to 0} \frac{S[\boldsymbol{\Omega}+s\delta\boldsymbol{\Omega}]-S[\boldsymbol{\Omega}]}{s} = \frac{d}{ds}\bigg|_{s=0} S[\boldsymbol{\Omega}+s\delta\boldsymbol{\Omega}] =: \Big\langle \frac{\delta S}{\delta \boldsymbol{\Omega}}\,, \delta\boldsymbol{\Omega}\Big\rangle,$$

where the new pairing, also denoted as $\langle\,\cdot\,,\,\cdot\,\rangle$, is between the space of body angular velocities and its dual, the space of body angular momenta.

Theorem 2.4.1 (Euler's rigid-body equations) *Euler's rigid-body equations are equivalent to Hamilton's principle*

$$\delta S(\boldsymbol{\Omega}) = \delta \int_a^b l(\boldsymbol{\Omega})\,dt = 0, \tag{2.4.3}$$

in which the Lagrangian $l(\boldsymbol{\Omega})$ appearing in the **action integral** *$S(\boldsymbol{\Omega}) = \int_a^b l(\boldsymbol{\Omega})\,dt$ is given by the kinetic energy in principal axis coordinates,*

$$l(\boldsymbol{\Omega}) = \frac{1}{2}\langle \mathbb{I}\boldsymbol{\Omega}, \boldsymbol{\Omega}\rangle = \frac{1}{2}\,\mathbb{I}\boldsymbol{\Omega}\cdot\boldsymbol{\Omega} = \frac{1}{2}\left(I_1\Omega_1^2 + I_2\Omega_2^2 + I_3\Omega_3^2\right), \tag{2.4.4}$$

and variations of $\boldsymbol{\Omega}$ are restricted to be of the form

$$\delta\boldsymbol{\Omega} = \dot{\boldsymbol{\Xi}} + \boldsymbol{\Omega}\times\boldsymbol{\Xi}\,, \tag{2.4.5}$$

where $\boldsymbol{\Xi}(t)$ is a curve in $\mathbb{R}^3$ that vanishes at the endpoints in time.

Proof. Since $l(\boldsymbol{\Omega}) = \frac{1}{2}\langle \mathbb{I}\boldsymbol{\Omega}, \boldsymbol{\Omega}\rangle$, and $\mathbb{I}$ is symmetric, one obtains

$$\begin{aligned}
\delta \int_a^b l(\boldsymbol{\Omega})\,dt &= \int_a^b \Big\langle \mathbb{I}\boldsymbol{\Omega}, \delta\boldsymbol{\Omega}\Big\rangle\,dt \\
&= \int_a^b \Big\langle \mathbb{I}\boldsymbol{\Omega}, \dot{\boldsymbol{\Xi}} + \boldsymbol{\Omega}\times\boldsymbol{\Xi}\Big\rangle\,dt \\
&= \int_a^b \left[\Big\langle -\frac{d}{dt}\mathbb{I}\boldsymbol{\Omega}, \boldsymbol{\Xi}\Big\rangle + \Big\langle \mathbb{I}\boldsymbol{\Omega}, \boldsymbol{\Omega}\times\boldsymbol{\Xi}\Big\rangle\right] dt \\
&= \int_a^b \Big\langle -\frac{d}{dt}\mathbb{I}\boldsymbol{\Omega} + \mathbb{I}\boldsymbol{\Omega}\times\boldsymbol{\Omega}, \boldsymbol{\Xi}\Big\rangle\,dt + \Big\langle \mathbb{I}\boldsymbol{\Omega}, \boldsymbol{\Xi}\Big\rangle\Big|_{t_a}^{t_b},
\end{aligned}$$

upon integrating by parts. The last term vanishes, upon using the endpoint conditions,

$$\Xi(a) = 0 = \Xi(b)\,.$$

Since Ξ is otherwise arbitrary, (2.4.3) is equivalent to

$$-\frac{d}{dt}(\mathbb{I}\boldsymbol{\Omega}) + \mathbb{I}\boldsymbol{\Omega} \times \boldsymbol{\Omega} = 0,$$

which recovers Euler's Equations (2.4.1) in vector form. ■

Proposition 2.4.1 (Derivation of the restricted variation) *The restricted variation in (2.4.5) arises via the following steps:*

(i) Vary the definition of body angular velocity, $\widehat{\Omega} = O^{-1}\dot{O}$.

(ii) Take the time derivative of the variation, $\widehat{\Xi} = O^{-1}O'$.

(iii) Use the equality of cross derivatives, $O^{\cdot\prime} = d^2O/dtds = O^{\prime\cdot}$.

(iv) Apply the hat map.

Proof. One computes directly that

$$\begin{aligned}
\widehat{\Omega}' &= (O^{-1}\dot{O})' = -O^{-1}O'O^{-1}\dot{O} + O^{-1}O^{\cdot\prime} = -\widehat{\Xi}\widehat{\Omega} + O^{-1}O^{\cdot\prime}, \\
\widehat{\Xi}^{\cdot} &= (O^{-1}O')^{\cdot} = -O^{-1}\dot{O}O^{-1}O' + O^{-1}O^{\prime\cdot} = -\widehat{\Omega}\widehat{\Xi} + O^{-1}O^{\prime\cdot}.
\end{aligned}$$

On taking the difference, the cross derivatives cancel and one finds a variational formula equivalent to (2.4.5),

$$\widehat{\Omega}' - \widehat{\Xi}^{\cdot} = \left[\widehat{\Omega}\,,\,\widehat{\Xi}\right] \quad \text{with} \quad [\widehat{\Omega}\,,\,\widehat{\Xi}] := \widehat{\Omega}\widehat{\Xi} - \widehat{\Xi}\widehat{\Omega}\,. \tag{2.4.6}$$

Under the bracket relation (2.1.12) for the hat map, this equation recovers the vector relation (2.4.5) in the form

$$\boldsymbol{\Omega}' - \dot{\boldsymbol{\Xi}} = \boldsymbol{\Omega} \times \boldsymbol{\Xi}\,. \tag{2.4.7}$$

Thus, Euler's equations for the rigid body in $T\mathbb{R}^3$,

$$\mathbb{I}\dot{\boldsymbol{\Omega}} = \mathbb{I}\boldsymbol{\Omega} \times \boldsymbol{\Omega}\,, \tag{2.4.8}$$

do follow from the variational principle (2.4.3) with variations of the form (2.4.5) derived from the definition of body angular velocity $\widehat{\Omega}$.

■

Remark 2.4.1 The body angular velocity is expressed in terms of the spatial angular velocity by $\boldsymbol{\Omega}(t) = O^{-1}(t)\boldsymbol{\omega}(t)$. Consequently, the kinetic energy Lagrangian in (2.4.4) transforms as

$$l(\boldsymbol{\Omega}) = \frac{1}{2}\boldsymbol{\Omega}\cdot\mathbb{I}\boldsymbol{\Omega} = \frac{1}{2}\boldsymbol{\omega}\cdot\mathbb{I}_{space}(t)\boldsymbol{\omega} =: l_{space}(\boldsymbol{\omega})\,,$$

where

$$\mathbb{I}_{space}(t) = O(t)\mathbb{I}O^{-1}(t)\,,$$

as in (2.1.31). □

Exercise. Show that Hamilton's principle for the action

$$S(\boldsymbol{\omega}) = \int_a^b l_{space}(\boldsymbol{\omega})\,dt$$

yields conservation of spatial angular momentum

$$\boldsymbol{\pi} = \mathbb{I}_{space}(t)\boldsymbol{\omega}(t)\,.$$

Hint: First derive $\delta\mathbb{I}_{space} = [\xi, \mathbb{I}_{space}]$ with right-invariant $\xi = \delta O O^{-1}$. ★

Exercise. (Noether's theorem for the rigid body) What conservation law does Theorem 2.3.1 (Noether's theorem) imply for the rigid-body Equations (2.4.2)?

Hint: Transform the endpoint terms arising on integrating the variation δS by parts in the proof of Theorem 2.4.1 into the spatial representation by setting $\boldsymbol{\Xi} = O^{-1}(t)\boldsymbol{\Gamma}$ and $\boldsymbol{\Omega} = O^{-1}(t)\boldsymbol{\omega}$. ★

Remark 2.4.2 (Reconstruction of $O(t) \in SO(3)$) The Euler solution is expressed in terms of the time-dependent angular velocity vector in the body, $\boldsymbol{\Omega}$. The body angular velocity vector $\boldsymbol{\Omega}(t)$ yields the tangent vector $\dot{O}(t) \in T_{O(t)}SO(3)$ along the integral curve in the rotation group $O(t) \in SO(3)$ by the relation

$$\dot{O}(t) = O(t)\widehat{\Omega}(t)\,, \tag{2.4.9}$$

where the left-invariant skew-symmetric 3×3 matrix $\widehat{\Omega}$ is defined by the hat map (2.1.27)

$$(O^{-1}\dot{O})_{jk} = \widehat{\Omega}_{jk} = -\,\Omega_i\epsilon_{ijk}\,. \tag{2.4.10}$$

Equation (2.4.9) is the ***reconstruction formula*** for $O(t) \in SO(3)$.

Once the time dependence of $\boldsymbol{\Omega}(t)$ and hence $\widehat{\Omega}(t)$ is determined from the Euler equations, solving formula (2.4.9) as a linear differential equation with time-dependent coefficients yields the integral curve $O(t) \in SO(3)$ for the orientation of the rigid body. □

2.4.1 Hamilton–Pontryagin constrained variations

Formula (2.4.6) for the variation $\widehat{\Omega}$ of the skew-symmetric matrix

$$\widehat{\Omega} = O^{-1}\dot{O}$$

may be imposed as a constraint in Hamilton's principle and thereby provide a variational derivation of Euler's Equations (2.1.33) for rigid-body motion in principal axis coordinates. This constraint is incorporated into the matrix Euler equations, as follows.

Proposition 2.4.2 (Matrix Euler equations) *Euler's rigid-body equation may be written in matrix form as*

$$\frac{d\Pi}{dt} = -\left[\,\widehat{\Omega}\,,\,\Pi\,\right] \quad \textit{with} \quad \Pi = \mathbb{I}\widehat{\Omega} = \frac{\delta l}{\delta\widehat{\Omega}}\,, \tag{2.4.11}$$

for the Lagrangian $l(\widehat{\Omega})$ given by

$$l = \frac{1}{2}\Big\langle\,\mathbb{I}\widehat{\Omega}\,,\,\widehat{\Omega}\,\Big\rangle\,. \tag{2.4.12}$$

Here, the bracket

$$\left[\,\widehat{\Omega}\,,\,\Pi\,\right] := \widehat{\Omega}\Pi - \Pi\widehat{\Omega} \tag{2.4.13}$$

denotes the commutator and $\langle\,\cdot\,,\,\cdot\,\rangle$ denotes the **trace pairing**, *e.g.,*

$$\Big\langle\,\Pi\,,\,\widehat{\Omega}\,\Big\rangle =: \frac{1}{2}\,\mathrm{trace}\left(\Pi^T\,\widehat{\Omega}\right). \tag{2.4.14}$$

Remark 2.4.3 Note that the symmetric part of Π does not contribute in the pairing and if set equal to zero initially, it will remain zero. □

Proposition 2.4.3 (Constrained variational principle) *The matrix Euler Equations (2.4.11) are equivalent to stationarity $\delta S = 0$ of the following* **constrained action:**

$$\begin{aligned} S(\widehat{\Omega}, O, \dot{O}, \Pi) &= \int_a^b l(\widehat{\Omega}, O, \dot{O}, \Pi)\,dt \qquad (2.4.15) \\ &= \int_a^b \Big[l(\widehat{\Omega}) + \langle\,\Pi\,,\,(O^{-1}\dot{O} - \widehat{\Omega}\,)\rangle\Big]\,dt\,. \end{aligned}$$

Remark 2.4.4 The integrand of the constrained action in (2.4.15) is similar to the formula for the Legendre transform, but its functional dependence is different. This variational approach is related to the classic ***Hamilton–Pontryagin principle*** for control theory in [YoMa2006]. It is also used in [BoMa2009] to develop algorithms for geometric numerical integrations of rotating motion. □

Proof. The variations of S in formula (2.4.15) are given by

$$\begin{aligned} \delta S &= \int_a^b \Big\{ \Big\langle \frac{\delta l}{\delta \widehat{\Omega}} - \Pi \,,\, \delta \widehat{\Omega} \Big\rangle \\ &\qquad + \Big\langle \delta \Pi \,,\, (O^{-1}\dot{O} - \Omega) \Big\rangle + \Big\langle \Pi \,,\, \delta (O^{-1}\dot{O}) \Big\rangle \Big\} \, dt \,, \end{aligned}$$

where

$$\delta(O^{-1}\dot{O}) = \widehat{\Xi}^{\cdot} + [\widehat{\Omega} \,,\, \widehat{\Xi}] \,, \tag{2.4.16}$$

and $\widehat{\Xi} = (O^{-1}\delta O)$ from Equation (2.4.6).

Substituting for $\delta(O^{-1}\dot{O})$ into the last term of δS produces

$$\begin{aligned} \int_a^b \Big\langle \Pi \,,\, \delta(O^{-1}\dot{O}) \Big\rangle \, dt &= \int_a^b \Big\langle \Pi \,,\, \widehat{\Xi}^{\cdot} + [\widehat{\Omega} \,,\, \widehat{\Xi}] \Big\rangle \, dt \\ &= \int_a^b \Big\langle - \Pi^{\cdot} - [\widehat{\Omega} \,,\, \Pi] \,,\, \widehat{\Xi} \Big\rangle \, dt \\ &\qquad + \Big\langle \Pi \,,\, \widehat{\Xi} \Big\rangle \Big|_a^b \,, \end{aligned} \tag{2.4.17}$$

where one uses the cyclic properties of the trace operation for matrices,

$$\text{trace}\Big(\Pi^T \, \widehat{\Xi} \, \widehat{\Omega} \Big) = \text{trace}\Big(\widehat{\Omega} \, \Pi^T \, \widehat{\Xi} \Big) \,. \tag{2.4.18}$$

Thus, stationarity of the Hamilton–Pontryagin variational principle for vanishing endpoint conditions $\widehat{\Xi}(a) = 0 = \widehat{\Xi}(b)$ implies the following set of equations:

$$\frac{\delta l}{\delta \widehat{\Omega}} = \Pi \,, \quad O^{-1}\dot{O} = \widehat{\Omega} \,, \quad \Pi^{\cdot} = -[\widehat{\Omega} \,,\, \Pi] \,. \tag{2.4.19}$$

■

Remark 2.4.5 (Interpreting the formulas in (2.4.19)) The first formula in (2.4.19) defines the angular momentum matrix Π as the ***fibre derivative*** of the Lagrangian with respect to the angular velocity matrix $\widehat{\Omega}$. The second formula is the reconstruction formula (2.4.9) for the solution curve $O(t) \in SO(3)$, given the solution $\widehat{\Omega}(t) = O^{-1}\dot{O}$. And the third formula is Euler's equation for rigid-body motion in matrix form. □

We transform the endpoint terms in (2.4.20), arising on integrating the variation δS by parts in the proof of Theorem 2.4.3, into the spatial representation by setting $\widehat{\Xi}(t) =: O(t)\,\widehat{\xi}\,O^{-1}(t)$ and $\widehat{\Pi}(t) =: O(t)\widehat{\pi}(t)O^{-1}(t)$, as follows:

$$\Big\langle \Pi\,,\,\widehat{\Xi}\Big\rangle = \text{trace}\left(\Pi^T\,\widehat{\Xi}\right) = \text{trace}\left(\pi^T\,\widehat{\xi}\right) = \Big\langle \pi\,,\,\widehat{\xi}\Big\rangle . \qquad (2.4.20)$$

Thus, the vanishing of both endpoints for a *constant* infinitesimal spatial rotation $\widehat{\xi} = (\delta O O^{-1}) = const$ implies

$$\pi(a) = \pi(b)\,. \qquad (2.4.21)$$

This is Noether's theorem for the rigid body.

Theorem 2.4.2 (Noether's theorem for the rigid body) *Invariance of the constrained Hamilton–Pontryagin action under spatial rotations implies conservation of spatial angular momentum,*

$$\pi = O^{-1}(t)\Pi(t)O(t) =: \text{Ad}^*_{O^{-1}(t)}\Pi(t). \qquad (2.4.22)$$

Proof.

$$\begin{aligned}
\frac{d}{dt}\Big\langle \pi\,,\,\widehat{\xi}\Big\rangle &= \frac{d}{dt}\Big\langle O^{-1}\Pi O\,,\,\widehat{\xi}\Big\rangle = \frac{d}{dt}\,\text{trace}\left(\Pi^T\,O^{-1}\widehat{\xi}O\right) \\
&= \Big\langle \frac{d}{dt}\Pi + [\,\widehat{\Omega}\,,\,\Pi\,]\,,\,O^{-1}\widehat{\xi}O\Big\rangle = 0 \\
&=: \Big\langle \frac{d}{dt}\Pi - \text{ad}^*_{\widehat{\Omega}}\Pi\,,\,\text{Ad}_{O^{-1}}\widehat{\xi}\Big\rangle, \\
\frac{d}{dt}\Big\langle \text{Ad}^*_{O^{-1}}\Pi\,,\,\widehat{\xi}\Big\rangle &= \Big\langle \text{Ad}^*_{O^{-1}}\Big(\frac{d}{dt}\Pi - \text{ad}^*_{\widehat{\Omega}}\Pi\Big)\,,\,\widehat{\xi}\Big\rangle. \qquad (2.4.23)
\end{aligned}$$

The proof of Noether's theorem for the rigid body is already on the second line. However, the last line gives a general result. ■

Remark 2.4.6 The proof of Noether's theorem for the rigid body when the constrained Hamilton–Pontryagin action is invariant under spatial rotations also proves a general result in Equation (2.4.23), with $\widehat{\Omega} = O^{-1}\dot{O}$ for a Lie group O, that

$$\frac{d}{dt}\Big(\mathrm{Ad}^*_{O^{-1}}\Pi\Big) = \mathrm{Ad}^*_{O^{-1}}\Big(\frac{d}{dt}\Pi - \mathrm{ad}^*_{\widehat{\Omega}}\Pi\Big). \tag{2.4.24}$$

This equation will be useful in the remainder of the text. In particular, it provides the solution of a differential equation defined on the dual of a Lie algebra. Namely, for a Lie group O with Lie algebra $\mathfrak{o}$, the equation for $\Pi \in \mathfrak{o}^*$ and $\widehat{\Omega} = O^{-1}\dot{O} \in \mathfrak{o}$

$$\frac{d}{dt}\Pi - \mathrm{ad}^*_{\widehat{\Omega}}\Pi = 0 \quad \text{has solution} \quad \Pi(t) = \mathrm{Ad}^*_{O(t)}\pi\,, \tag{2.4.25}$$

in which the constant $\pi \in \mathfrak{o}^*$ is obtained from the initial conditions. □

2.4.2 Manakov's formulation of the $SO(n)$ rigid body

Proposition 2.4.4 (Manakov [Man1976]) *Euler's equations for a rigid body on $SO(n)$ take the matrix commutator form,*

$$\frac{dM}{dt} = [\,M\,,\,\Omega\,] \quad \textit{with} \quad M = \mathbb{A}\Omega + \Omega\mathbb{A}\,, \tag{2.4.26}$$

where the $n \times n$ matrices M, Ω are skew-symmetric (forgoing superfluous hats) and $\mathbb{A}$ is symmetric.

Proof. Manakov's commutator form of the $SO(n)$ rigid-body Equations (2.4.26) follows as the Euler–Lagrange equations for Hamilton's principle $\delta S = 0$ with $S = \int l\,dt$ for the Lagrangian

$$l = -\frac{1}{2}\mathrm{tr}(\Omega\mathbb{A}\Omega)\,,$$

where $\Omega = O^{-1}\dot{O} \in so(n)$ and the $n \times n$ matrix $\mathbb{A}$ is symmetric. Taking matrix variations in Hamilton's principle yields

$$\delta S = -\frac{1}{2}\int_a^b \mathrm{tr}\big(\delta\Omega\,(\mathbb{A}\Omega + \Omega\mathbb{A})\big)\,dt = -\frac{1}{2}\int_a^b \mathrm{tr}\big(\delta\Omega\,M\big)\,dt\,,$$

after cyclically permuting the order of matrix multiplication under the trace and substituting $M := \mathbb{A}\Omega + \Omega\mathbb{A}$. Using the variational formula (2.4.16) for $\delta\Omega$ now leads to

$$\delta S = -\frac{1}{2}\int_a^b \operatorname{tr}\big((\Xi^{\cdot} + \Omega\Xi - \Xi\Omega)M\big)\, dt\,.$$

Integrating by parts and permuting under the trace then yields the equation

$$\delta S = \frac{1}{2}\int_a^b \operatorname{tr}\big(\Xi\,(\,\dot{M} + \Omega M - M\Omega\,)\big)\, dt\,.$$

Finally, invoking stationarity for arbitrary Ξ implies the commutator form (2.4.26). ■

2.4.3 Matrix Euler–Poincaré equations

Manakov's commutator form of the rigid-body equations recalls much earlier work by Poincaré [Po1901], who also noticed that the matrix commutator form of Euler's rigid-body equations suggests an additional mathematical structure going back to Lie's theory of groups of transformations depending continuously on parameters. In particular, Poincaré [Po1901] remarked that the commutator form of Euler's rigid-body equations would make sense for any Lie algebra, not just for $so(3)$. The proof of Manakov's commutator form (2.4.26) by Hamilton's principle is essentially the same as Poincaré's proof in [Po1901], which is translated into English in Appendix D.

Theorem 2.4.3 (Matrix Euler–Poincaré equations) *The Euler–Lagrange equations for Hamilton's principle $\delta S = 0$ with $S = \int l(\Omega)\, dt$ may be expressed in matrix commutator form,*

$$\frac{dM}{dt} = [\,M\,,\,\Omega\,] \quad with \quad M = \frac{\delta l}{\delta\Omega}\,, \tag{2.4.27}$$

for any Lagrangian $l(\Omega)$, where $\Omega = g^{-1}\dot{g} \in \mathfrak{g}$ and $\mathfrak{g}$ is the matrix Lie algebra of any matrix Lie group G.

Proof. The proof here is the same as the proof of Manakov's commutator formula via Hamilton's principle, modulo replacing $O^{-1}\dot{O} \in so(n)$ with $g^{-1}\dot{g} \in \mathfrak{g}$. ■

Remark 2.4.7 Poincaré's observation leading to the matrix Euler–Poincaré Equation (2.4.27) was reported in two pages with no references [Po1901]. The proof above shows that the matrix Euler–Poincaré equations possess a natural variational principle. Note that if $\Omega = g^{-1}\dot{g} \in \mathfrak{g}$, then $M = \delta l/\delta\Omega \in \mathfrak{g}^*$, where the dual is defined in terms of the matrix trace pairing. □

Exercise. Retrace the proof of the variational principle for the Euler–Poincaré equation, replacing the left-invariant quantity $g^{-1}\dot{g}$ with the right-invariant quantity $\dot{g}g^{-1}$. ★

2.4.4 An isospectral eigenvalue problem for the $SO(n)$ rigid body

The solution of the $SO(n)$ rigid-body dynamics

$$\frac{dM}{dt} = [\,M\,,\,\Omega\,] \quad \text{with} \quad M = \mathbb{A}\Omega + \Omega\mathbb{A}\,,$$

for the evolution of the $n \times n$ skew-symmetric matrices M, Ω, with constant symmetric $\mathbb{A}$, is given by a similarity transformation (later to be identified as coadjoint motion),

$$M(t) = O(t)^{-1}M(0)O(t) =: \mathrm{Ad}^*_{O(t)}M(0)\,,$$

with $O(t) \in SO(n)$ and $\Omega := O^{-1}\dot{O}(t)$. Consequently, the evolution of $M(t)$ is ***isospectral***. This means that

- The initial eigenvalues of the matrix $M(0)$ are preserved by the motion; that is, $d\lambda/dt = 0$ in

$$M(t)\psi(t) = \lambda\psi(t)\,,$$

provided its eigenvectors $\psi \in \mathbb{R}^n$ evolve according to

$$\psi(t) = O(t)^{-1}\psi(0)\,.$$

The proof of this statement follows from the corresponding property of similarity transformations.

- Its matrix invariants are preserved:

$$\frac{d}{dt}\mathrm{tr}(M - \lambda \mathrm{Id})^K = 0\,,$$

for every non-negative integer power K.

This is clear because the invariants of the matrix M may be expressed in terms of its eigenvalues; but these are invariant under a similarity transformation.

Proposition 2.4.5 *Isospectrality allows the quadratic rigid-body dynamics (2.4.26) on $SO(n)$ to be rephrased as a system of two coupled linear equations: the eigenvalue problem for M and an evolution equation for its eigenvectors ψ, as follows:*

$$M\psi = \lambda\psi \quad \textit{and} \quad \dot{\psi} = -\,\Omega\psi\,, \quad \textit{with} \quad \Omega = O^{-1}\dot{O}(t)\,.$$

Proof. Applying isospectrality in the time derivative of the first equation yields

$$(\,\dot{M} + [\,\Omega, M\,]\,)\psi + (M - \lambda \mathrm{Id})(\dot{\psi} + \Omega\psi) = 0\,.$$

Now substitute the second equation to recover (2.4.26). ∎

2.4.5 Manakov's integration of the $SO(n)$ rigid body

Manakov [Man1976] observed that Equations (2.4.26) may be "deformed" into

$$\frac{d}{dt}(M + \lambda A) = [(M + \lambda A), (\Omega + \lambda B)]\,, \tag{2.4.28}$$

where A, B are also $n \times n$ matrices and λ is a scalar constant parameter. For these deformed rigid-body equations on $SO(n)$ to hold for any value of λ, the coefficient of each power must vanish.

- The coefficent of λ^2 is

$$0 = [A, B]\,.$$

  Therefore, A and B must commute. For this, let them be constant and diagonal:

$$A_{ij} = \mathrm{diag}(a_i)\delta_{ij}\,, \quad B_{ij} = \mathrm{diag}(b_i)\delta_{ij} \quad \text{(no sum)}.$$

- The coefficent of λ is

$$0 = \frac{dA}{dt} = [A, \Omega] + [M, B]\,.$$

  Therefore, by antisymmetry of M and Ω,

$$(a_i - a_j)\Omega_{ij} = (b_i - b_j)M_{ij}\,,$$

  which implies that

$$\Omega_{ij} = \frac{b_i - b_j}{a_i - a_j} M_{ij} \quad \text{(no sum)}.$$

  Hence, angular velocity Ω is a linear function of angular momentum, M.

- Finally, the coefficent of λ^0 recovers the Euler equation

$$\frac{dM}{dt} = [M, \Omega]\,,$$

but now with the restriction that the moments of inertia are of the form

$$\Omega_{ij} = \frac{b_i - b_j}{a_i - a_j} M_{ij} \quad \text{(no sum).}$$

This relation turns out to possess only five free parameters for $n = 4$.

Under these conditions, Manakov's deformation of the $SO(n)$ rigid-body equation into the commutator form (2.4.28) implies for every non-negative integer power K that

$$\frac{d}{dt}(M + \lambda A)^K = [(M + \lambda A)^K, (\Omega + \lambda B)]\,.$$

Since the commutator is antisymmetric, its trace vanishes and K conservation laws emerge, as

$$\frac{d}{dt}\mathrm{tr}(M + \lambda A)^K = 0\,,$$

after commuting the trace operation with the time derivative. Consequently,

$$\mathrm{tr}(M + \lambda A)^K = \mathrm{constant}\,,$$

for each power of λ. That is, all the coefficients of each power of λ are constant in time for the $SO(n)$ rigid body. Manakov [Man1976] proved that these constants of motion are sufficient to completely determine the solution for $n = 4$.

Remark 2.4.8 This result generalises considerably. For example, Manakov's method determines the solution for all the algebraically solvable rigid bodies on $SO(n)$. The moments of inertia of these bodies possess only $2n - 3$ parameters. (Recall that in Manakov's case for $SO(4)$ the moment of inertia possesses only five parameters.) □

Exercise. Try computing the constants of motion $\mathrm{tr}(M + \lambda A)^K$ for the values $K = 2, 3, 4$.

Hint: Keep in mind that M is a skew-symmetric matrix, $M^T = -M$, so the trace of the product of any diagonal matrix times an odd power of M vanishes. ★

Answer. The traces of the powers $\mathrm{trace}(M + \lambda A)^n$ are given by

$$
\begin{aligned}
\boxed{n = 2} &: \mathrm{tr}\, M^2 + 2\lambda \mathrm{tr}(AM) + \lambda^2 \mathrm{tr}\, A^2 \,, \\
\boxed{n = 3} &: \mathrm{tr}\, M^3 + 3\lambda \mathrm{tr}(AM^2) + 3\lambda^2 \mathrm{tr}\, A^2 M + \lambda^3 \mathrm{tr}\, A^3 \,, \\
\boxed{n = 4} &: \mathrm{tr}\, M^4 + 4\lambda \mathrm{tr}(AM^3) \\
&\qquad + \lambda^2 (2\mathrm{tr}\, A^2 M^2 + 4\mathrm{tr}\, AMAM) \\
&\qquad + \lambda^3 \mathrm{tr}\, A^3 M + \lambda^4 \mathrm{tr}\, A^4 \,.
\end{aligned}
$$

The number of conserved quantities for $n = 2, 3, 4$ are, respectively, one ($C_2 = \mathrm{tr}\, M^2$), one ($C_3 = \mathrm{tr}\, AM^2$) and two ($C_4 = \mathrm{tr}\, M^4$ and $I_4 = 2\mathrm{tr}\, A^2 M^2 + 4\mathrm{tr}\, AMAM$). ▲

Exercise. How do the Euler equations look on $so(4)^*$ as a matrix equation? Is there an analogue of the hat map for $so(4)$?

Hint: The Lie algebra $so(4)$ is locally isomorphic to $so(3) \times so(3)$. ★

2.5 Hamilton

The ***Legendre transform*** of the Lagrangian (2.4.4) in the variational principle (2.4.3) for Euler's rigid-body dynamics (2.4.8) on $\mathbb{R}^3$ will reveal its well-known Hamiltonian formulation.

Definition 2.5.1 (Legendre transformation) *The Legendre transformation* $\mathbb{F}l : \mathbb{R}^3 \to \mathbb{R}^{3*} \simeq \mathbb{R}^3$ *is defined by the* ***fibre derivative,***

$$\mathbb{F}l(\Omega) = \frac{\delta l}{\delta \Omega} = \Pi \,.$$

The Legendre transformation defines the ***body angular momentum*** by the variations of the rigid body's reduced Lagrangian with respect to the body angular velocity. For the Lagrangian in (2.4.3), the $\mathbb{R}^3$ components of the body angular momentum are

$$\Pi_i = I_i \Omega_i = \frac{\partial l}{\partial \Omega_i}, \quad i = 1, 2, 3. \tag{2.5.1}$$

Remark 2.5.1 This is also how body angular momentum was defined in the Newtonian setting. See Definition 2.1.10. □

Exercise. Express the Lagrangian (2.4.4) in terms of the matrices $O(t)$ and $\dot{O}(t)$. Show that this Lagrangian is left-invariant under $(O, \dot{O}) \mapsto (RO, R\dot{O})$ for any orthogonal matrix $R^T = R^{-1}$. Compute the Euler–Lagrange equations for this Lagrangian in geodesic form (2.2.4). ★

Exercise. Compute the Legendre transformation and pass to the canonical Hamiltonian formulation using the Lagrangian $l(\Omega) = L(O\,, \dot{O})$ and the following definitions of the canonical momentum and Hamiltonian,

$$P = \frac{\partial L(O\,, \dot{O})}{\partial \dot{O}} \quad \text{and} \quad H(P\,, O) = \left\langle P\,, \dot{O} \right\rangle - L(O\,, \dot{O})\,,$$

in combination with the chain rule for $\Omega = O^{-1}\dot{O}$. ★

2.5.1 Hamiltonian form of rigid-body motion

Definition 2.5.2 (Dynamical systems in Hamiltonian form) *A dynamical system on a manifold M*

$$\dot{\mathbf{x}}(t) = \mathbf{F}(\mathbf{x}), \quad \mathbf{x} \in M,$$

is said to be in ***Hamiltonian form****, if it can be expressed as*

$$\dot{\mathbf{x}}(t) = \{\mathbf{x}, H\}, \quad \textit{for} \quad H : M \mapsto \mathbb{R},$$

in terms of a Poisson bracket operation $\{\cdot\,,\,\cdot\}$ *among smooth real functions* $\mathcal{F}(M) : M \mapsto \mathbb{R}$ *on the manifold* M,

$$\{\cdot\,,\,\cdot\} : \mathcal{F}(M) \times \mathcal{F}(M) \mapsto \mathcal{F}(M),$$

so that $\dot{F} = \{F\,,\,H\}$ *for any* $F \in \mathcal{F}(M)$.

Definition 2.5.3 (Poisson bracket) *A* ***Poisson bracket operation*** $\{\cdot\,,\,\cdot\}$ *is defined as possessing the following properties:*

- *It is* ***bilinear.***
- *It is* ***skew-symmetric,*** $\{F\,,\,H\} = -\{H\,,\,F\}$.
- *It satisfies the* ***Leibniz rule*** *(product rule),*

$$\{FG\,,\,H\} = \{F\,,\,H\}G + F\{G\,,\,H\},$$

for the product of any two functions F and G on M.

- *It satisfies the* ***Jacobi identity,***

$$\{F\,,\,\{G\,,\,H\}\} + \{G\,,\,\{H\,,\,F\}\} + \{H\,,\,\{F\,,\,G\}\} = 0, \quad (2.5.2)$$

for any three functions F, G and H on M.

Remark 2.5.2 This definition of a Poisson bracket does not require it to be the standard canonical bracket in position q and conjugate momentum p, although it does include that case as well. □

2.5.2 Lie–Poisson Hamiltonian rigid-body dynamics

Let

$$h(\mathbf{\Pi}) := \langle \Pi, \Omega \rangle - l(\mathbf{\Omega}), \tag{2.5.3}$$

where the pairing $\langle \cdot, \cdot \rangle : \mathbb{R}^{3*} \times \mathbb{R}^3 \to \mathbb{R}$ denotes the vector dot product on $\mathbb{R}^3$,

$$\langle \Pi, \Omega \rangle := \mathbf{\Pi} \cdot \mathbf{\Omega}, \tag{2.5.4}$$

in which indices are suppressed within the brackets. Hence, one finds the expected expression for the rigid-body Hamiltonian

$$h = \frac{1}{2} \langle \Pi, \mathbb{I}^{-1} \Pi \rangle = \frac{1}{2} \mathbf{\Pi} \cdot \mathbb{I}^{-1} \mathbf{\Pi} := \frac{\Pi_1^2}{2I_1} + \frac{\Pi_2^2}{2I_2} + \frac{\Pi_3^2}{2I_3}. \tag{2.5.5}$$

The Legendre transform $\mathbb{F}l$ for this case is a diffeomorphism, so one may solve for

$$\frac{\partial h}{\partial \mathbf{\Pi}} = \mathbf{\Omega} + \left\langle \Pi, \frac{\partial \Omega}{\partial \mathbf{\Pi}} \right\rangle - \left\langle \frac{\partial l}{\partial \Omega}, \frac{\partial \Omega}{\partial \mathbf{\Pi}} \right\rangle = \mathbf{\Omega} \tag{2.5.6}$$

upon using the definition of angular momentum $\Pi = \partial l / \partial \Omega$ in (2.5.1). In $\mathbb{R}^3$ coordinates, the relation (2.5.6) expresses the body angular velocity as the derivative of the reduced Hamiltonian with respect to the body angular momentum, namely,

$$\frac{\partial h}{\partial \mathbf{\Pi}} = \mathbf{\Omega}. \tag{2.5.7}$$

Hence, the reduced Euler–Lagrange equations for l may be expressed equivalently in angular momentum vector components in $\mathbb{R}^3$ and Hamiltonian h as

$$\frac{d}{dt}(\mathbb{I}\mathbf{\Omega}) = \mathbb{I}\mathbf{\Omega} \times \mathbf{\Omega} \Longleftrightarrow \dot{\mathbf{\Pi}} = \mathbf{\Pi} \times \frac{\partial h}{\partial \mathbf{\Pi}} := \{\mathbf{\Pi}, h\}.$$

This expression suggests we introduce the following ***rigid-body Poisson bracket*** on functions of the $\mathbf{\Pi}$'s:

$$\{f, h\}(\mathbf{\Pi}) := -\mathbf{\Pi} \cdot \left(\frac{\partial f}{\partial \mathbf{\Pi}} \times \frac{\partial h}{\partial \mathbf{\Pi}} \right). \tag{2.5.8}$$

For the Hamiltonian (2.5.5), one checks that the Euler equations in terms of the rigid-body angular momenta,

$$\begin{aligned}\dot{\Pi}_1 &= \left(\frac{1}{I_3} - \frac{1}{I_2}\right)\Pi_2\Pi_3\,,\\ \dot{\Pi}_2 &= \left(\frac{1}{I_1} - \frac{1}{I_3}\right)\Pi_3\Pi_1\,,\\ \dot{\Pi}_3 &= \left(\frac{1}{I_2} - \frac{1}{I_1}\right)\Pi_1\Pi_2\,,\end{aligned} \tag{2.5.9}$$

that is, the equations

$$\dot{\mathbf{\Pi}} = \mathbf{\Pi} \times \mathbb{I}^{-1}\mathbf{\Pi}\,, \tag{2.5.10}$$

are equivalent to

$$\dot{f} = \{f, h\}\,, \quad \text{with} \quad f = \mathbf{\Pi}\,.$$

2.5.3 Lie–Poisson bracket

The Poisson bracket proposed in (2.5.8) is an example of a ***Lie–Poisson bracket***.

It satisfies the defining relations of a Poisson bracket for a number of reasons, not least because it is the hat map to $\mathbb{R}^3$ of the following bracket defined by the general form in Equation (2.4.23) in terms of the $so(3)^* \times so(3)$ pairing $\langle\,\cdot\,,\,\cdot\,\rangle$ in Equation (2.4.20). Namely,

$$\begin{aligned}\frac{dF}{dt} &= \left\langle \frac{d}{dt}\Pi,\, \frac{\partial F}{\partial \Pi}\right\rangle = \left\langle \mathrm{ad}^*_{\widehat{\Omega}}\Pi,\, \frac{\partial F}{\partial \Pi}\right\rangle\\ &= \left\langle \Pi,\, \mathrm{ad}_{\widehat{\Omega}}\frac{\partial F}{\partial \Pi}\right\rangle = \left\langle \Pi,\, \left[\widehat{\Omega},\, \frac{\partial F}{\partial \Pi}\right]\right\rangle\\ &= -\left\langle \Pi,\, \left[\frac{\partial F}{\partial \Pi},\, \frac{\partial H}{\partial \Pi}\right]\right\rangle,\end{aligned} \tag{2.5.11}$$

where we have used the equation corresponding to (2.5.7) under the inverse of the hat map

$$\widehat{\Omega} = \frac{\partial H}{\partial \Pi}$$

and applied antisymmetry of the matrix commutator. Writing Equation (2.5.11) as

$$\frac{dF}{dt} = -\left\langle \Pi, \left[\frac{\partial F}{\partial \Pi}, \frac{\partial H}{\partial \Pi}\right]\right\rangle =: \Big\{F, H\Big\} \tag{2.5.12}$$

defines the Lie–Poisson bracket $\{\cdot, \cdot\}$ on smooth functions $(F, H) : so(3)^* \to \mathbb{R}$. This bracket satisfies the defining relations of a Poisson bracket because it is a linear functional of the commutator product of skew-symmetric matrices, which is bilinear, skew-symmetric, satisfies the Leibniz rule (because of the partial derivatives) and also satisfies the Jacobi identity.

These Lie–Poisson brackets may be written in tabular form as

$\{\Pi_i, \Pi_j\} =$ (2.5.13)

$\{\cdot,\cdot\}$	Π_1	Π_2	Π_3
Π_1	0	$-\Pi_3$	Π_2
Π_2	Π_3	0	$-\Pi_1$
Π_3	$-\Pi_2$	Π_1	0

or, in index notation,

$$\{\Pi_i\, , \, \Pi_j\} = -\epsilon_{ijk}\Pi_k = \widehat{\Pi}_{ij}\, . \tag{2.5.14}$$

Remark 2.5.3 The Lie–Poisson bracket in the form (2.5.12) would apply to any Lie algebra. This Lie–Poisson Hamiltonian form of the rigid-body dynamics substantiates Poincaré's observation in [Po1901] that the corresponding equations could have been written on the dual of any Lie algebra by using the ad^* operation for that Lie algebra. □

The corresponding Poisson bracket in (2.5.8) in $\mathbb{R}^3$-vector form also satisfies the defining relations of a Poisson bracket because it is an example of a ***Nambu bracket***, to be discussed next.

2.5.4 Nambu's $\mathbb{R}^3$ Poisson bracket

The rigid-body Poisson bracket (2.5.8) is a special case of the Poisson bracket for functions of $\mathbf{x} \in \mathbb{R}^3$ introduced in [Na1973],

$$\{f, h\} = -\nabla c \cdot \nabla f \times \nabla h\, . \tag{2.5.15}$$

This bracket generates the motion

$$\dot{\mathbf{x}} = \{\mathbf{x}, h\} = \nabla c \times \nabla h \,. \tag{2.5.16}$$

For this bracket the motion takes place along the intersections of level surfaces of the functions c and h in $\mathbb{R}^3$. In particular, for the rigid body, the motion takes place along intersections of angular momentum spheres $c = |\mathbf{x}|^2/2$ and energy ellipsoids $h = \mathbf{x} \cdot \mathbb{I}\mathbf{x}$. (See the cover illustration of [MaRa1994].)

Exercise. Consider the Nambu $\mathbb{R}^3$ bracket

$$\{f, h\} = -\nabla c \cdot \nabla f \times \nabla h \,. \tag{2.5.17}$$

Let $c = \mathbf{x}^T \cdot \mathbb{C}\mathbf{x}/2$ be a quadratic form on $\mathbb{R}^3$, and let $\mathbb{C}$ be the associated symmetric 3×3 matrix. Show by direct computation that this Nambu bracket satisfies the Jacobi identity. ★

Exercise. Find the general conditions on the function $\mathbf{c}(\mathbf{x})$ so that the $\mathbb{R}^3$ bracket

$$\{f, h\} = -\nabla c \cdot \nabla f \times \nabla h$$

satisfies the defining properties of a Poisson bracket. Is this $\mathbb{R}^3$ bracket also a derivation satisfying the Leibniz relation for a product of functions on $\mathbb{R}^3$? If so, why? ★

Answer. The bilinear skew-symmetric Nambu $\mathbb{R}^3$ bracket yields the divergenceless vector field

$$X_{c,h} = \{\,\cdot\,, h\} = (\nabla c \times \nabla h) \cdot \nabla \quad \text{with} \quad \operatorname{div}(\nabla c \times \nabla h) = 0 \,.$$

Divergenceless vector fields are derivative operators that satisfy the Leibniz product rule. They also satisfy the Jacobi identity for any choice of C^2 functions c and h. Hence, the Nambu $\mathbb{R}^3$ bracket is a bilinear skew-symmetric operation satisfying the defining properties of a Poisson bracket. ▲

Theorem 2.5.1 (Jacobi identity) *The Nambu $\mathbb{R}^3$ bracket (2.5.17) satisfies the Jacobi identity.*

Proof. The isomorphism $X_H = \{\,\cdot\,,\, H\}$ between the Lie algebra of divergenceless vector fields and functions under the $\mathbb{R}^3$ bracket is the key to proving this theorem. The Lie derivative among vector fields is identified with the Nambu bracket by

$$\mathcal{L}_{X_G} X_H = [X_G, X_H] = -X_{\{G,H\}}\,.$$

Repeating the Lie derivative produces

$$\mathcal{L}_{X_F}(\mathcal{L}_{X_G} X_H) = [X_F, [X_G, X_H]] = X_{\{F,\{G,H\}\}}\,.$$

The result follows because both the left- and right-hand sides in this equation satisfy the Jacobi identity. ■

Exercise. How is the $\mathbb{R}^3$ bracket related to the canonical Poisson bracket?

Hint: Restrict to level surfaces of the function $c(\mathbf{x})$. ★

Exercise. (Casimirs of the $\mathbb{R}^3$ bracket) The Casimirs (or distinguished functions, as Lie called them) of a Poisson bracket satisfy

$$\{c, h\}(\mathbf{x}) = 0\,, \quad \text{for all } h(\mathbf{x})\,.$$

Suppose the function $c(\mathbf{x})$ is chosen so that the $\mathbb{R}^3$ bracket (2.5.15) defines a proper Poisson bracket. What are the Casimirs for the $\mathbb{R}^3$ bracket (2.5.15)? Why? ★

Exercise. (Geometric interpretation of Nambu motion)

- Show that the Nambu motion equation (2.5.16)

$$\dot{\mathbf{x}} = \{\mathbf{x}, h\} = \nabla c \times \nabla h$$

  for the $\mathbb{R}^3$ bracket (2.5.15) is invariant under a certain linear combination of the functions c and h. Interpret this invariance geometrically.

- Show that the rigid-body equations for

$$\mathbb{I} = \text{diag}(1,\ 1/2,\ 1/3)$$

  may be interpreted as intersections in $\mathbb{R}^3$ of the spheres $x_1^2 + x_2^2 + x_3^2 = \text{constant}$ and the hyperbolic cylinders $x_1^2 - x_3^2 = \text{constant}$. See [HoMa1991] for more discussions of this geometric interpretation of solutions under the $\mathbb{R}^3$ bracket.

- A special case of the equations for three-wave interactions is [AlLuMaRo1998]

$$\dot{x}_1 = s_1\gamma_1 x_2 x_3\,,\ \dot{x}_2 = s_2\gamma_2 x_3 x_1\,,\ \dot{x}_3 = s_3\gamma_3 x_1 x_2\,,$$

  for a set of constants $\gamma_1 + \gamma_2 + \gamma_3 = 0$ and signs $s_1, s_2, s_3 = \pm 1$. Write these equations as a Nambu motion equation on $\mathbb{R}^3$ of the form (2.5.16). Interpret their solutions geometrically as intersections of level surfaces of quadratic functions for various values and signs of the γ's. ★

2.5.5 Clebsch variational principle for the rigid body

Proposition 2.5.1 (Clebsch variational principle) *The Euler rigid-body Equations (2.4.2) on $T\mathbb{R}^3$ are equivalent to the* ***constrained variational principle,***

$$\delta S(\mathbf{\Omega}, \mathbf{Q}, \dot{\mathbf{Q}}; \mathbf{P}) = \delta \int_a^b l(\mathbf{\Omega}, \mathbf{Q}, \dot{\mathbf{Q}}; \mathbf{P})\, dt = 0, \tag{2.5.18}$$

for a ***constrained action integral***

$$\begin{aligned} S(\mathbf{\Omega}, \mathbf{Q}, \dot{\mathbf{Q}}) &= \int_a^b l(\mathbf{\Omega}, \mathbf{Q}, \dot{\mathbf{Q}})\, dt \\ &= \int_a^b \frac{1}{2}\mathbf{\Omega} \cdot \mathbb{I}\mathbf{\Omega} + \mathbf{P} \cdot \left(\dot{\mathbf{Q}} + \mathbf{\Omega} \times \mathbf{Q}\right) dt\,. \end{aligned} \tag{2.5.19}$$

Remark 2.5.4 (Reconstruction as constraint)

- The first term in the Lagrangian (2.5.19),

$$l(\mathbf{\Omega}) = \frac{1}{2}(I_1\Omega_1^2 + I_2\Omega_2^2 + I_3\Omega_3^2) = \frac{1}{2}\mathbf{\Omega}^T\mathbb{I}\mathbf{\Omega}\,, \tag{2.5.20}$$

  is again the (rotational) kinetic energy of the rigid body in (2.1.23).

- The second term in the Lagrangian (2.5.19) introduces the Lagrange multiplier **P** which imposes the constraint

$$\dot{\mathbf{Q}} + \mathbf{\Omega} \times \mathbf{Q} = 0\,.$$

  This *reconstruction formula* has the solution

$$\mathbf{Q}(t) = O^{-1}(t)\mathbf{Q}(0)\,,$$

  which satisfies

$$\begin{aligned} \dot{\mathbf{Q}}(t) &= -(O^{-1}\dot{O})O^{-1}(t)\mathbf{Q}(0) \\ &= -\hat{\Omega}(t)\mathbf{Q}(t) = -\,\mathbf{\Omega}(t) \times \mathbf{Q}(t)\,. \end{aligned} \tag{2.5.21}$$

□

Proof. The variations of S are given by

$$\begin{aligned}\delta S &= \int_a^b \left(\frac{\delta l}{\delta \boldsymbol{\Omega}} \cdot \delta \boldsymbol{\Omega} + \frac{\delta l}{\delta \mathbf{P}} \cdot \delta \mathbf{P} + \frac{\delta l}{\delta \mathbf{Q}} \cdot \delta \mathbf{Q} \right) dt \\ &= \int_a^b \Big[\Big(\mathbb{I}\boldsymbol{\Omega} - \mathbf{P} \times \mathbf{Q} \Big) \cdot \delta \boldsymbol{\Omega} \\ &\qquad + \delta \mathbf{P} \cdot \Big(\dot{\mathbf{Q}} + \boldsymbol{\Omega} \times \mathbf{Q} \Big) - \delta \mathbf{Q} \cdot \Big(\dot{\mathbf{P}} + \boldsymbol{\Omega} \times \mathbf{P} \Big) \Big] dt \,.\end{aligned}$$

Thus, stationarity of this ***implicit variational principle*** implies the following set of equations:

$$\mathbb{I}\boldsymbol{\Omega} = \mathbf{P} \times \mathbf{Q} \,, \quad \dot{\mathbf{Q}} = -\,\boldsymbol{\Omega} \times \mathbf{Q} \,, \quad \dot{\mathbf{P}} = -\,\boldsymbol{\Omega} \times \mathbf{P} \,. \qquad (2.5.22)$$

These ***symmetric equations*** for the rigid body first appeared in the theory of optimal control of rigid bodies [BlCrMaRa1998]. Euler's form of the rigid-body equations emerges from these, upon elimination of $\mathbf{Q}$ and $\mathbf{P}$, as

$$\begin{aligned}\mathbb{I}\dot{\boldsymbol{\Omega}} &= \dot{\mathbf{P}} \times \mathbf{Q} + \mathbf{P} \times \dot{\mathbf{Q}} \\ &= \mathbf{Q} \times (\boldsymbol{\Omega} \times \mathbf{P}) + \mathbf{P} \times (\mathbf{Q} \times \boldsymbol{\Omega}) \\ &= -\,\boldsymbol{\Omega} \times (\mathbf{P} \times \mathbf{Q}) = -\,\boldsymbol{\Omega} \times \mathbb{I}\boldsymbol{\Omega} \,,\end{aligned}$$

which are Euler's equations for the rigid body in $T\mathbb{R}^3$. ■

Remark 2.5.5 The Clebsch variational principle for the rigid body is a natural approach in developing geometric algorithms for numerical integrations of rotating motion. Geometric integrators for rotations are derived using the Clebsch approach in [CoHo2007]. □

Remark 2.5.6 The Clebsch approach is also a natural path across to the Hamiltonian formulation of the rigid-body equations. This becomes clear in the course of the following exercise. □

Exercise. Given that the canonical Poisson brackets in Hamilton's approach are

$$\{Q_i, P_j\} = \delta_{ij} \quad \text{and} \quad \{Q_i, Q_j\} = 0 = \{P_i, P_j\} \,,$$

what are the Poisson brackets for $\mathbf{\Pi} = \mathbf{P} \times \mathbf{Q} \in \mathbb{R}^3$ in (2.5.22)? Show that these Poisson brackets recover the rigid-body Poisson bracket (2.5.8). ★

Answer. The components of the angular momentum $\mathbf{\Pi} = \mathbb{I}\mathbf{\Omega}$ in (2.5.22) are

$$\Pi_a = \epsilon_{abc} P_b Q_c \,,$$

and their canonical Poisson brackets are (noting the similarity with the hat map)

$$\{\Pi_a, \Pi_i\} = \{\epsilon_{abc} P_b Q_c \,,\, \epsilon_{ijk} P_j Q_k\} = -\,\epsilon_{ail}\Pi_l \,.$$

Consequently, the derivative property of the canonical Poisson bracket yields

$$\{f, h\}(\mathbf{\Pi}) = \frac{\partial f}{\partial \Pi_a}\{\Pi_a, \Pi_i\}\frac{\partial h}{\partial \Pi_b} = -\,\epsilon_{abc}\Pi_c \frac{\partial f}{\partial \Pi_a}\frac{\partial h}{\partial \Pi_b} \,, \tag{2.5.23}$$

which is indeed the Lie–Poisson bracket in (2.5.8) on functions of the $\mathbf{\Pi}$'s. The correspondence with the hat map noted above shows that this Poisson bracket satisfies the Jacobi identity as a result of the Jacobi identity for the vector cross product on $\mathbb{R}^3$. ▲

Remark 2.5.7 This exercise proves that the map $T^*\mathbb{R}^3 \to \mathbb{R}^3$ given by $\mathbf{\Pi} = \mathbf{P} \times \mathbf{Q} \in \mathbb{R}^3$ in (2.5.22) is Poisson. That is, the map takes Poisson brackets on one manifold into Poisson brackets on another manifold. Later we will recognise such an occurrence as one of the properties of a ***momentum map***. □

Exercise. The Euler–Lagrange equations in matrix commutator form of Manakov's formulation of the rigid body on $SO(n)$ are

$$\frac{dM}{dt} = [\, M \,, \Omega \,] \,,$$

where the $n \times n$ matrices M, Ω are skew-symmetric. Show that these equations may be derived from Hamilton's principle $\delta S = 0$ with constrained action integral

$$S(\Omega, Q, P) = \int_a^b l(\Omega) + \mathrm{tr}\Big(P^T \left(\dot{Q} - Q\Omega \right) \Big) \, dt \,,$$

for which $M = \delta l / \delta \Omega = P^T Q - Q^T P$ and $Q, P \in SO(n)$ satisfy the following symmetric equations reminiscent of those in (2.5.22),

$$\dot{Q} = Q\Omega \quad \text{and} \quad \dot{P} = P\Omega \,, \tag{2.5.24}$$

as a result of the constraints. ★

Exercise. Write Manakov's deformation of the rigid-body Equations (2.4.28) in the symmetric form (2.5.24). ★

2.5.6 Rotating motion with potential energy

Manakov's method for showing the integrability of the n-dimensional rigid body illustrates the conditions necessary to prove isospectral integrability for any Lie–Poisson system. For example, consider the problem of a rigid body in a quadratic potential, first studied in [Bo1985].

The Lagrangian of an arbitrary rigid body rotating about a fixed point at the origin of spatial coordinates $x \in \mathbb{R}^n$ in a field with a quadratic potential

$$\phi(x) = \frac{1}{2}\mathrm{tr}\big(x^T \mathbb{S}_0 x\big)$$

is defined in the body coordinates by the difference between its kinetic and potential energies in the form

$$l = \underbrace{\frac{1}{2}\mathrm{tr}(\Omega^T \mathbb{A}\Omega)}_{\text{kinetic}} - \underbrace{\frac{1}{2}\mathrm{tr}(\mathbb{S}\mathbb{A})}_{\text{potential}} .$$

Here, $\Omega(t) = O^{-1}(t)\dot{O}(t) \in so(n)$, the $n \times n$ constant matrices $\mathbb{A}$ and $\mathbb{S}_0$ are symmetric, and $\mathbb{S}(t) = O^{-1}(t)\mathbb{S}_0 O(t)$.

The reduced Euler–Lagrange equations for this Lagrangian are computed by taking matrix variations in its Hamilton's principle $\delta S = 0$ with $S = \int l \, dt$, to find

$$\delta S = \frac{1}{2}\int_a^b \mathrm{tr}(\delta\Omega \, M)\, dt + \frac{1}{2}\int_a^b \mathrm{tr}\big(\Xi\,[\,\mathbb{S}\,,\,\mathbb{A}\,]\big)\, dt ,$$

with matrix commutator $[\mathbb{S}, \mathbb{A}] := \mathbb{S}\mathbb{A} - \mathbb{A}\mathbb{S}$, variation $\Xi := O^{-1}\delta O \in so(n)$ and variational derivative $M := \partial l/\partial\Omega = \mathbb{A}\Omega + \Omega\mathbb{A}$.

Integrating by parts, invoking homogeneous endpoint conditions, then rearranging as in the proof of Proposition 2.4.2 and using the variational relation (2.4.16), rewritten here as

$$\delta\Omega = \frac{d\Xi}{dt} + [\,\Omega\,,\,\Xi\,],$$

finally yields the following formula for the variation,

$$\delta S = -\frac{1}{2}\int_a^b \mathrm{tr}\bigg(\Big(\frac{dM}{dt} - [\,M\,,\,\Omega\,] - [\,\mathbb{S}\,,\,\mathbb{A}\,]\Big)\,\Xi\bigg)\, dt .$$

Hence, Hamilton's principle for $\delta S = 0$ with arbitrary Ξ implies an equation for the evolution of M given by

$$\frac{dM}{dt} = [\,M\,,\,\Omega\,] + [\,\mathbb{S}(t)\,,\,\mathbb{A}\,] . \tag{2.5.25}$$

A differential equation for $\mathbb{S}(t)$ follows from the time derivative of its definition $\mathbb{S}(t) := O^{-1}(t)\mathbb{S}_0 O(t)$, as

$$\frac{d\mathbb{S}}{dt} = \big[\,\mathbb{S}\,,\,\Omega\,\big]\,. \tag{2.5.26}$$

The last two equations constitute a closed dynamical system for $M(t)$ and $\mathbb{S}(t)$, with initial conditions specified by the values of $\Omega(0)$ and $\mathbb{S}(0) = \mathbb{S}_0$ for $O(0) = \mathrm{Id}$ at time $t = 0$.

Following Manakov's idea [Man1976], these equations may be combined into a commutator of polynomials,

$$\frac{d}{dt}\big(\mathbb{S} + \lambda M + \lambda^2\mathbb{A}^2\big) = \big[\,\mathbb{S} + \lambda M + \lambda^2\mathbb{A}^2\,,\,\Omega + \lambda\mathbb{A}\,\big]\,. \tag{2.5.27}$$

The commutator form (2.5.27) implies for every non-negative integer power K that

$$\frac{d}{dt}(\mathbb{S} + \lambda M + \lambda^2\mathbb{A}^2)^K = [(\mathbb{S} + \lambda M + \lambda^2\mathbb{A}^2)^K, (\Omega + \lambda\mathbb{A})]\,.$$

Since the commutator is antisymmetric, its trace vanishes and K conservation laws emerge, as

$$\frac{d}{dt}\mathrm{tr}(\mathbb{S} + \lambda M + \lambda^2\mathbb{A}^2)^K = 0\,,$$

after commuting the trace operation with the time derivative. Consequently,

$$\mathrm{tr}(\mathbb{S} + \lambda M + \lambda^2\mathbb{A}^2)^K = \text{constant}\,, \tag{2.5.28}$$

for each power of λ. That is, all the coefficients of each power of λ are constant in time for the motion of a rigid body in a quadratic field.

Exercise. Show that the Hamiltonian formulation of this system is Lie–Poisson, with Hamiltonian function

$$H(M, \mathbb{S}) = \frac{1}{2}\mathrm{tr}(\Omega^T M) + \frac{1}{2}\mathrm{tr}\big(\mathbb{S}, \mathbb{A}\,\big)\,.$$

Determine the Lie algebra involved. ★

Exercise. Explicitly compute the conservation laws in (2.5.28) for $n = 4$. ★

Exercise. What is the dimension of the generic solution of the system of equations (2.5.25) and (2.5.26)? That is, what is the sum of the dimensions of $so(n)$ and the symmetric $n \times n$ matrices, minus the number of conservation laws? ★

Exercise. Write the equations of motion and their Lie–Poisson Hamiltonian formulation in $\mathbb{R}^3$-vector form for the case when $\Omega(t) = O^{-1}(t)\dot{O}(t) \in so(3)$ by using the hat map. List the conservation laws in this case. ★

Exercise. How would the variational calculation of the system (2.5.25) and (2.5.26) have changed if the Lie group had been unitary instead of orthogonal and the matrices $\mathbb{S}_0$, $\mathbb{A}$ and $\mathbb{S}(t)$ were Hermitian, rather than symmetric? ★

3 QUATERNIONS

Contents

3.1 Operating with quaternions — 78
- 3.1.1 Multiplying quaternions using Pauli matrices — 79
- 3.1.2 Quaternionic conjugate — 82
- 3.1.3 Decomposition of three-vectors — 85
- 3.1.4 Alignment dynamics for Newton's second law — 86
- 3.1.5 Quaternionic dynamics of Kepler's problem — 90

3.2 Quaternionic conjugation — 93
- 3.2.1 Cayley–Klein parameters — 93
- 3.2.2 Pure quaternions, Pauli matrices and $SU(2)$ — 99
- 3.2.3 Tilde map: $\mathbb{R}^3 \simeq su(2) \simeq so(3)$ — 102
- 3.2.4 Dual of the tilde map: $\mathbb{R}^{3^*} \simeq su(2)^* \simeq so(3)^*$ — 103
- 3.2.5 Pauli matrices and Poincaré's sphere $\mathbb{C}^2 \to S^2$ — 103
- 3.2.6 Poincaré's sphere and Hopf's fibration — 105
- 3.2.7 Coquaternions — 108

3.1 Operating with quaternions

William Rowan Hamilton

Hamilton had great hopes for the utility of quaternions, although he was not entirely sure how their utility would emerge. This was evidenced by his appeal to the following quotation of John Wallis [Wa1685] in the preface of Hamilton's book on quaternions:

> We find therefore that in Equations, whether Lateral or Quadratick, which in the strict Sense, and first Prospect, appear Impossible; some mitigation may be allowed to make them Possible; and in such a mitigated interpretation they may yet be useful.
> – Wallis, 1685

However, not all of his peers had such great hopes for quaternions and history treated them rather unkindly. Decades later, Lord Kelvin condemned them harshly [OcoRo1998]:

> Quaternions came from Hamilton after his best work had been done, and though beautifully ingenious, they have been an unmixed evil to those who have touched them in any way. – Lord Kelvin (William Thomson), 1890

Hamilton's hope that quaternions "may yet be useful" was eventually redeemed by their broad modern applications. The relation between quaternions and vectors is now understood, as we shall explain, and quaternions are used for their special advantages in the robotics and avionics industries to track objects moving continuously along a curve of tumbling rotations. They are also heavily used in graphics [Ha2006, Ku1999].

Hamilton was correct: quaternions are special. For example, they form the only associative division ring containing both real and complex numbers. For us, they also form a natural introduction to geometric mechanics. In particular, quaternions will introduce us to ***mechanics on Lie groups***; namely, mechanics on the Lie group $SU(2)$ of 2×2 special unitary matrices.

3.1.1 Multiplying quaternions using Pauli matrices

Every quaternion $\mathfrak{q} \in \mathbb{H}$ is a real linear combination of the ***basis quaternions***, denoted as $(\mathbb{J}_0, \mathbb{J}_1, \mathbb{J}_2, \mathbb{J}_3)$. The ***multiplication rules*** for their basis are given by the triple product

$$\mathbb{J}_1\mathbb{J}_2\mathbb{J}_3 = -\mathbb{J}_0\,, \tag{3.1.1}$$

and the squares

$$\mathbb{J}_1^2 = \mathbb{J}_2^2 = \mathbb{J}_3^2 = -\mathbb{J}_0\,, \tag{3.1.2}$$

where $\mathbb{J}_0$ is the identity element. Thus, $\mathbb{J}_1\mathbb{J}_2 = \mathbb{J}_3$ holds, with cyclic permutations of $(1, 2, 3)$. According to a famous story, Hamilton inscribed a version of their triple product formula on Brougham (pronounced "Broom") bridge in Dublin [OcoRo1998].

Quaternions combine a real scalar $q \in \mathbb{R}$ and a real three-vector $\boldsymbol{q} \in \mathbb{R}^3$ with components q_a $a = 1, 2, 3$, into a ***tetrad***

$$\mathfrak{q} = [q_0, \boldsymbol{q}] = q_0\mathbb{J}_0 + q_1\mathbb{J}_1 + q_2\mathbb{J}_2 + q_3\mathbb{J}_3 \in \mathbb{H}\,. \tag{3.1.3}$$

The multiplication table of the quaternion basis elements may be expressed as

	$\mathbb{J}_0$	$\mathbb{J}_1$	$\mathbb{J}_2$	$\mathbb{J}_3$
$\mathbb{J}_0$	$\mathbb{J}_0$	$\mathbb{J}_1$	$\mathbb{J}_2$	$\mathbb{J}_3$
$\mathbb{J}_1$	$\mathbb{J}_1$	$-\mathbb{J}_0$	$\mathbb{J}_3$	$-\mathbb{J}_2$
$\mathbb{J}_2$	$\mathbb{J}_2$	$-\mathbb{J}_3$	$-\mathbb{J}_0$	$\mathbb{J}_1$
$\mathbb{J}_3$	$\mathbb{J}_3$	$\mathbb{J}_2$	$-\mathbb{J}_1$	$-\mathbb{J}_0$

. (3.1.4)

Definition 3.1.1 (Multiplication of quaternions) *The* ***multiplication rule*** *for two quaternions,*

$$\mathfrak{q} = [q_0, \mathbf{q}] \quad and \quad \mathfrak{r} = [r_0, \mathbf{r}] \in \mathbb{H}\,,$$

may be defined in vector notation as

$$\mathfrak{q}\mathfrak{r} = [q_0,\, \mathbf{q}][r_0,\, \mathbf{r}] = [\, q_0 r_0 - \mathbf{q} \cdot \mathbf{r},\, q_0\mathbf{r} + r_0\mathbf{q} + \mathbf{q} \times \mathbf{r}\,]\,. \quad (3.1.5)$$

Remark 3.1.1 The antisymmetric and symmetric parts of the quaternionic product correspond to ***vector operations***[1]:

$$\frac{1}{2}\Big(\mathfrak{q}\mathfrak{r} - \mathfrak{r}\mathfrak{q}\Big) = [0\,,\, \boldsymbol{q} \times \boldsymbol{r}]\,, \quad (3.1.6)$$

$$\frac{1}{2}\Big(\mathfrak{q}\mathfrak{r} + \mathfrak{r}\mathfrak{q}\Big) = [q_0 r_0 - \boldsymbol{q} \cdot \boldsymbol{r},\, q_0\boldsymbol{r} + r_0\boldsymbol{q}\,]\,. \quad (3.1.7)$$

The product of quaternions is not commutative. (It has a nonzero antisymmetric part.) □

Theorem 3.1.1 (Isomorphism with Pauli matrix product) *The multiplication rule (3.1.5) may be represented in a 2×2 matrix basis as*

$$\mathfrak{q} = [q_0,\, \mathbf{q}] = q_0\sigma_0 - i\mathbf{q} \cdot \boldsymbol{\sigma}\,, \quad \textit{with } \mathbf{q} \cdot \boldsymbol{\sigma} := \sum_{a=1}^{3} q_a \sigma_a\,, \quad (3.1.8)$$

where σ_0 is the 2×2 identity matrix and σ_a, with $a = 1, 2, 3$, are the Hermitian ***Pauli spin matrices****,*

$$\sigma_0 = \begin{bmatrix} 1 & 0 \\ 0 & 1 \end{bmatrix}, \quad \sigma_1 = \begin{bmatrix} 0 & 1 \\ 1 & 0 \end{bmatrix},$$

$$\sigma_2 = \begin{bmatrix} 0 & -i \\ i & 0 \end{bmatrix}, \quad \sigma_3 = \begin{bmatrix} 1 & 0 \\ 0 & -1 \end{bmatrix}. \quad (3.1.9)$$

Proof. The isomorphism is implied by the product relation for the Pauli matrices

$$\sigma_a\sigma_b = \delta_{ab}\,\sigma_0 + i\epsilon_{abc}\sigma_c \quad \text{for} \quad a, b, c = 1, 2, 3, \quad (3.1.10)$$

[1]Hamilton introduced the word ***vector*** in 1846 as a synonym for a ***pure quaternion,*** whose scalar part vanishes.

where ϵ_{abc} is the totally antisymmetric tensor density with $\epsilon_{123} = 1$. The Pauli matrices also satisfy $\sigma_1^2 = \sigma_2^2 = \sigma_3^2 = \sigma_0$ and one has $\sigma_1\sigma_2\sigma_3 = i\,\sigma_0$ as well as cyclic permutations of $\{1, 2, 3\}$. Identifying $\mathbb{J}_0 = \sigma_0$ and $\mathbb{J}_a = -i\sigma_a$, with $a = 1, 2, 3$, provides the basic quaternionic properties. ■

Exercise. Verify by antisymmetry of ϵ_{abc} the ***commutator relation*** for the Pauli matrices

$$[\,\sigma_a\,,\,\sigma_b\,] := \sigma_a\sigma_b - \sigma_b\sigma_a = 2i\epsilon_{abc}\sigma_c \quad \text{for} \quad a, b, c = 1, 2, 3, \tag{3.1.11}$$

and their ***anticommutator relation***

$$\{\,\sigma_a\,,\,\sigma_b\,\}_+ := \sigma_a\sigma_b + \sigma_b\sigma_a = 2\delta_{ab}\sigma_0 \quad \text{for} \quad a, b = 1, 2, 3. \tag{3.1.12}$$

The corresponding relations among quaternions are given in (3.1.6) and (3.1.7), respectively. ★

Exercise. Verify the quaternionic multiplication rule expressed in the tetrad-bracket notation in (3.1.5) by using the isomorphism (3.1.8) and the product relation for the Pauli matrices in Equation (3.1.10). ★

Answer.

$$\begin{aligned}
\mathfrak{q}\mathfrak{r} &= (q_0\sigma_0 - iq_a\sigma_a)(r_0\sigma_0 - ir_b\sigma_b) \\
&= (q_0r_0 - \boldsymbol{q}\cdot\boldsymbol{r})\,\sigma_0 - i(q_0\boldsymbol{r} + r_0\boldsymbol{q} + \boldsymbol{q}\times\boldsymbol{r})\cdot\boldsymbol{\sigma}\,.
\end{aligned}$$

▲

Exercise. Use Equations (3.1.11), (3.1.12) and isomorphism (3.1.8) to verify relations (3.1.6) and (3.1.7). ★

Exercise. Use formula (3.1.10) to verify the decomposition of a vector in Pauli matrices

$$\mathbf{q}\sigma_0 = (\mathbf{q}\cdot\boldsymbol{\sigma})\boldsymbol{\sigma} - i\mathbf{q}\times\boldsymbol{\sigma}\,, \tag{3.1.13}$$

which is valid for three-vectors $\mathbf{q} \in \mathbb{R}^3$. Verify also that

$$-|\mathbf{q}\times\boldsymbol{\sigma}|^2 = 2|\mathbf{q}|^2\sigma_0 = 2(\mathbf{q}\cdot\boldsymbol{\sigma})^2.$$

★

Exercise. Use Equations (3.1.11) to verify the commutation relation

$$[\,\mathbf{p}\cdot\boldsymbol{\sigma},\, \mathbf{q}\cdot\boldsymbol{\sigma}\,] = 2i\mathbf{p}\times\mathbf{q}\cdot\boldsymbol{\sigma}$$

for three-vectors $\mathbf{p}, \mathbf{q} \in \mathbb{R}^3$. ★

3.1.2 Quaternionic conjugate

Remark 3.1.2 (Quaternionic product is associative) The quaternionic product is ***associative***:

$$\mathfrak{p}(\mathfrak{q}\mathfrak{r}) = (\mathfrak{p}\mathfrak{q})\mathfrak{r}\,. \tag{3.1.14}$$

However, the quaternionic product is not commutative,

$$[\mathfrak{p},\, \mathfrak{q}] := \mathfrak{p}\mathfrak{q} - \mathfrak{q}\mathfrak{p} = [0,\, 2\boldsymbol{p}\times\boldsymbol{q}\,]\,, \tag{3.1.15}$$

as we saw earlier in (3.1.6). □

Definition 3.1.2 (Quaternionic conjugate) *One defines the* ***conjugate*** *of* $\mathfrak{q} := [q_0\,,\,\mathbf{q}]$ *in analogy to complex variables as*

$$\mathfrak{q}^* = [q_0\,,\,-\,\mathbf{q}]\,. \tag{3.1.16}$$

Following this analogy, the scalar and vector parts of a quaternion are defined as

$$\mathrm{Re}\,\mathfrak{q} \;:=\; \frac{1}{2}(\mathfrak{q}+\mathfrak{q}^*) = [\,q_0,\,0\,]\,, \tag{3.1.17}$$

$$\mathrm{Im}\,\mathfrak{q} \;:=\; \frac{1}{2}(\mathfrak{q}-\mathfrak{q}^*) = [\,0\,,\,\mathbf{q}\,]\,. \tag{3.1.18}$$

Lemma 3.1.1 (Properties of quaternionic conjugation) *Two important properties of quaternionic conjugation are easily demonstrated. Namely,*

$$(\mathfrak{p}\mathfrak{q})^* \;=\; \mathfrak{q}^*\mathfrak{p}^* \quad \textit{(note reversed order)}, \tag{3.1.19}$$

$$\begin{aligned}\mathrm{Re}(\mathfrak{p}\mathfrak{q}^*) \;&:=\; \frac{1}{2}(\mathfrak{p}\mathfrak{q}^* + \mathfrak{q}\mathfrak{p}^*)\\ &=\; [p_0q_0 + \mathbf{p}\cdot\mathbf{q}\,,\,0] \quad \textit{(yields real part)}.\end{aligned} \tag{3.1.20}$$

Note that conjugation reverses the order in the product of two quaternions.

Definition 3.1.3 (Dot product of quaternions) *The quaternionic* ***dot product****, or* ***inner product****, is defined as*

$$\begin{aligned}\mathfrak{p}\cdot\mathfrak{q} \;&=\; [p_0\,,\,\mathbf{p}]\cdot[q_0\,,\,\mathbf{q}]\\ &:=\; [p_0q_0 + \mathbf{p}\cdot\mathbf{q}\,,\,0] = \mathrm{Re}(\mathfrak{p}\mathfrak{q}^*)\,.\end{aligned} \tag{3.1.21}$$

Definition 3.1.4 (Pairing of quaternions) *The quaternionic dot product (3.1.21) defines a real symmetric pairing* $\langle\,\cdot\,,\,\cdot\,\rangle : \mathbb{H}\times\mathbb{H}\mapsto\mathbb{R}$*, denoted as*

$$\langle\,\mathfrak{p}\,,\,\mathfrak{q}\,\rangle = \mathrm{Re}(\mathfrak{p}\mathfrak{q}^*) := \mathrm{Re}(\mathfrak{q}\mathfrak{p}^*) = \langle\,\mathfrak{q}\,,\,\mathfrak{p}\,\rangle\,. \tag{3.1.22}$$

In particular, $\langle\,\mathfrak{q}\,,\,\mathfrak{q}\,\rangle = \mathrm{Re}(\mathfrak{q}\mathfrak{q}^*) =: |\mathfrak{q}|^2$ *is a positive real number.*

Definition 3.1.5 (Magnitude of a quaternion) *The* ***magnitude*** *of a quaternion* $\mathfrak{q}$ *may be defined by*

$$|\mathfrak{q}| := (\mathfrak{q} \cdot \mathfrak{q})^{1/2} = ({q_0}^2 + \mathbf{q} \cdot \mathbf{q})^{1/2} . \tag{3.1.23}$$

Remark 3.1.3 A level set of $|\mathfrak{q}|$ defines a three-sphere S^3. □

Definition 3.1.6 (Quaternionic inverse) *We have the product*

$$|\mathfrak{q}|^2 := \mathfrak{q}\mathfrak{q}^* = (\mathfrak{q} \cdot \mathfrak{q})\mathfrak{e} , \tag{3.1.24}$$

where $\mathfrak{e} = [1, 0]$ *is the* ***identity quaternion****. Hence, one may define*

$$\mathfrak{q}^{-1} := \mathfrak{q}^* / |\mathfrak{q}|^2 \tag{3.1.25}$$

to be the ***inverse*** *of quaternion* $\mathfrak{q}$.

Exercise. Does a quaternion $\mathfrak{q}$ have a square root? Prove it. ★

Exercise. Show that the magnitude of the product of two quaternions is the product of their magnitudes. ★

Answer. From the definitions of the quaternionic multiplication rule (3.1.5), inner product (3.1.21) and magnitude (3.1.23), one verifies that

$$\begin{aligned} |\mathfrak{p}\mathfrak{q}|^2 &= (p_0 q_0 - \mathbf{p} \cdot \mathbf{q})^2 + |p_0 \mathbf{q} + \mathbf{p} q_0 + \mathbf{p} \times \mathbf{q}|^2 \\ &= ({p_0}^2 + |\mathbf{p}|^2)({q_0}^2 + |\mathbf{q}|^2) = |\mathfrak{p}|^2 |\mathfrak{q}|^2 . \end{aligned}$$

▲

Definition 3.1.7 *A quaternion* $\mathfrak{q}$ *with magnitude* $|\mathfrak{q}| = 1$ *is called a* ***unit quaternion****.*

Definition 3.1.8 *A quaternion with no scalar (or real) part* $\mathfrak{q} = [0, \mathbf{q}]$ *is called a* ***pure quaternion***, *or equivalently a* ***vector*** *(a term introduced by Hamilton in 1846 [Ne1997]).*

Exercise. Show that the antisymmetric and symmetric parts of the product of two pure quaternions $\mathfrak{v} = [0, \boldsymbol{v}]$ and $\mathfrak{w} = [0, \boldsymbol{w}]$ yield, respectively, the cross product and (minus) the scalar product of the two corresponding vectors $\boldsymbol{v}$, $\boldsymbol{w}$. ★

Answer. The quaternionic product of pure quaternions $\mathfrak{v} = [0, \boldsymbol{v}]$ and $\mathfrak{w} = [0, \boldsymbol{w}]$ is defined as

$$\mathfrak{v}\mathfrak{w} = \big[-\boldsymbol{v} \cdot \boldsymbol{w},\, \boldsymbol{v} \times \boldsymbol{w} \big].$$

Its antisymmetric (vector) part yields the cross product of the corresponding vectors:

$$\mathrm{Im}(\mathfrak{v}\mathfrak{w}) = \frac{1}{2}\Big(\mathfrak{v}\mathfrak{w} - \mathfrak{w}\mathfrak{v}\Big) = [0\,,\, \boldsymbol{v} \times \boldsymbol{w}]$$

(vanishes for $\boldsymbol{v} \| \boldsymbol{w}$).

Its symmetric (or real) part yields minus the scalar product of the vectors:

$$\mathrm{Re}(\mathfrak{v}\mathfrak{w}) = \frac{1}{2}\Big(\mathfrak{v}\mathfrak{w} + \mathfrak{w}\mathfrak{v}\Big) = [-\boldsymbol{v} \cdot \boldsymbol{w}\,,\, 0]$$

(vanishes for $\boldsymbol{v} \perp \boldsymbol{w}$). ▲

Remark 3.1.4 ($\mathbb{H}_0 \simeq \mathbb{R}^3$) Being three-dimensional linear spaces possessing the same vector and scalar products, pure quaternions in $\mathbb{H}_0$ (with no real part) are equivalent to vectors in $\mathbb{R}^3$. □

3.1.3 Decomposition of three-vectors

Pure quaternions have been identified with vectors in $\mathbb{R}^3$. Under this identification, the two types of products of pure quaternions

$[0,\,\mathbf{v}]$ and $[0,\,\mathbf{w}]$ are given by

$$[0,\,\mathbf{v}]\cdot[0,\,\mathbf{w}] = [\mathbf{v}\cdot\mathbf{w},\,0] \quad \text{and} \quad [0,\,\mathbf{v}][0,\,\mathbf{w}] = [-\mathbf{v}\cdot\mathbf{w},\,\mathbf{v}\times\mathbf{w}].$$

Thus, the dot $(\cdot)$ and cross $(\times)$ products of three-vectors may be identified with these two products of pure quaternions. The product of an arbitrary quaternion $[\alpha,\,\boldsymbol{\chi}]$ with a pure unit quaternion $[0,\,\hat{\boldsymbol{\omega}}]$ produces another pure quaternion, provided $\boldsymbol{\chi}\cdot\hat{\boldsymbol{\omega}} = 0$. In this case, one computes

$$[\alpha,\,\boldsymbol{\chi}][0,\,\hat{\boldsymbol{\omega}}] = [-\boldsymbol{\chi}\cdot\hat{\boldsymbol{\omega}},\,\alpha\,\hat{\boldsymbol{\omega}}+\boldsymbol{\chi}\times\hat{\boldsymbol{\omega}}] =: [0,\,\mathbf{v}],\ \text{for}\ \boldsymbol{\chi}\cdot\hat{\boldsymbol{\omega}} = 0\,. \quad (3.1.26)$$

Remark 3.1.5 Quaternions are summoned whenever a three-vector $\mathbf{v}$ is decomposed into its components parallel $(\|)$ and perpendicular $(\perp)$ to a unit three-vector direction $\hat{\boldsymbol{\omega}}$, according to

$$\mathbf{v} = \alpha\,\hat{\boldsymbol{\omega}} + \boldsymbol{\chi}\times\hat{\boldsymbol{\omega}} = [\alpha,\,\boldsymbol{\chi}][0,\,\hat{\boldsymbol{\omega}}] = \mathbf{v}_{\|} + \mathbf{v}_{\perp}\,. \quad (3.1.27)$$

Here $\alpha = \hat{\boldsymbol{\omega}}\cdot\mathbf{v}$ and $\boldsymbol{\chi} = \hat{\boldsymbol{\omega}}\times\mathbf{v}$ so that $\boldsymbol{\chi}\cdot\hat{\boldsymbol{\omega}} = 0$ and one uses $\hat{\boldsymbol{\omega}}\cdot\hat{\boldsymbol{\omega}} = 1$ to find $\mathbf{v}\cdot\mathbf{v} = \alpha^2 + \chi^2$ with $\chi := |\boldsymbol{\chi}|$. The vector decomposition (3.1.27) is precisely the quaternionic product (3.1.26), in which the vectors $\mathbf{v}$ and $\hat{\boldsymbol{\omega}}$ are treated as pure quaternions. □

This remark may be summarised by the following.

Proposition 3.1.1 (Vector decomposition) *Quaternionic left multiplication of* $[0,\,\hat{\boldsymbol{\omega}}]$ *by* $[\alpha,\,\boldsymbol{\chi}] = [\hat{\boldsymbol{\omega}}\cdot\mathbf{v},\,\hat{\boldsymbol{\omega}}\times\mathbf{v}]$ *decomposes the pure quaternion* $[0\,,\mathbf{v}]$ *into components that are* $\|$ *and* $\perp$ *to the pure unit quaternion* $[0\,,\hat{\boldsymbol{\omega}}]$.

3.1.4 Alignment dynamics for Newton's second law

Newton's second law of motion is a set of ordinary differential equations for vectors of position and velocity $(\mathbf{r},\,\mathbf{v}) \in \mathbb{R}^3\times\mathbb{R}^3$. Namely,

$$\frac{d\mathbf{r}}{dt} = \mathbf{v} \quad \text{and} \quad \frac{d\mathbf{v}}{dt} = \mathbf{f}\,,$$

where $\mathbf{f}$ is the ***force per unit mass***. Quaternions $[\alpha_v,\,\boldsymbol{\chi}_v]$ and $[\alpha_f,\,\boldsymbol{\chi}_f]$ are defined by the dot $(\cdot)$ and cross $(\times)$ products of the three-vectors of velocity and force with the radial unit vector $\hat{\mathbf{r}}$ as

$$[\alpha_v, \chi_v] = \left[\hat{\mathbf{r}} \cdot \frac{d\mathbf{r}}{dt}, \hat{\mathbf{r}} \times \frac{d\mathbf{r}}{dt}\right] = [\hat{\mathbf{r}} \cdot \mathbf{v}, \hat{\mathbf{r}} \times \mathbf{v}], \text{ with } \chi_v \cdot \hat{\mathbf{r}} = 0,$$

$$[\alpha_f, \chi_f] = \left[\hat{\mathbf{r}} \cdot \frac{d\mathbf{v}}{dt}, \hat{\mathbf{r}} \times \frac{d\mathbf{v}}{dt}\right] = [\hat{\mathbf{r}} \cdot \mathbf{f}, \hat{\mathbf{r}} \times \mathbf{f}], \text{ with } \chi_f \cdot \hat{\mathbf{r}} = 0.$$

In these equations, $\alpha_v = \hat{\mathbf{r}} \cdot \mathbf{v} = dr/dt$ represents radial velocity, $\chi_v = \hat{\mathbf{r}} \times \mathbf{v}$ represents angular velocity, $\alpha_f = \hat{\mathbf{r}} \cdot \mathbf{f}$ represents radial force per unit mass and $\chi_f = \hat{\mathbf{r}} \times \mathbf{f}$ represents ***twist***, or torque per unit (mass $\times$ length), which vanishes for central forces.

Upon decomposing into components that are parallel ($\|$) and perpendicular ($\perp$) to the position unit vector $\hat{\mathbf{r}}$, Newton's second law $(d\mathbf{r}/dt, d\mathbf{v}/dt) = (\mathbf{v}, \mathbf{f})$ may be expressed as a pair of quaternionic equations,

$$\frac{d\mathbf{r}}{dt} = \mathbf{v} = \alpha_v \hat{\mathbf{r}} + \chi_v \times \hat{\mathbf{r}} = [\alpha_v, \chi_v][0, \hat{\mathbf{r}}], \quad \text{using} \quad \chi_v \cdot \hat{\mathbf{r}} = 0,$$

$$\frac{d\mathbf{v}}{dt} = \mathbf{f} = \alpha_f \hat{\mathbf{r}} + \chi_f \times \hat{\mathbf{r}} = [\alpha_f, \chi_f][0, \hat{\mathbf{r}}], \quad \text{using} \quad \chi_f \cdot \hat{\mathbf{r}} = 0.$$

In this representation the alignment parameters for force $[\alpha_f, \chi_f]$ drive those for velocity $[\alpha_v, \chi_v]$. That is, upon using

$$d[0, \hat{\mathbf{r}}]/dt = [0, \chi_v/r][0, \hat{\mathbf{r}}],$$

one finds

$$\left(\frac{d}{dt}[\alpha_v, \chi_v] + [\alpha_v, \chi_v][0, \chi_v/r] - [\alpha_f, \chi_f]\right)[0, \hat{\mathbf{r}}] = [0, \mathbf{0}]. \quad (3.1.28)$$

Since $\chi_v \cdot \hat{\mathbf{r}} = 0 = \chi_f \cdot \hat{\mathbf{r}}$, the term in parentheses vanishes.

The force vector $\mathbf{f}$ in the orthonormal frame $\mathbf{e}_a = (\hat{\mathbf{r}}, \hat{\mathbf{r}} \times \hat{\chi}_v, \hat{\chi}_v)$ may be expanded as

$$\mathbf{f} = \alpha_f \hat{\mathbf{r}} + \beta_f \hat{\mathbf{r}} \times \hat{\chi}_v + \gamma_f \hat{\chi}_v, \quad (3.1.29)$$

in which the coefficients β_f and γ_f are defined as

$$\beta_f = \mathbf{f} \cdot \hat{\mathbf{r}} \times \hat{\chi}_v \quad \text{and} \quad \gamma_f = \mathbf{f} \cdot \hat{\chi}_v. \quad (3.1.30)$$

In this orthonormal frame the twist vector becomes

$$\boldsymbol{\chi}_f := \hat{\mathbf{r}} \times \mathbf{f} = -\beta_f \hat{\boldsymbol{\chi}}_v + \gamma_f \hat{\mathbf{r}} \times \hat{\boldsymbol{\chi}}_v ,$$

in which β_f and γ_f represent components of the twist that are parallel and perpendicular to the direction of angular frequency $\hat{\boldsymbol{\chi}}_v$. As a pure quaternion, the twist vector $[0, \boldsymbol{\chi}_f]$ may be decomposed as

$$[0, \boldsymbol{\chi}_f] = [-\beta_f, \gamma_f \hat{\mathbf{r}}][0, \hat{\boldsymbol{\chi}}_v] .$$

Expanding the quaternionic representation of Newton's second law in Equation (3.1.28) provides the ***alignment dynamics,***

$$\frac{d}{dt}[\alpha_v, \boldsymbol{\chi}_v] + [-\chi_v^2/r, \alpha_v \boldsymbol{\chi}_v/r] = [\alpha_f, -\beta_f \hat{\boldsymbol{\chi}}_v + \gamma_f \hat{\mathbf{r}} \times \hat{\boldsymbol{\chi}}_v] . \quad (3.1.31)$$

The scalar part of this equation and the magnitude of its vector part yield

$$\frac{d\alpha_v}{dt} = \frac{d^2 r}{dt^2} = \frac{\chi_v^2}{r} + \alpha_f \quad \text{and} \quad \frac{d\chi_v}{dt} = -\frac{\alpha_v \chi_v}{r} - \beta_f , \quad (3.1.32)$$

with $\alpha_v = \hat{\mathbf{r}} \cdot \mathbf{v}$, $\chi_v = |\hat{\mathbf{r}} \times \mathbf{v}|$ and force components α_f, β_f defined in (3.1.29). Its unit-vector parts yield an evolution equation for the alignment dynamics of the following ***orthonormal frame,***

$$\frac{d}{dt}\begin{pmatrix} \hat{\mathbf{r}} \\ \hat{\mathbf{r}} \times \hat{\boldsymbol{\chi}}_v \\ \hat{\boldsymbol{\chi}}_v \end{pmatrix} = \begin{pmatrix} 0 & -\chi_v/r & 0 \\ \chi_v/r & 0 & -\gamma_f/\chi_v \\ 0 & \gamma_f/\chi_v & 0 \end{pmatrix} \begin{pmatrix} \hat{\mathbf{r}} \\ \hat{\mathbf{r}} \times \hat{\boldsymbol{\chi}}_v \\ \hat{\boldsymbol{\chi}}_v \end{pmatrix} , \quad (3.1.33)$$

with $\gamma_f = \mathbf{f} \cdot \hat{\boldsymbol{\chi}}_v = \hat{\mathbf{v}} \cdot \mathbf{f} \times \hat{\mathbf{r}}$. Thus, the alignment dynamics of Newton's second law is expressed as rotation of the orthonormal frame $\mathbf{e}_a = (\hat{\mathbf{r}}, \hat{\mathbf{r}} \times \hat{\boldsymbol{\chi}}_v, \hat{\boldsymbol{\chi}}_v)$, $a = 1, 2, 3$ given by

$$\frac{d\mathbf{e}_a}{dt} = \widehat{\mathcal{D}}\,\mathbf{e}_a = \mathbf{D} \times \mathbf{e}_a\,, \tag{3.1.34}$$

with angular frequency vector

$$\mathbf{D} = \frac{\chi_v}{r}\hat{\boldsymbol{\chi}}_v + \frac{\gamma_f}{\chi_v}\hat{\mathbf{r}} \tag{3.1.35}$$

lying in the ($\hat{\mathbf{r}}$, $\hat{\boldsymbol{\chi}}_v$) plane. The components of the antisymmetric matrix $\widehat{\mathcal{D}}$ in Equation (3.1.34) are related to the components of the vector $\mathbf{D}$ by the hat map $\widehat{\mathcal{D}}_{ij} = -\,\epsilon_{ijk}D_k$.

The three unit vectors in the orthonormal frame

$$\mathbf{e}_a = (\hat{\mathbf{r}},\, \hat{\mathbf{r}} \times \hat{\boldsymbol{\chi}}_v,\, \hat{\boldsymbol{\chi}}_v)^T$$

point along the position vector, opposite the nonradial component of the velocity vector, and along the angular velocity vector, respectively. For central forces, there are no torques and $\beta_f = 0 = \gamma_f$ in Equations (3.1.32) and (3.1.33). In that case, Equation (3.1.33) implies that the direction of the angular velocity $\hat{\boldsymbol{\chi}}_v = \hat{\mathbf{r}} \times \hat{\mathbf{v}}$ remains fixed. The other two unit vectors in the orthonormal frame $\mathbf{e}_a$ then rotate around $\hat{\boldsymbol{\chi}}_v$ at angular frequency χ_v/r. For $\chi_v \neq 0$, one may assume $r(t) \neq 0$ for $r(0) \neq 0$. (One may have $r = 0$ for $\chi_v \equiv 0$, but in that case the motions are only one-dimensional and rotations are not defined.)

Remark 3.1.6 The alignment Equation (3.1.33) governs the evolution of an orthonormal frame for any application of Newton's second law. However, this idea goes much further than particle dynamics. For example, in ideal incompressible fluid dynamics, the alignment of the vorticity vector with the velocity shear tensor in ideal incompressible fluids may be analysed by following similar steps to those used here for Newtonian mechanics [GiHoKeRo2006]. □

3.1.5 Quaternionic dynamics of Kepler's problem

Johannes Kepler

Newton's dynamical equation for the reduced Kepler problem is

$$\ddot{\mathbf{r}} + \frac{\mu \mathbf{r}}{r^3} = 0\,, \tag{3.1.36}$$

in which μ is the gravitational constant times the reduced mass of the system of two particles in the centre of mass frame. Scale invariance of this equation under the changes $R \to s^2 R$ and $T \to s^3 T$ in the units of space R and time T for any constant s means that it admits families of solutions whose space and time scales are related by $T^2/R^3 = const$ (Kepler's third law).

The reduced Kepler problem conserves the quantities

$$\begin{aligned} E &= \frac{1}{2}|\dot{\mathbf{r}}|^2 - \frac{\mu}{r} \quad \text{(energy)}\,, \\ \mathbf{L} &= \mathbf{r} \times \dot{\mathbf{r}} \quad \text{(specific angular momentum)}\,. \end{aligned}$$

Constancy of magnitude L means the orbit sweeps out equal areas in equal times (Kepler's second law). Constancy of direction $\hat{\mathbf{L}}$ means the orbital position $\mathbf{r}$ and velocity $\dot{\mathbf{r}}$ span a plane with unit normal vector $\hat{\mathbf{L}}$. In that orbital plane one may specify plane polar coordinates (r, θ) with unit vectors $(\hat{\mathbf{r}}, \hat{\boldsymbol{\theta}})$ in the plane and $\hat{\mathbf{r}} \times \hat{\boldsymbol{\theta}} = \hat{\mathbf{L}}$ normal to it.

The radial quaternionic variables in Equation (3.1.31) for Kepler's problem are $\alpha_v = \dot{r}$ and $\alpha_f = -\mu/r^2$. The angular velocity is $\chi_v = \mathbf{L}/r$ and the angular frequency vector in (3.1.35) is

$$\mathbf{D} = \chi_v/r = \mathbf{L}/r^2\,,$$

whose magnitude is $\chi_v/r = L/r^2$. The quantity χ_f and its components γ_f and β_f all vanish in (3.1.29) because gravity is a central force.

Substituting these values into the scalar alignment equations in (3.1.32) yields

$$\frac{d\alpha_v}{dt} = \frac{d^2 r}{dt^2} = \frac{L}{r^3} - \frac{\mu}{r^2} \quad \text{and} \quad \frac{d}{dt}\Big(\frac{L}{r}\Big) = -\frac{\dot{r}L}{r^2}\,. \tag{3.1.37}$$

The former equation is the balance of radial centrifugal and gravitational forces. The latter implies $dL/dt = 0$.

The unit vectors for polar coordinates in the orbital plane are

$$\hat{\mathbf{r}}\,, \quad \hat{\chi}_v = \hat{\mathbf{L}} = \hat{\mathbf{r}} \times \hat{\boldsymbol{\theta}} \quad \text{and} \quad \hat{\mathbf{r}} \times \hat{\chi}_v = -\,\hat{\boldsymbol{\theta}}\,. \tag{3.1.38}$$

These orthogonal unit vectors form an orthonormal frame, whose alignment dynamics is governed by Equation (3.1.33) as

$$\frac{d}{dt}\begin{pmatrix} \hat{\mathbf{r}} \\ -\,\hat{\boldsymbol{\theta}} \\ \hat{\mathbf{r}} \times \hat{\boldsymbol{\theta}} \end{pmatrix} = \begin{pmatrix} 0 & -\,L/r^2 & 0 \\ L/r^2 & 0 & 0 \\ 0 & 0 & 0 \end{pmatrix}\begin{pmatrix} \hat{\mathbf{r}} \\ -\,\hat{\boldsymbol{\theta}} \\ \hat{\mathbf{r}} \times \hat{\boldsymbol{\theta}} \end{pmatrix} = \begin{pmatrix} L/r^2\,\hat{\boldsymbol{\theta}} \\ L/r^2\,\hat{\mathbf{r}} \\ 0 \end{pmatrix}.$$

That is, the normal to the orbital plane is the constant unit vector $\hat{\mathbf{r}} \times \hat{\boldsymbol{\theta}} = \hat{\mathbf{L}}$, while

$$\frac{d\hat{\mathbf{r}}}{dt} = \dot{\theta}\,\hat{\boldsymbol{\theta}}\,, \quad \frac{d\hat{\boldsymbol{\theta}}}{dt} = -\,\dot{\theta}\,\hat{\mathbf{r}} \quad \text{and} \quad \dot{\theta} = \frac{L}{r^2}\,.$$

Newton's equation of motion (3.1.36) for the Kepler problem may now be written equivalently as

$$0 = \ddot{\mathbf{r}} + \frac{\mu\mathbf{r}}{r^3} = \ddot{\mathbf{r}} + \frac{\mu}{L}\dot{\theta}\,\hat{\mathbf{r}} = \frac{d}{dt}\Big(\dot{\mathbf{r}} - \frac{\mu}{L}\hat{\boldsymbol{\theta}}\Big)\,.$$

This equation implies conservation of the following vector in the plane of motion:

$$\mathbf{K} = \dot{\mathbf{r}} - \frac{\mu}{L}\hat{\boldsymbol{\theta}} \quad \textit{(Hamilton's vector)}\,.$$

The vector in the plane given by the cross product of the two conserved vectors **K** and **L**,

$$\mathbf{J} = \mathbf{K} \times \mathbf{L} = \dot{\mathbf{r}} \times \mathbf{L} - \mu\hat{\mathbf{r}} \quad \textit{(Laplace–Runge–Lenz vector)}\,,$$

is also conserved. From their definitions, these conserved quantities are related by

$$K^2 = 2E + \frac{\mu^2}{L^2} = \frac{J^2}{L^2}\,.$$

Choose the conserved Laplace–Runge–Lenz vector $\mathbf{J}$ in the plane of the orbit as the reference line for the measurement of the polar angle θ. The scalar product of $\mathbf{r}$ and $\mathbf{J}$ then yields an elegant result for the Kepler orbit in plane polar coordinates:

$$\mathbf{r}\cdot\mathbf{J} = rJ\cos\theta = \mathbf{r}\cdot(\dot{\mathbf{r}}\times\mathbf{L}) - \mu\,\mathbf{r}\cdot\hat{\mathbf{r}}\,,$$

which implies

$$r(\theta) = \frac{L^2}{\mu + J\cos\theta} = \frac{l_\perp}{1 + e\cos\theta}\,.$$

As expected, the orbit is a ***conic section*** whose origin is at one of the two foci. This is ***Kepler's first law.***

The Laplace–Runge–Lenz vector $\mathbf{J}$ is directed from the focus of the orbit to its perihelion (point of closest approach). The eccentricity of the conic section is $e = J/\mu = KL/\mu$ and its semi-latus rectum (normal distance from the line through the foci to the orbit) is $l_\perp = L^2/\mu$. The eccentricity vanishes ($e = 0$) for a circle and correspondingly $K = 0$ implies that $\dot{\mathbf{r}} = \mu\hat{\boldsymbol{\theta}}/L$. The eccentricity takes values $0 < e < 1$ for an ellipse, $e = 1$ for a parabola and $e > 1$ for a hyperbola.

Exercise. (Monopole Kepler problem [LeFl2003]) Consider the Kepler problem with a magnetic monopole, whose dynamical equation is, for constants λ and μ,

$$\ddot{\mathbf{r}} + \frac{\lambda}{r^3}\mathbf{L} + \left(\frac{\mu}{r^3} - \frac{\lambda^2}{r^4}\right)\mathbf{r} = 0\,. \qquad (3.1.39)$$

Take vector cross products of this equation with $\mathbf{r}$ and $\mathbf{L} = \mathbf{r}\times\dot{\mathbf{r}}$ to find its conserved Hamilton's vector and

Laplace–Runge–Lenz vector. Are energy and angular momentum conserved? Are negative energy orbits ellipses? Do Kepler's three laws still hold when $\lambda \neq 0$? ★

Exercise. Derive Equation (3.1.39) from Hamilton's principle.

Hint: Assume there exists a generalised function $\mathbf{A}(\mathbf{r})$ whose curl satisfies $\text{curl}\,\mathbf{A} = \mathbf{r}/r^3$. ★

Exercise. Write the Hamiltonian formulation and Poisson brackets for Equation (3.1.39). Compute the Poisson brackets for $\{\dot{r}_i, \dot{r}_j\}$. ★

Exercise. Write the alignment Equations (3.1.33) for the dynamics of Equation (3.1.39). ★

3.2 Quaternionic conjugation

3.2.1 Cayley–Klein parameters

Definition 3.2.1 (Quaternionic conjugation) *Quaternionic conjugation is defined as the map under the quaternionic product (recalling that it is associative),*

$$\mathfrak{r} \to \mathfrak{r}' = \hat{\mathfrak{q}}\, \mathfrak{r}\, \hat{\mathfrak{q}}^*\,, \tag{3.2.1}$$

where $\hat{\mathfrak{q}} = [q_0\,,\,\mathbf{q}]$ *is a unit quaternion,* $\hat{\mathfrak{q}} \cdot \hat{\mathfrak{q}} = 1$, *so* $\hat{\mathfrak{q}}\hat{\mathfrak{q}}^* = \mathfrak{e} = [1, 0]$. *The inverse map is*

$$\mathfrak{r} = \hat{\mathfrak{q}}^* \mathfrak{r}' \hat{\mathfrak{q}}\,.$$

Exercise. Show that the product of a quaternion $\mathfrak{r} = [r_0, \boldsymbol{r}]$ with a unit quaternion $\hat{\mathfrak{q}} = [q_0, \boldsymbol{q}]$, whose inverse is $\hat{\mathfrak{q}}^* = [q_0, -\boldsymbol{q}]$, satisfies

$$\mathfrak{r}\hat{\mathfrak{q}}^* = \left[\mathfrak{r}\cdot\hat{\mathfrak{q}}, -r_0\boldsymbol{q} + q_0\boldsymbol{r} + \boldsymbol{q}\times\boldsymbol{r}\right],$$

$$\hat{\mathfrak{q}}\,\mathfrak{r}\,\hat{\mathfrak{q}}^* = \left[r_0|\hat{\mathfrak{q}}|^2, \boldsymbol{r} + 2q_0\boldsymbol{q}\times\boldsymbol{r} + 2\boldsymbol{q}\times(\boldsymbol{q}\times\boldsymbol{r})\right],$$

where $\mathfrak{r}\cdot\hat{\mathfrak{q}} = r_0q_0 + \boldsymbol{r}\cdot\boldsymbol{q}$ and $|\hat{\mathfrak{q}}|^2 = \hat{\mathfrak{q}}\cdot\hat{\mathfrak{q}} = q_0{}^2 + \boldsymbol{q}\cdot\boldsymbol{q} = 1$ according to the definitions of the dot product in (3.1.21) and magnitude in (3.1.23). ★

Remark 3.2.1 The same products using the pure unit quaternion $\hat{\mathfrak{z}} = [0, \hat{\mathbf{z}}]$ with $\hat{\mathbf{z}} = (0, 0, 1)^T$ and the unit quaternion $\hat{\mathfrak{q}} = [q_0, \boldsymbol{q}]$ satisfy

$$\hat{\mathfrak{z}}\hat{\mathfrak{q}}^* = \left[q_3, q_0\hat{\mathbf{z}} + \boldsymbol{q}\times\hat{\mathbf{z}}\right],$$

$$\hat{\mathfrak{q}}\,\hat{\mathfrak{z}}\,\hat{\mathfrak{q}}^* = \left[0, \hat{\mathbf{z}} + 2q_0\boldsymbol{q}\times\hat{\mathbf{z}} + 2\boldsymbol{q}\times(\boldsymbol{q}\times\hat{\mathbf{z}})\right],$$

which produces a complete set of unit vectors. □

Remark 3.2.2 Conjugation $\hat{\mathfrak{q}}\,\mathfrak{r}\,\hat{\mathfrak{q}}^*$ is a wise choice, as opposed to, say, choosing the apparently less meaningful triple product

$$\hat{\mathfrak{q}}\,\mathfrak{r}\,\hat{\mathfrak{q}} = [0, \boldsymbol{r}] + (r_0q_0 - \boldsymbol{r}\cdot\boldsymbol{q})[q_0, \boldsymbol{q}]$$

for quaternions $\mathfrak{r} = [r_0, \boldsymbol{r}]$ and $\hat{\mathfrak{q}} = [q_0, \boldsymbol{q}]$ with $|\mathfrak{q}|^2 = q_0{}^2 + \boldsymbol{q}\cdot\boldsymbol{q} = 1$. □

Exercise. For $\mathfrak{q}^* = [q_0, -\boldsymbol{q}]$, such that $\mathfrak{q}^*\mathfrak{q} = \mathbb{J}_0|\mathfrak{q}|^2$, verify that

$$2\mathfrak{q}^* = -\mathbb{J}_0\mathfrak{q}\mathbb{J}_0^* + \mathbb{J}_1\mathfrak{q}\mathbb{J}_1^* + \mathbb{J}_2\mathfrak{q}\mathbb{J}_2^* + \mathbb{J}_3\mathfrak{q}\mathbb{J}_3^* .$$

What does this identity mean geometrically? Does the complex conjugate z^* for $z \in \mathbb{C}$ satisfy such an identity? Prove it. ★

Lemma 3.2.1 *As a consequence of Remark 3.2.1 and the Exercise just before it, one finds that conjugation $\hat{\mathfrak{q}}\,\mathfrak{r}\,\hat{\mathfrak{q}}^*$ of a quaternion $\mathfrak{r}$ by a unit quaternion $\hat{\mathfrak{q}}$ preserves the sphere $S^3_{|\mathfrak{r}|}$ given by any level set of $|\mathfrak{r}|$. That is, the value of $|\mathfrak{r}|^2$ is invariant under conjugation by a unit quaternion:*

$$|\hat{\mathfrak{q}}\,\mathfrak{r}\,\hat{\mathfrak{q}}^*|^2 = |\mathfrak{r}|^2 = {r_0}^2 + \mathbf{r}\cdot\mathbf{r}\,. \qquad (3.2.2)$$

Definition 3.2.2 (Conjugacy classes) *The set*

$$C(\mathfrak{r}) := \Big\{\mathfrak{r}' \in \mathbb{H} \Big| \, \mathfrak{r}' = \hat{\mathfrak{q}}\,\mathfrak{r}\,\hat{\mathfrak{q}}^* \Big\} \qquad (3.2.3)$$

is called the ***conjugacy class*** *of the quaternion $\mathfrak{r}$.*

Corollary 3.2.1 *The conjugacy classes of the three-sphere $S^3_{|\mathfrak{r}|}$ under conjugation by a unit quaternion $\hat{\mathfrak{q}}$ are the two-spheres given by*

$$\Big\{\mathbf{r} \in \mathbb{R}^3 \Big| \, \mathbf{r}\cdot\mathbf{r} = |\mathfrak{r}|^2 - {r_0}^2\Big\}. \qquad (3.2.4)$$

Proof. The proof is a straightforward exercise. ■

Remark 3.2.3 The expressions in Remark 3.2.1 correspond to ***spatial rotations*** when $r_0 = 0$ so that $\mathfrak{r} = [0, \boldsymbol{r}]$. □

Lemma 3.2.2 (Euler–Rodrigues formula) *If* $\mathfrak{r} = [0, \mathbf{r}]$ *is a pure quaternion and* $\hat{\mathfrak{q}} = [q_0, \mathbf{q}]$ *is a unit quaternion, then under quaternionic conjugation,*

$$\begin{aligned} \mathfrak{r}' &= \hat{\mathfrak{q}}\,\mathfrak{r}\,\hat{\mathfrak{q}}^* = [0, \mathbf{r}'] \\ &= [0, \mathbf{r} + 2q_0(\mathbf{q}\times\mathbf{r}) + 2\mathbf{q}\times(\mathbf{q}\times\mathbf{r})] . \end{aligned} \tag{3.2.5}$$

For $\hat{\mathfrak{q}} := \pm[\cos\frac{\theta}{2}, \sin\frac{\theta}{2}\,\hat{\mathbf{n}}]$, *we have*

$$[0, \mathbf{r}'] = \left[\cos\frac{\theta}{2}, \sin\frac{\theta}{2}\,\hat{\mathbf{n}}\right] [0, \mathbf{r}] \left[\cos\frac{\theta}{2}, -\sin\frac{\theta}{2}\,\hat{\mathbf{n}}\right],$$

so that

$$\begin{aligned} \mathbf{r}' &= \mathbf{r} + 2\cos\frac{\theta}{2}\sin\frac{\theta}{2}(\hat{\mathbf{n}}\times\mathbf{r}) + 2\sin^2\frac{\theta}{2}(\hat{\mathbf{n}}\times(\hat{\mathbf{n}}\times\mathbf{r})) \\ &= \mathbf{r} + \sin\theta\,(\hat{\mathbf{n}}\times\mathbf{r}) + (1-\cos\theta)\big(\hat{\mathbf{n}}\times(\hat{\mathbf{n}}\times\mathbf{r})\big) \\ &=: O^{\theta}_{\hat{\mathbf{n}}}\,\mathbf{r} . \end{aligned} \tag{3.2.6}$$

This is the famous ***Euler–Rodrigues formula*** *for the rotation* $O^{\theta}_{\hat{\mathbf{n}}}\,\mathbf{r}$ *of a vector* $\mathbf{r}$ *by an angle* θ *about the unit vector* $\hat{\mathbf{n}}$.

Exercise. Verify the Euler–Rodrigues formula (3.2.6) by a direct computation using quaternionic multiplication. ★

Exercise. Write formula (3.2.5) for conjugation of a pure quaternion by a unit quaternion $q_0{}^2 + \mathbf{q}\cdot\mathbf{q} = 1$ as a 3×3 matrix operation acting on a vector. ★

Answer. As a 3×3 matrix operation acting on a vector, $\boldsymbol{r}' = O_{3\times3}\boldsymbol{r}$, formula (3.2.5) becomes

$$\begin{aligned}\boldsymbol{r}' &= \boldsymbol{r} + 2q_0(\boldsymbol{q}\times\boldsymbol{r}) + 2\boldsymbol{q}\times(\boldsymbol{q}\times\boldsymbol{r}) \\ &= \Big[(2{q_0}^2 - 1)Id + 2q_0\widehat{q} + 2\mathbf{q}\mathbf{q}^T\Big]\boldsymbol{r} =: O_{3\times3}\boldsymbol{r},\end{aligned}$$

where $\widehat{q} = \boldsymbol{q}\times$, or in components $\widehat{q}_{lm} = -q^k\epsilon_{klm}$ by the hat map in (2.1.10) and (2.1.11). When $\mathfrak{q} = [q_0, \boldsymbol{q}]$ is a unit quaternion, the Euler–Rodrigues formula implies $O_{3\times3} \in SO(3)$. ▲

Definition 3.2.3 (Euler parameters) *In the Euler–Rodrigues formula (3.2.6) for the rotation of vector* $\mathbf{r}$ *by angle* θ *about* $\hat{\mathbf{n}}$, *the quantities* θ, $\hat{\mathbf{n}}$ *are called the* ***Euler parameters.***

Definition 3.2.4 (Cayley–Klein parameters) *The unit quaternion* $\hat{\mathfrak{q}} = [q_0\,, \mathbf{q}]$ *corresponding to the rotation of a pure quaternion* $\mathfrak{r} = [0\,, \mathbf{r}]$ *by angle* θ *about* $\hat{\mathbf{n}}$ *using quaternionic conjugation is*

$$\hat{\mathfrak{q}} := \pm\Big[\cos\frac{\theta}{2}\,,\ \sin\frac{\theta}{2}\,\hat{\mathbf{n}}\Big]. \tag{3.2.7}$$

The quantities $q_0 = \pm\cos\frac{\theta}{2}$ *and* $\mathbf{q} = \pm\sin\frac{\theta}{2}\,\hat{\mathbf{n}}$ *in (3.2.7) are called the* ***Cayley–Klein parameters.***

Remark 3.2.4 (Cayley–Klein coordinates of a quaternion) An arbitrary quaternion may be written in terms of its magnitude and its Cayley–Klein parameters as

$$\mathfrak{q} = |\mathfrak{q}|\hat{\mathfrak{q}} = |\mathfrak{q}|\Big[\cos\frac{\theta}{2},\ \sin\frac{\theta}{2}\,\hat{\boldsymbol{n}}\Big]. \tag{3.2.8}$$

□

The calculation of the Euler–Rodrigues formula (3.2.6) shows the equivalence of quaternionic conjugation and rotations of vectors. Moreover, compositions of quaternionic products imply the following.

Corollary 3.2.2 *Composition of rotations*

$$O^{\theta'}_{\hat{\mathbf{n}}'} O^{\theta}_{\hat{\mathbf{n}}} \mathbf{r} = \hat{\mathfrak{q}}'(\hat{\mathfrak{q}}\, \mathfrak{r}\, \hat{\mathfrak{q}}^*)\hat{\mathfrak{q}}'^*$$

is equivalent to multiplication of (±) *unit quaternions.*

Exercise. Show directly by quaternionic multiplication that

$$O^{\pi/2}_{\hat{\mathbf{y}}} O^{\pi/2}_{\hat{\mathbf{x}}} = O^{2\pi/3}_{\hat{n}} \quad \text{with} \quad \hat{n} = (\hat{\mathbf{x}} + \hat{\mathbf{y}} - \hat{\mathbf{z}})/\sqrt{3}\,.$$

★

Answer. One multiplies the corresponding unit quaternions, yielding

$$\begin{aligned}
\left[\frac{1}{\sqrt{2}}, \hat{\mathbf{y}}\frac{1}{\sqrt{2}}\right]\left[\frac{1}{\sqrt{2}}, \hat{\mathbf{x}}\frac{1}{\sqrt{2}}\right] &= \left[\frac{1}{2}, \frac{1}{2}(\hat{\mathbf{x}} + \hat{\mathbf{y}} - \hat{\mathbf{z}})\right] \\
&= \left[\frac{1}{2}, \frac{1}{\sqrt{3}}(\hat{\mathbf{x}} + \hat{\mathbf{y}} - \hat{\mathbf{z}})\frac{\sqrt{3}}{2}\right] \\
&= \left[\cos\frac{\pi}{3}, \frac{1}{\sqrt{3}}(\hat{\mathbf{x}} + \hat{\mathbf{y}} - \hat{\mathbf{z}})\sin\frac{\pi}{3}\right] \\
&= \left[\cos\frac{\theta}{2}, \sin\frac{\theta}{2}\,\hat{n}\right],
\end{aligned}$$

which is the Cayley–Klein form for a rotation of

$$\theta = 2\pi/3 \quad \text{about} \quad \hat{n} = (\hat{\mathbf{x}} + \hat{\mathbf{y}} - \hat{\mathbf{z}})/\sqrt{3}\,.$$

▲

Exercise. Compute $O^{\pi}_{\hat{\mathbf{y}}} O^{\pi}_{\hat{\mathbf{x}}} - O^{\pi}_{\hat{\mathbf{x}}} O^{\pi}_{\hat{\mathbf{y}}}$ by quaternionic multiplication. Does it vanish? Prove it. ★

Remark 3.2.5 (Cayley–Klein parameters for three-vectors) Consider the unit Cayley–Klein quaternion, $\hat{\mathfrak{p}} := \pm[\cos\frac{\theta}{2}\,,\,\sin\frac{\theta}{2}\,\hat{\chi}]$. Then the decompositions for quaternions (3.1.26) and for vectors (3.1.27) may be set equal to find

$$
\begin{aligned}
[0,\,\hat{\mathbf{v}}] &:= |\mathbf{v}|^{-1}[0,\,\mathbf{v}] = \hat{\mathfrak{p}}[0,\,\hat{\omega}]\hat{\mathfrak{p}}^* \\
&= [0, \cos\theta\,\hat{\omega} + \sin\theta\,\hat{\chi}\times\hat{\omega}] \\
&= |\mathbf{v}|^{-1}[\alpha,\,\chi][0,\hat{\omega}] \\
&= (\alpha^2+\chi^2)^{-1/2}\,[0, \alpha\,\hat{\omega} + \chi\times\hat{\omega}]\,.
\end{aligned}
$$

Thus, the unit vector $\hat{\mathbf{v}} = |\mathbf{v}|^{-1}\mathbf{v}$ is a rotation of $\hat{\omega}$ by angle θ around $\hat{\chi}$ with

$$
\cos\theta = \frac{\alpha}{(\alpha^2+\chi^2)^{1/2}} \quad \text{and} \quad \sin\theta = \frac{\chi}{(\alpha^2+\chi^2)^{1/2}}\,.
$$

Hence, the alignment parameters α and χ in (3.1.26) and (3.1.27) define the three-vector $\mathbf{v}$ in $[0,\,\mathbf{v}] = [\alpha,\,\chi][0,\hat{\omega}]$ as a stretching of $\hat{\omega}$ by $(\alpha^2+\chi^2)^{1/2}$ and a rotation of $\hat{\omega}$ by $\theta = \tan^{-1}\chi/\alpha$ about $\hat{\chi}$. The Cayley–Klein angle θ is the ***relative angle*** between the directions $\hat{\mathbf{v}}$ and $\hat{\omega}$. □

3.2.2 Pure quaternions, Pauli matrices and $SU(2)$

Exercise. Write the product of two pure unit quaternions as a multiplication of Pauli matrices. ★

Answer. By the quaternionic multiplication rule (3.1.5), one finds

$$
[0\,,\,\hat{\mathbf{v}}][0\,,\,\hat{\mathbf{w}}] = [-\,\hat{\mathbf{v}}\cdot\hat{\mathbf{w}}\,,\,\hat{\mathbf{v}}\times\hat{\mathbf{w}}] =: [\cos\theta\,,\,\hat{\boldsymbol{n}}\sin\theta]\,. \quad (3.2.9)
$$

Here $\hat{\mathrm{v}} \cdot \hat{\mathrm{w}} = -\cos\theta$, so that θ is the relative angle between the unit three-vectors $\hat{\mathrm{v}}$ and $\hat{\mathrm{w}}$, and $\hat{\mathrm{v}} \times \hat{\mathrm{w}} = \hat{\boldsymbol{n}} \sin\theta$ is their cross product, satisfying

$$|\hat{\mathrm{v}} \times \hat{\mathrm{w}}|^2 = |\hat{\mathrm{v}}|^2 |\hat{\mathrm{w}}|^2 - (\hat{\mathrm{v}} \cdot \hat{\mathrm{w}})^2 = 1 - \cos^2\theta = \sin^2\theta\,.$$

This is equivalent to following the multiplication of Pauli matrices,

$$\begin{aligned}(-i\hat{\mathrm{v}} \cdot \boldsymbol{\sigma})(-i\hat{\mathrm{w}} \cdot \boldsymbol{\sigma}) &= -\hat{\mathrm{v}} \cdot \hat{\mathrm{w}}\, \sigma_0 - i\, \hat{\mathrm{v}} \times \hat{\mathrm{w}} \cdot \boldsymbol{\sigma} \\ &= -\left(\cos\theta\, \sigma_0 + i\, \sin\theta\, \hat{\boldsymbol{n}} \cdot \boldsymbol{\sigma}\right),\end{aligned} \tag{3.2.10}$$

with, e.g., $\hat{\boldsymbol{n}} \cdot \boldsymbol{\sigma} = \sum_{a=1}^{3} \hat{n}_a \sigma_a$. ▲

Proposition 3.2.1 (De Moivre's theorem for quaternions) *De Moivre's theorem for complex numbers of unit modulus is*

$$(\cos\theta + i \sin\theta)^m = (\cos m\theta + i \sin m\theta)\,.$$

The analogue of De Moivre's theorem for unit quaternions is

$$[\cos\theta,\ \sin\theta\hat{\mathbf{n}}]^m = [\cos m\theta,\ \sin m\theta\hat{\mathbf{n}}]\,.$$

Proof. The proof follows immediately from the Cayley–Klein representation of a unit quaternion. ■

Theorem 3.2.1 *The unit quaternions form a representation of the matrix Lie group $SU(2)$.*

Proof. The matrix representation of a unit quaternion is given in (3.1.8). Let $\hat{\mathrm{q}} = [q_0,\, \boldsymbol{q}]$ be a unit quaternion ($|\hat{\mathrm{q}}|^2 = q_0^2 + \boldsymbol{q} \cdot \boldsymbol{q} = 1$) and define the matrix Q by

$$\begin{aligned} Q &= q_0\sigma_0 - i\boldsymbol{q} \cdot \boldsymbol{\sigma} \\ &= \begin{bmatrix} q_0 - iq_3 & -iq_1 - q_2 \\ -iq_1 + q_2 & q_0 + iq_3 \end{bmatrix}. \end{aligned} \tag{3.2.11}$$

The matrix Q is a unitary 2×2 matrix ($QQ^\dagger = Id$) with unit determinant ($\det Q = 1$). That is, $Q \in SU(2)$. In fact, we may rewrite the map (3.2.1) for quaternionic conjugation of a vector $\mathfrak{r} = [0, \boldsymbol{r}]$ by a unit quaternion equivalently in terms of unitary conjugation of the Hermitian Pauli spin matrices as

$$\mathfrak{r}' = \hat{\mathfrak{q}}\,\mathfrak{r}\,\hat{\mathfrak{q}}^* \iff \mathbf{r}' \cdot \boldsymbol{\sigma} = Q\,\mathbf{r} \cdot \boldsymbol{\sigma} Q^\dagger \,, \tag{3.2.12}$$

with

$$\mathbf{r} \cdot \boldsymbol{\sigma} = \begin{bmatrix} r_3 & r_1 - ir_2 \\ r_1 + ir_2 & -r_3 \end{bmatrix} . \tag{3.2.13}$$

This is the standard representation of $SO(3)$ rotations as a double covering ($\pm J$) by $SU(2)$ matrices, which is now seen to be equivalent to quaternionic multiplication. ■

Remark 3.2.6 A variant of the map (3.2.11), known as the Kustaanheimo–Stiefel map [KuSt1965], establishes a relation between the solutions of a constrained isotropic harmonic oscillator in four dimensions and those of the Kepler problem in three dimensions. The KS map is beyond our present scope. However, for an interesting discussion of it, see [Co2003]. □

Remark 3.2.7 Composition of $SU(2)$ matrices by matrix multiplication forms a Lie subgroup of the Lie group of 2×2 complex matrices $GL(2, \mathbb{C})$, see, e.g., [MaRa1994]. □

Exercise. Check that the matrix Q in (3.2.11) is a special unitary matrix so that $Q \in SU(2)$. That is, show that Q is unitary and has unit determinant. ★

Exercise. Verify the conjugacy formula (3.2.12) arising from the isomorphism between unit quaternions and $SU(2)$. ★

Remark 3.2.8 The $(\pm)$ in the Cayley–Klein parameters reflects the 2:1 covering of the map $SU(2) \to SO(3)$. □

3.2.3 Tilde map: $\mathbb{R}^3 \simeq su(2) \simeq so(3)$

The following ***tilde map*** may be defined by considering the isomorphism (3.1.8) for a pure quaternion $[0,\, \boldsymbol{q}\,]$. Namely,

$$\boldsymbol{q} \in \mathbb{R}^3 \mapsto -i\,\boldsymbol{q} \cdot \boldsymbol{\sigma} \;=\; -\,i \sum_{j=1}^{3} q_j \sigma_j \tag{3.2.14}$$

$$= \begin{bmatrix} -iq_3 & -iq_1 - q_2 \\ -iq_1 + q_2 & iq_3 \end{bmatrix} =: \tilde{\boldsymbol{q}} \in su(2)\,.$$

The tilde map (3.2.14) is a Lie algebra isomorphism between $\mathbb{R}^3$ with the cross product of vectors and the Lie algebra $su(2)$ of 2×2 skew-Hermitian traceless matrices. Just as in the hat map one writes

$$JJ^\dagger(t) = Id \quad \Longrightarrow \quad \dot{J}J^\dagger + (\dot{J}J^\dagger)^\dagger = 0\,,$$

so the tangent space at the identity for the $SU(2)$ matrices comprises 2×2 skew-Hermitian traceless matrices, whose basis is $-i\,\boldsymbol{\sigma}$, the imaginary number $(-i)$ times the three Pauli matrices. This completes the ***circle of the isomorphisms*** between Pauli matrices and quaternions, and between *pure* quaternions and vectors in $\mathbb{R}^3$. In particular, their Lie products are all isomorphic. That is,

$$\boxed{\mathrm{Im}(\mathfrak{pq}) = \frac{1}{2}\Big(\mathfrak{pq} - \mathfrak{qp}\Big) = [\tilde{\boldsymbol{p}},\, \tilde{\boldsymbol{q}}] = (\boldsymbol{p} \times \boldsymbol{q})\tilde{}\,.} \tag{3.2.15}$$

In addition, recalling that $\mathrm{Re}(\mathfrak{p}\mathfrak{q}^*) = [\boldsymbol{p}\cdot\boldsymbol{q}\,,\,0]$ helps prove the following identities:

$$\det(\boldsymbol{q}\cdot\boldsymbol{\sigma}) = |\boldsymbol{q}|^2\,, \quad (\tilde{\boldsymbol{p}}\tilde{\boldsymbol{q}}) = -\,\boldsymbol{p}\cdot\boldsymbol{q}\,.$$

3.2.4 Dual of the tilde map: $\mathbb{R}^{3*} \simeq su(2)^* \simeq so(3)^*$

One may identify $su(2)^*$ with $\mathbb{R}^3$ via the map $\mu \in su(2)^* \to \breve{\boldsymbol{\mu}} \in \mathbb{R}^3$ defined by

$$\breve{\boldsymbol{\mu}}\cdot\mathbf{q} := \big\langle \mu, \tilde{\mathbf{q}}\big\rangle_{su(2)^*\times su(2)}$$

for any $\mathbf{q} \in \mathbb{R}^3$.

Then, for example,

$$\breve{\boldsymbol{\mu}}\cdot(\boldsymbol{p}\times\boldsymbol{q}) := \big\langle \mu, [\tilde{\boldsymbol{p}},\,\tilde{\boldsymbol{q}}]\big\rangle_{su(2)^*\times su(2)},$$

which foreshadows the adjoint and coadjoint actions of $SU(2)$ to appear in our discussions of rigid-body dynamics in Chapter 5 and momentum maps in Chapter 11.

3.2.5 Pauli matrices and Poincaré's sphere $\mathbb{C}^2 \to S^2$

The Lie algebra isomorphisms given by the Pauli matrix representation of the quaternions (3.1.8) and the tilde map (3.2.14) are related to a map $\mathbb{C}^2 \mapsto S^2$ first introduced by Poincaré [Po1892] and later studied by Hopf [Ho1931]. Consider for $a_k \in \mathbb{C}^2$, with $k = 1, 2$ the four real combinations written in terms of the Pauli matrices

$$n_\alpha = \sum_{k,l=1}^{2} a_k^*\{\sigma_\alpha\}_{kl}\, a_l \quad \text{with} \quad \alpha = 0, 1, 2, 3\,. \tag{3.2.16}$$

The $n_\alpha \in \mathbb{R}^4$ have components

$$\begin{aligned} n_0 &= |a_1|^2 + |a_2|^2\,, \\ n_3 &= |a_1|^2 - |a_2|^2\,, \\ n_1 + i\,n_2 &= 2a_1^* a_2\,. \end{aligned} \tag{3.2.17}$$

Remark 3.2.9 One may motivate the definition of $n_\alpha \in \mathbb{R}^4$ in (3.2.16) by introducing the following Hermitian matrix,

$$\rho = \mathbf{a} \otimes \mathbf{a}^* = \frac{1}{2}\Big(n_0\sigma_0 + \mathbf{n}\cdot\boldsymbol{\sigma}\Big)\,, \tag{3.2.18}$$

in which the vector $\mathbf{n}$ is defined as

$$\mathbf{n} = \operatorname{tr}\rho\,\boldsymbol{\sigma} = a_l a_k^* \boldsymbol{\sigma}_{kl}\,. \tag{3.2.19}$$

The last equation recovers (3.2.16). We will return to the interpretation of this map when we discuss momentum maps in Chapter 11. For now, we simply observe that the components of the singular Hermitian matrix ($\det\rho = 0$)

$$\rho = \mathbf{a} \otimes \mathbf{a}^* = \frac{1}{2}\begin{bmatrix} n_0 + n_3 & n_1 - in_2 \\ n_1 + in_2 & n_0 - n_3 \end{bmatrix}$$

are all invariant under the diagonal action

$$S^1 : \mathbf{a} \to e^{i\phi}\mathbf{a},\ \mathbf{a}^* \to e^{-i\phi}\mathbf{a}^*.$$

□

A fixed value $n_0 = const$ defines a three-sphere $S^3 \in \mathbb{R}^4$. Moreover, because $\det\rho = 0$ the remaining three components satisfy an additional relation which defines the ***Poincaré sphere*** $S^2 \in S^3$ as

$$n_0^2 = n_1^2 + n_2^2 + n_3^2 = |\mathbf{n}|^2\,. \tag{3.2.20}$$

Each point on this sphere defines a direction introduced by Poincaré to represent polarised light. The north (resp. south) pole represents right (resp. left) circular polarisation and the equator represents the various inclinations of linear polarisation. Off the equator and the poles the remaining directions in the upper and lower hemispheres represent right- and left-handed elliptical polarisations, respectively. Opposing directions $\pm\mathbf{n}$ correspond to orthogonal polarisations. See [BoWo1965] for details of the physical interpretation of the Poincaré sphere for polarised ray optics.

3.2.6 Poincaré's sphere and Hopf's fibration

The same map $S^3 \mapsto S^2$ given by (3.2.16) from the $n_0 = const$ S^3 to the Poincaré sphere S^2 was later studied by Hopf [Ho1931], who found it to be a ***fibration*** of S^3 over S^2. That is, $S^3 \simeq S^2 \times S^1$ locally, where S^1 is the fibre. A fibre bundle structure is defined descriptively, as follows.

Definition 3.2.5 (Fibre bundle) *In topology, a* ***fibre bundle*** *is a space which locally looks like a product of two spaces but may possess a different global structure. Every fibre bundle consists of a continuous surjective map $\pi : E \mapsto B$, where small regions in the total space E look like small regions in the product space $B \times F$, of the* ***base space*** *B with the* ***fibre space*** *F (Figure 3.1). Fibre bundles comprise a rich mathematical subject that is explained more completely in, e.g., [Is1999, La1999, Wa1983, Sp1979]. We shall confine our attention here to the one particular case leading to the Poincaré sphere.*

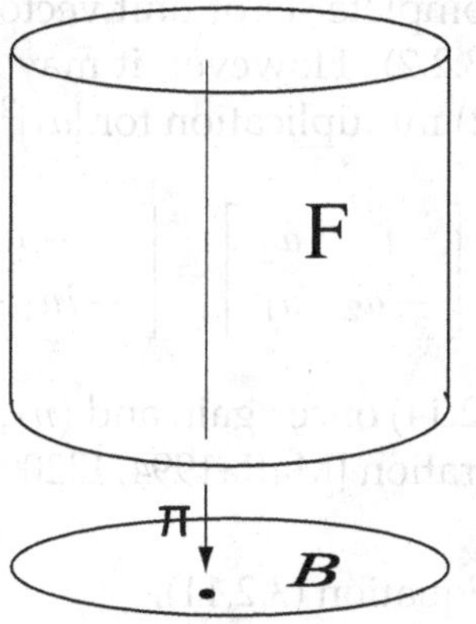

Figure 3.1. A fibre bundle E looks locally like the product space $B \times F$, of the base space B with the fibre space F. The map $\pi : E \approx B \times F \mapsto B$ projects E onto the base space B.

Remark 3.2.10 The ***Hopf fibration,*** or fibre bundle, $S^3 \simeq S^2 \times S^1$ has spheres as its total space, base space and fibre, respectively. In terms of the Poincaré sphere one may think of the Hopf fibration locally as a sphere S^2 which has a great circle S^1 attached at every point. The phase on the great circles at opposite points are orthogonal (rotated by $\pi/2$, not π); so passing once around the Poincaré sphere along a great circle rotates the S^1 phase only by π, not 2π. One must pass twice around a great circle on the Poincaré sphere to return to the original phase. Thus, the relation $S^3 \simeq S^2 \times S^1$ only holds locally, not globally. □

Remark 3.2.11 The conjugacy classes of S^3 by unit quaternions yield the family of two-spheres S^2 in formula (3.2.4) of Corollary 3.2.1. These also produce a version of the Hopf fibration $S^3 \simeq S^2 \times S^1$, obtained by identifying the Poincaré sphere (3.2.20) from the definitions (3.2.17). □

Remark 3.2.12 (Hopf fibration/quaternionic conjugation) Conjugating the pure unit quaternion along the z-axis $[0, \hat{\mathbf{z}}]$ by the other unit quaternions yields the entire unit two-sphere S^2. This is to be expected from the complete set of unit vectors found by quaternionic conjugation in (3.2.2). However, it may be shown explicitly by computing the $SU(2)$ multiplication for $|a_1|^2 + |a_2|^2 = 1$,

$$\begin{bmatrix} a_1 & -a_2^* \\ a_2 & a_1^* \end{bmatrix} \begin{bmatrix} -i & 0 \\ 0 & i \end{bmatrix} \begin{bmatrix} a_1^* & a_2^* \\ -a_2 & a_1 \end{bmatrix} = \begin{bmatrix} -in_3 & -in_1 + n_2 \\ -in_1 - n_2 & in_3 \end{bmatrix}.$$

This is the tilde map (3.2.14) once again and (n_1, n_2, n_3) are the components of the Hopf fibration [MaRa1994, El2007]. □

In other words, cf. Equation (3.2.11),

$$-ig\sigma_3 g^\dagger = -i\mathbf{n} \cdot \boldsymbol{\sigma},$$

in which $g^\dagger = g^{-1} \in SU(2)$ and $|\mathbf{n}|^2 = 1$. This is the tilde map (3.2.14) once again and (n_1, n_2, n_3) are the spatial components of the Hopf fibration in (3.2.17).

Thus, the isomorphism given in (3.1.8) and (3.2.14) between the unit quaternions and $SU(2)$ expressed in terms of the Pauli spin matrices connects the quaternions to the mathematics of Poincaré's sphere $\mathbb{C}^2 \mapsto S^2$, Hopf's fibration $S^3 \simeq S^2 \times S^1$ and the geometry of fibre bundles. This is a very deep network of connections that will amply reward the efforts of further study.

Exercise. Show that the Hopf fibration is a decomposition law for the group $SU(2)$.

Hint: Write the Hopf fibration in quaternionic form. ★

Exercise. Write the quaternionic version of unitary transformations of Hermitian matrices.

Hint: The Pauli spin matrices defined in (3.1.9) are Hermitian. To get started, you may want to take a look at Equation (3.2.12). ★

Exercise. Write the quaternionic version of orthogonal transformations of symmetric matrices as in Section 2.5.6. ★

Remark 3.2.13 In quantum mechanics, the quantity corresponding to the Hermitian matrix $\rho = \mathbf{a} \otimes \mathbf{a}^*$ in Equation (3.2.18) is called the ***density matrix*** and the Poincaré sphere in Equation (3.2.20) is called the ***Bloch sphere***. □

3.2.7 Coquaternions

Not long after Hamilton discovered quaternions, James Cockle [Co1848] proposed an alternative field of numbers called ***co-quaternions***. A coquaternion $\mathfrak{c} = [w, x, y, z]$ is defined by four real numbers, as follows:

$$\mathfrak{c} = w\mathbb{1} + x\mathsf{i} + y\mathsf{j} + z\mathsf{k}, \quad \text{with} \quad (w, x, y, z) \in \mathbb{R}^4,$$

where $\mathbb{1}$ is the identity element and the other three coquaternion basis elements $(\mathsf{i}, \mathsf{j}, \mathsf{k})$ satisfy the multiplication rules,

$$\mathsf{ij} = \mathsf{k} = -\mathsf{ji}, \quad \mathsf{jk} = -\mathsf{i} = -\mathsf{kj}, \quad \mathsf{ki} = \mathsf{j} = -\mathsf{ik},$$
$$\mathbb{1}^2 = \mathbb{1}, \quad \mathsf{i}^2 = -\mathbb{1}, \quad \mathsf{j}^2 = \mathbb{1}, \quad \mathsf{k}^2 = \mathbb{1}, \quad \mathsf{ijk} = \mathbb{1}\,.$$

The multiplication table of the coquaternion basis may be expressed as

	$\mathbb{1}$	i	j	k
$\mathbb{1}$	$\mathbb{1}$	i	j	k
i	i	$-\mathbb{1}$	k	$-\mathsf{j}$
j	j	$-\mathsf{k}$	$\mathbb{1}$	$-\mathsf{i}$
k	k	j	i	$\mathbb{1}$

. (3.2.21)

The multiplication rules for coquaternions may be represented in a basis of real 2×2 matrices,

$$\mathbb{1} = \begin{bmatrix} 1 & 0 \\ 0 & 1 \end{bmatrix}, \quad \mathsf{i} = \begin{bmatrix} 0 & 1 \\ -1 & 0 \end{bmatrix},$$
$$\mathsf{j} = \begin{bmatrix} 0 & 1 \\ 1 & 0 \end{bmatrix}, \quad \mathsf{k} = \begin{bmatrix} 1 & 0 \\ 0 & -1 \end{bmatrix}. \tag{3.2.22}$$

This means the coquaternion $\mathfrak{c} = [w, x, y, z]$ may be represented as a *real matrix*

$$C = \begin{bmatrix} w+z & y+x \\ y-x & w-z \end{bmatrix}, \tag{3.2.23}$$

with determinant $\det(C) = w^2 + x^2 - y^2 - z^2$.

One defines the ***conjugate*** of the coquaternion $\mathfrak{c} = [w, x, y, z]$ as $\mathfrak{c}^* = [w, -x, -y, -z]$,

$$\mathfrak{c}^* = w\mathbb{1} - x\mathsf{i} - y\mathsf{j} - z\mathsf{k}.$$

Thus, in a matrix representation, the conjugate coquaternion is expressed as

$$C^* = \begin{bmatrix} w - z & -y - x \\ -y + x & w + z \end{bmatrix}. \tag{3.2.24}$$

An ***inner product*** for coquaternions based on the matrix inner product may be defined as

$$(\mathfrak{c}_1,\, \mathfrak{c}_2) = \mathfrak{c}_1^* \cdot \mathfrak{c}_2 := \frac{1}{2}\mathrm{tr}(C_1^* C_2) = w_1 w_2 + x_1 x_2 - y_1 y_2 - z_1 z_2.$$

The inner product of a coquaternion with its conjugate then defines its squared magnitude

$$|\mathfrak{c}|^2 = \mathfrak{c}^* \cdot \mathfrak{c} := \frac{1}{2}\mathrm{tr}(C^* C) = w^2 + x^2 - y^2 - z^2 = \det(C),$$

which is indefinite in sign.

Exercise. Show that the unit coquaternions with $|\mathfrak{c}|^2 = 1$ form a representation of the Lie group of 2×2 symplectic matrices $Sp(2, \mathbb{R})$.

Hint: A symplectic matrix $M \in Sp(2, \mathbb{R})$ satisfies

$$MJM^T = J \quad \text{for} \quad J = \begin{bmatrix} 0 & -1 \\ 1 & 0 \end{bmatrix}. \tag{3.2.25}$$

★

Remark 3.2.14 For a discussion of symplectic matrices in the context of the geometric mechanics approach to ray optics via Fermat's principle, see, e.g., [Ho2008]. □

Exercise. Follow the developments in the earlier part of this chapter for quaternions far enough to define conjugacy classes for the action of unit coquaternions on vectors in $\mathbb{R}^3$. ★

Exercise. Compute the Euler–Rodrigues formula for the coquaternions. ★

Remark 3.2.15 For a recent discussion of coquaternions in the study of complexified mechanics, see [BrGr2011]. □

4 ADJOINT AND COADJOINT ACTIONS

Contents

4.1 Cayley–Klein dynamics for the rigid body 112
- 4.1.1 Cayley–Klein parameters, rigid-body dynamics 112
- 4.1.2 Body angular frequency 113
- 4.1.3 Cayley–Klein parameters 115

4.2 Actions of quaternions, Lie groups and Lie algebras 116
- 4.2.1 AD, Ad, ad, Ad* and ad* actions of quaternions 117
- 4.2.2 AD, Ad, and ad for Lie algebras and groups 118

4.3 Example: The Heisenberg Lie group 124
- 4.3.1 Definitions for the Heisenberg group 124
- 4.3.2 Adjoint actions: AD, Ad and ad 126
- 4.3.3 Coadjoint actions: Ad* and ad* 127
- 4.3.4 Coadjoint motion and harmonic oscillations 129

4.1 Cayley–Klein dynamics for the rigid body

4.1.1 Cayley–Klein parameters, rigid-body dynamics

Recall that a time-dependent rotation by angle $\theta(t)$ about direction $\hat{\boldsymbol{n}}(t)$ is given by the unit quaternion (3.2.7) in Cayley–Klein parameters,

$$\hat{\mathfrak{q}}(t) = [q_0(t), \mathbf{q}(t)] = \pm\Big[\cos\frac{\theta(t)}{2}\,,\ \sin\frac{\theta(t)}{2}\,\hat{\boldsymbol{n}}(t)\Big]. \tag{4.1.1}$$

The ***operation*** of the unit quaternion $\hat{\mathfrak{q}}(t)$ on a vector $\mathbf{X} = [0,\, \mathbf{X}]$ is given by quaternionic multiplication as

$$\begin{aligned}\hat{\mathfrak{q}}(t)\mathbf{X} &= [q_0, \mathbf{q}][0,\, \mathbf{X}] = [-\mathbf{q}\cdot\mathbf{X}\,,\ q_0\mathbf{X} + \mathbf{q}\times\mathbf{X}] \qquad (4.1.2)\\ &= \pm\Big[-\sin\frac{\theta(t)}{2}\,\hat{\boldsymbol{n}}(t)\cdot\mathbf{X},\ \cos\frac{\theta(t)}{2}\,\mathbf{X} + \sin\frac{\theta(t)}{2}\,\hat{\boldsymbol{n}}(t)\times\mathbf{X}\Big].\end{aligned}$$

The corresponding time-dependent rotation is given by the Euler–Rodrigues formula (3.2.6) as

$$\mathbf{x}(t) = \hat{\mathfrak{q}}(t)\mathbf{X}\hat{\mathfrak{q}}^*(t) \quad \text{so that} \quad \mathbf{X} = \hat{\mathfrak{q}}^*(t)\mathbf{x}(t)\hat{\mathfrak{q}}(t) \tag{4.1.3}$$

in terms of the unit quaternion $\hat{\mathfrak{q}}(t)$. Its time derivative is given by

$$\begin{aligned}\dot{\mathbf{x}}(t) &= \dot{\hat{\mathfrak{q}}}\hat{\mathfrak{q}}^*\mathbf{x}\hat{\mathfrak{q}}\hat{\mathfrak{q}}^* + \hat{\mathfrak{q}}\hat{\mathfrak{q}}^*\mathbf{x}\hat{\mathfrak{q}}\dot{\hat{\mathfrak{q}}}^* = \dot{\hat{\mathfrak{q}}}\hat{\mathfrak{q}}^*\mathbf{x} + \mathbf{x}\hat{\mathfrak{q}}\dot{\hat{\mathfrak{q}}}^*\\ &= \dot{\hat{\mathfrak{q}}}\hat{\mathfrak{q}}^*\mathbf{x} + \mathbf{x}(\dot{\hat{\mathfrak{q}}}\hat{\mathfrak{q}}^*)^* = \dot{\hat{\mathfrak{q}}}\hat{\mathfrak{q}}^*\mathbf{x} - \mathbf{x}(\dot{\hat{\mathfrak{q}}}\hat{\mathfrak{q}}^*)\\ &= \dot{\hat{\mathfrak{q}}}\hat{\mathfrak{q}}^*\mathbf{x} - \big((\dot{\hat{\mathfrak{q}}}\hat{\mathfrak{q}}^*)\mathbf{x}\big)^*\\ &= 2\mathrm{Im}\big((\dot{\hat{\mathfrak{q}}}\hat{\mathfrak{q}}^*)\mathbf{x}\big)\\ &= 2(\dot{\hat{\mathfrak{q}}}\hat{\mathfrak{q}}^*)\mathbf{x}.\end{aligned} \tag{4.1.4}$$

In quaternion components, this equation may be rewritten using

$$2\dot{\hat{\mathfrak{q}}}\hat{\mathfrak{q}}^* = [\,0\,,\ \dot{\theta}\hat{\boldsymbol{n}} + \sin\theta\,\dot{\hat{\boldsymbol{n}}} + (1-\cos\theta)\hat{\boldsymbol{n}}\times\dot{\hat{\boldsymbol{n}}}\,]. \tag{4.1.5}$$

The quantity $2\dot{\hat{\mathfrak{q}}}\hat{\mathfrak{q}}^* = [0,\, 2(\dot{\hat{\mathfrak{q}}}\hat{\mathfrak{q}}^*)]$ is a pure quaternion whose vector component is denoted for the moment by enclosing it in parentheses as

$$2(\dot{\hat{\mathfrak{q}}}\hat{\mathfrak{q}}^*) = \dot{\theta}\hat{\boldsymbol{n}} + \sin\theta\,\dot{\hat{\boldsymbol{n}}} + (1-\cos\theta)\hat{\boldsymbol{n}}\times\dot{\hat{\boldsymbol{n}}}\,.$$

As a consequence, the vector component of the quaternion equation (4.1.4) becomes

$$[0,\, \dot{\mathbf{x}}(t)] = 2[0,\, (\dot{\hat{\mathfrak{q}}}\hat{\mathfrak{q}}^*)]\,[0,\, \mathbf{x}] = 2[\,0\,,\, (\dot{\hat{\mathfrak{q}}}\hat{\mathfrak{q}}^*) \times \mathbf{x}\,]\,. \tag{4.1.6}$$

Spatial angular frequency

Upon recalling the isomorphism provided by the Euler–Rodrigues formula (3.2.6) for finite rotations,

$$\mathbf{x}(t) = O(t)\mathbf{X} = \hat{\mathfrak{q}}(t)\mathbf{X}\hat{\mathfrak{q}}^*(t)\,, \tag{4.1.7}$$

the vector component of (4.1.6) yields a series of isomorphisms for the angular frequency,

$$\begin{aligned} \dot{\mathbf{x}}(t) &= \dot{O}O^{-1}(t)\mathbf{x} \\ &= \hat{\omega}(t)\mathbf{x} \\ &= \boldsymbol{\omega}(t) \times \mathbf{x} \\ &= 2(\dot{\hat{\mathfrak{q}}}\hat{\mathfrak{q}}^*)(t) \times \mathbf{x}\,. \end{aligned} \tag{4.1.8}$$

Since the quaternion $2\dot{\hat{\mathfrak{q}}}\hat{\mathfrak{q}}^*(t) = [\,0\,,\, 2(\dot{\hat{\mathfrak{q}}}\hat{\mathfrak{q}}^*)(t)\,]$ is equivalent to a vector in $\mathbb{R}^3$, we may simply use vector notation for it and equate the spatial angular frequencies as ***vectors***. That is, we shall write

$$\boldsymbol{\omega}(t) = 2\dot{\hat{\mathfrak{q}}}\hat{\mathfrak{q}}^*(t)\,, \tag{4.1.9}$$

and drop the parentheses $(\,\cdot\,)$ when identifying pure quaternions with angular velocity vectors.

Remark 4.1.1 Pure quaternions of the form $\dot{\hat{\mathfrak{q}}}\hat{\mathfrak{q}}^*(t)$ may be identified with the tangent space of the *unit* quaternions at the identity. □

4.1.2 Body angular frequency

The quaternion for the body angular frequency will have the corresponding vector expression,

$$\boldsymbol{\Omega}(t) = 2\hat{\mathfrak{q}}^*\dot{\hat{\mathfrak{q}}}(t)\,. \tag{4.1.10}$$

Thus, only the vector parts enter the quaternionic descriptions of the spatial and body angular frequencies. The resulting isomorphisms are entirely sufficient to express the quaternionic versions of the rigid-body equations of motion in their Newtonian, Lagrangian and Hamiltonian forms, including the Lie–Poisson brackets. In particular, the kinetic energy for the rigid body is given by

$$K = \frac{1}{2}\boldsymbol{\Omega}(t) \cdot \mathbb{I}\,\boldsymbol{\Omega}(t) = 2\Big\langle\, \hat{\mathfrak{q}}^*\dot{\hat{\mathfrak{q}}}(t)\,,\, \mathbb{I}\,\hat{\mathfrak{q}}^*\dot{\hat{\mathfrak{q}}}(t)\,\Big\rangle\,. \tag{4.1.11}$$

So the quaternionic description of rigid-body dynamics reduces to the equivalent description in $\mathbb{R}^3$.

This equivalence in the two descriptions of rigid-body dynamics means that the relations for angular momentum, Hamilton's principle and the Lie–Poisson brackets in terms of vector quantities all have identical expressions in the quaternionic picture. Likewise, the reconstruction of the Cayley–Klein parameters from the solution for the body angular velocity vector may be accomplished by integrating the linear quaternionic equation

$$\dot{\hat{\mathfrak{q}}}(t) = \hat{\mathfrak{q}}\boldsymbol{\Omega}(t)/2\,, \tag{4.1.12}$$

or explicitly,

$$\frac{d}{dt}\Big[\cos\frac{\theta}{2}\,,\, \sin\frac{\theta}{2}\,\hat{\boldsymbol{n}}\Big] = \Big[\cos\frac{\theta}{2}\,,\, \sin\frac{\theta}{2}\,\hat{\boldsymbol{n}}\Big]\big[0,\, \boldsymbol{\Omega}(t)/2\big]\,. \tag{4.1.13}$$

This is the linear reconstruction formula for the Cayley–Klein parameters.

Remark 4.1.2 Expanding this linear equation for the Cayley–Klein parameters leads to a quaternionic equation for the Euler parameters that is linear in $\boldsymbol{\Omega}(t)$, but is nonlinear in θ and $\hat{\boldsymbol{n}}$, namely

$$\big[\dot{\theta},\, \dot{\hat{\boldsymbol{n}}}\big] = \big[0,\, \hat{\boldsymbol{n}}\big]\Big[0,\, \boldsymbol{\Omega}(t) + \boldsymbol{\Omega}(t)\times\hat{\boldsymbol{n}}\cot\frac{\theta}{2}\Big]\,. \tag{4.1.14}$$

□

4.1.3 Cayley–Klein parameters

The series of isomorphisms in Equation (4.1.8) holds the key for rewriting Hamilton's principle in Proposition 2.4.1 for Euler's rigid-body equations in quaternionic form using Cayley–Klein parameters. The key step in proving Proposition 2.4.1 was deriving formula (2.4.7) for the variation of the body angular velocity. For this, one invokes equality of cross derivatives with respect to time t and variational parameter s. The hat map in that case then led to the key variational formula (2.4.7). The corresponding step for the quaternionic form of Hamilton's principle in Cayley–Klein parameters also produces the key formula needed in this case.

Proposition 4.1.1 (Cayley–Klein variational formula) *The variation of the pure quaternion* $\boldsymbol{\Omega} = 2\hat{\mathfrak{q}}^*\dot{\hat{\mathfrak{q}}}$ *corresponding to body angular velocity in Cayley–Klein parameters satisfies the identity*

$$\boldsymbol{\Omega}' - \dot{\boldsymbol{\Xi}} = (\boldsymbol{\Omega}\,\boldsymbol{\Xi} - \boldsymbol{\Xi}\,\boldsymbol{\Omega})/2 = \mathrm{Im}(\boldsymbol{\Omega}\,\boldsymbol{\Xi})\,, \tag{4.1.15}$$

where $\boldsymbol{\Xi} := 2\hat{\mathfrak{q}}^*\hat{\mathfrak{q}}'$ *and* $(\,\cdot\,)'$ *denotes variation.*

Proof. The body angular velocity is defined as $\boldsymbol{\Omega} = 2\hat{\mathfrak{q}}^*\dot{\hat{\mathfrak{q}}}$ in (4.1.10). Its variational derivative is found to be

$$\delta\boldsymbol{\Omega} := \frac{d}{ds}\boldsymbol{\Omega}(s)\Big|_{s=0} =: \boldsymbol{\Omega}'\,. \tag{4.1.16}$$

Thus, the variation of $\boldsymbol{\Omega}$ may be expressed as

$$\boldsymbol{\Omega}'/2 = (\hat{\mathfrak{q}}^*)'\dot{\hat{\mathfrak{q}}} + \hat{\mathfrak{q}}^*\dot{\hat{\mathfrak{q}}}'\,. \tag{4.1.17}$$

Now $\mathfrak{e} = \hat{\mathfrak{q}}^*\hat{\mathfrak{q}}$ so that

$$\mathfrak{e}' = 0 = (\hat{\mathfrak{q}}^*)'\hat{\mathfrak{q}} + \hat{\mathfrak{q}}^*\hat{\mathfrak{q}}' \quad \text{and} \quad (\hat{\mathfrak{q}}^*)' = -\,\hat{\mathfrak{q}}^*\hat{\mathfrak{q}}'\hat{\mathfrak{q}}^*\,. \tag{4.1.18}$$

Hence, the variation of the angular frequency becomes

$$\begin{aligned}\delta\boldsymbol{\Omega}/2 = \boldsymbol{\Omega}'/2 &= -\,\hat{\mathfrak{q}}^*\hat{\mathfrak{q}}'\hat{\mathfrak{q}}^*\dot{\hat{\mathfrak{q}}} + \hat{\mathfrak{q}}^*\dot{\hat{\mathfrak{q}}}' \\ &= -\,\boldsymbol{\Xi}\boldsymbol{\Omega}/4 + \hat{\mathfrak{q}}^*\dot{\hat{\mathfrak{q}}}'\,,\end{aligned} \tag{4.1.19}$$

where we have defined $\Xi := 2\hat{\mathfrak{q}}^*\hat{\mathfrak{q}}'$, which satisfies a similar relation,

$$\dot{\Xi}/2 \;=\; -\boldsymbol{\Omega}\,\Xi/4 + \hat{\mathfrak{q}}^*\dot{\hat{\mathfrak{q}}}'\,. \tag{4.1.20}$$

Taking the difference of (4.1.19) and (4.1.20) yields

$$\boldsymbol{\Omega}' - \dot{\Xi} = (\boldsymbol{\Omega}\,\Xi - \Xi\,\boldsymbol{\Omega})/2 = \mathrm{Im}(\boldsymbol{\Omega}\,\Xi)\,. \tag{4.1.21}$$

In quaternion components this formula becomes

$$\begin{aligned}\left[0\,,\,\boldsymbol{\Omega}'\right] - \left[0\,,\,\dot{\Xi}\right] &= \frac{1}{2}\Big(\left[0\,,\,\boldsymbol{\Omega}\right]\left[0\,,\,\Xi\right] - \left[0\,,\,\Xi\right]\left[0\,,\,\boldsymbol{\Omega}\right]\Big)\\ &= \left[0\,,\,\boldsymbol{\Omega}\times\Xi\right], \end{aligned}\tag{4.1.22}$$

or, in the equivalent vector form,

$$\boldsymbol{\Omega}' - \dot{\Xi} = \boldsymbol{\Omega}\times\Xi\,. \tag{4.1.23}$$

This recovers the vector Equation (2.4.7), which was the key formula needed for writing Hamilton's principle in vector form, now reproduced in its pure quaternionic form for the Cayley–Klein parameters. ■

Remark 4.1.3 Having expressed the key vector variational formula (2.4.7) in quaternionic form (4.1.21), the path for deriving Hamilton's principle for the rigid body in the quaternionic picture proceeds in parallel with the vector case. □

Exercise. State and prove Hamilton's principle for the rigid body in quaternionic form. ★

4.2 Actions of quaternions, Lie groups and Lie algebras

Quaternionic operations are isomorphic to the actions of Lie groups and Lie algebras. This isomorphism will allow us to develop Hamilton's principle for mechanics on Lie groups by following a path that parallels the one taken for quaternions.

4.2.1 AD, Ad, ad, Ad* and ad* actions of quaternions

We introduce the following notation for how the quaternions act among themselves and on the vectors in their left- and right-invariant tangent spaces at the identity. (These left- and right-invariant vectors are the body and space angular frequencies, respectively.)

- AD (conjugacy of quaternions),

$$\mathrm{AD}_{\hat{\mathfrak{q}}}\, \mathfrak{r} := \hat{\mathfrak{q}}\, \mathfrak{r}\, \hat{\mathfrak{q}}^* ,$$

- Ad (conjugacy of angular velocities),

$$\mathrm{Ad}_{\hat{\mathfrak{q}}} \boldsymbol{\Omega} = \hat{\mathfrak{q}}\, \boldsymbol{\Omega}\, \hat{\mathfrak{q}}^* =: \boldsymbol{\omega} ,$$

- ad (commutator of angular velocities),

$$\mathrm{ad}_{\boldsymbol{\Omega}} \Xi = \mathrm{Im}(\boldsymbol{\Omega}\, \Xi) := (\boldsymbol{\Omega}\, \Xi - \Xi\, \boldsymbol{\Omega})/2 .$$

The pairing $\langle\, \cdot\, , \cdot\, \rangle : \mathbb{H} \times \mathbb{H} \mapsto \mathbb{R}$ in formula (3.1.22) also allows one to define the corresponding dual operations. These are

- coAD $\langle\, \mathrm{AD}^*_{\hat{\mathfrak{q}}}\, \mathfrak{s} , \mathfrak{r} \,\rangle = \langle\, \mathfrak{s} \,,\, \mathrm{AD}_{\hat{\mathfrak{q}}} \mathfrak{r} \,\rangle$,
- coAd $\langle\, \mathrm{Ad}^*_{\hat{\mathfrak{q}}}\, \Xi , \boldsymbol{\Omega} \,\rangle = \langle\, \Xi \,,\, \mathrm{Ad}_{\hat{\mathfrak{q}}} \boldsymbol{\Omega} \,\rangle$,
- coad $\langle\, \mathrm{ad}^*_{\boldsymbol{\Omega}} \Upsilon \,,\, \Xi \,\rangle = \langle\, \Upsilon \,,\, \mathrm{ad}_{\boldsymbol{\Omega}}\, \Xi \,\rangle$.

Exercise. Prove that any pure quaternion is in the conjugacy class of $[\,0,\, \hat{\mathbf{k}}\,]$ with $\hat{\mathbf{k}} = (0, 0, 1)^T$ under the Ad action of a unit quaternion.

Hint: Compare with the formula in Remark 3.2.12. ★

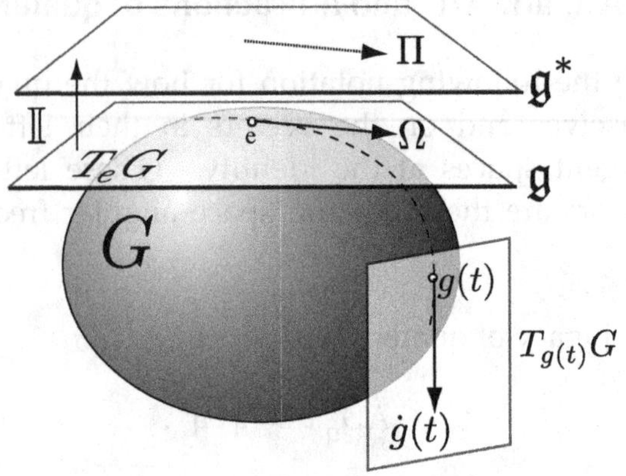

Figure 4.1. The tangent space at the identity e of the group G is its Lie algebra $\mathfrak{g}$, a vector space represented here as a plane. The moment of inertia $\mathbb{I}$ maps the vector $\Omega \in \mathfrak{g}$ into the dual vector $\Pi = \mathbb{I}\Omega \in \mathfrak{g}^*$. The dual Lie algebra $\mathfrak{g}^*$ is another vector space, also represented as a plane in the figure. A group orbit in G has tangent vector $\dot{g}(t)$ at point $g(t)$ which may be transported back to the identity by acting with $g^{-1}(t) \in G$ from either the left as $\Omega = g^{-1}(t)\dot{g}(t)$ or the right as $\omega = \dot{g}(t)g^{-1}(t)$.

4.2.2 AD, Ad, and ad for Lie algebras and groups

The notation for the conjugacy relations among the quaternions in Section 4.2.1 follows the standard notation for the corresponding actions of a Lie group on itself, on its Lie algebra (its tangent space at the identity), the action of the Lie algebra on itself, and their dual actions. By the isomorphism between the quaternions and the matrix Lie group $G = SU(2)$, one may define these corresponding operations for other ***matrix Lie groups***.

ADjoint, Adjoint and adjoint for matrix Lie groups

- AD (conjugacy classes of a matrix Lie group): The map $I_g : G \to G$ given by $I_g(h) \to ghg^{-1}$ for matrix Lie group elements $g, h \in G$ is the ***inner automorphism*** associated with g. Orbits of this action are called ***conjugacy classes***.

$$\mathrm{AD} : G \times G \to G : \quad \mathrm{AD}_g h := ghg^{-1} .$$

- Differentiate $I_g(h)$ with respect to h at $h = e$ to produce the ***Adjoint operation***,

$$\mathrm{Ad} : G \times \mathfrak{g} \to \mathfrak{g} : \quad \mathrm{Ad}_g \, \eta = T_e I_g \, \eta =: g\eta g^{-1} ,$$

with $\eta = h'(0)$.

- Differentiate $\mathrm{Ad}_g \, \eta$ with respect to g at $g = e$ in the direction ξ to produce the ***adjoint operation***,

$$\mathrm{ad} : \mathfrak{g} \times \mathfrak{g} \to \mathfrak{g} : \quad T_e(\mathrm{Ad}_g \, \eta) \, \xi = [\xi, \eta] = \mathrm{ad}_\xi \, \eta .$$

Explicitly, one computes the ad operation by differentiating the Ad operation directly as

$$\begin{aligned} \frac{d}{dt}\Big|_{t=0} \mathrm{Ad}_{g(t)} \, \eta &= \frac{d}{dt}\Big|_{t=0} \Big(g(t)\eta g^{-1}(t) \Big) \\ &= \dot{g}(0)\eta g^{-1}(0) - g(0)\eta g^{-1}(0)\dot{g}(0)g^{-1}(0) \\ &= \xi\eta - \eta\xi = [\xi, \eta] = \mathrm{ad}_\xi \, \eta , \end{aligned} \tag{4.2.1}$$

where $g(0) = Id$, $\xi = \dot{g}(0)$ and the ***Lie bracket***

$$[\xi, \eta] : \mathfrak{g} \times \mathfrak{g} \to \mathfrak{g} ,$$

is the matrix commutator for a matrix Lie algebra.

Remark 4.2.1 (Adjoint action) Composition of the Adjoint action of $G \times \mathfrak{g} \to \mathfrak{g}$ of a Lie group on its Lie algebra represents the group composition law as

$$\mathrm{Ad}_g \mathrm{Ad}_h \eta = g(h\eta h^{-1})g^{-1} = (gh)\eta(gh)^{-1} = \mathrm{Ad}_{gh}\eta ,$$

for any $\eta \in \mathfrak{g}$. □

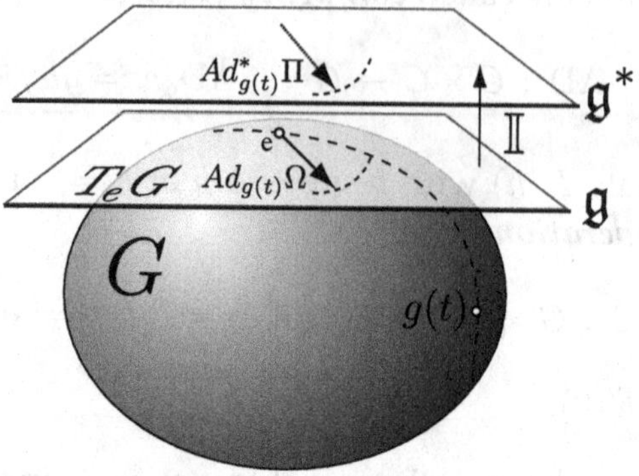

Figure 4.2. The Ad and Ad* operations of $g(t)$ act, respectively, on the Lie algebra $\mathrm{Ad} : G \times \mathfrak{g} \to \mathfrak{g}$ and on its dual $\mathrm{Ad}^* : G \times \mathfrak{g}^* \to \mathfrak{g}^*$.

Exercise. Verify that (note the minus sign)

$$\frac{d}{dt}\bigg|_{t=0} \mathrm{Ad}_{g^{-1}(t)}\, \eta = -\, \mathrm{ad}_\xi\, \eta\,,$$

for any fixed $\eta \in \mathfrak{g}$. ★

Proposition 4.2.1 (Adjoint motion equation) *Let $g(t)$ be a path in a Lie group G and $\eta(t)$ be a path in its Lie algebra $\mathfrak{g}$. Then*

$$\frac{d}{dt}\mathrm{Ad}_{g(t)}\eta(t) = \mathrm{Ad}_{g(t)}\left[\frac{d\eta}{dt} + \mathrm{ad}_{\xi(t)}\eta(t)\right],$$

where $\xi(t) = g(t)^{-1}\dot{g}(t)$.

Proof. By Equation (4.2.1), for a curve $\eta(t) \in \mathfrak{g}$,

$$\begin{aligned}
\frac{d}{dt}\Big|_{t=t_0} \mathrm{Ad}_{g(t)}\, \eta(t) &= \frac{d}{dt}\Big|_{t=t_0} \Big(g(t)\eta(t)g^{-1}(t) \Big) \\
&= g(t_0)\Big(\dot{\eta}(t_0) + g^{-1}(t_0)\dot{g}(t_0)\eta(t_0) \\
&\quad - \eta(t_0)g^{-1}(t_0)\dot{g}(t_0) \Big) g^{-1}(t_0) \\
&= \left[\mathrm{Ad}_{g(t)} \left(\frac{d\eta}{dt} + \mathrm{ad}_\xi \eta \right) \right]_{t=t_0} . \qquad (4.2.2)
\end{aligned}$$

■

Exercise. (Inverse Adjoint motion relation) Verify that

$$\frac{d}{dt}\mathrm{Ad}_{g(t)^{-1}}\eta = -\mathrm{ad}_\xi \mathrm{Ad}_{g(t)^{-1}}\eta \,, \qquad (4.2.3)$$

for any fixed $\eta \in \mathfrak{g}$. Note the placement of $\mathrm{Ad}_{g(t)^{-1}}$ and compare with Exercise on page 120. ★

Compute the coAdjoint and coadjoint operations by taking duals

The pairing

$$\langle \cdot , \cdot \rangle : \mathfrak{g}^* \times \mathfrak{g} \mapsto \mathbb{R} \qquad (4.2.4)$$

(which is assumed to be nondegenerate) between a Lie algebra $\mathfrak{g}$ and its dual vector space $\mathfrak{g}^*$ allows one to define the following dual operations:

- The *coAdjoint operation* of a Lie group on the dual of its Lie algebra is defined by the pairing with the Ad operation,

$$\mathrm{Ad}^* : G \times \mathfrak{g}^* \to \mathfrak{g}^* : \quad \langle \mathrm{Ad}^*_g \mu \,,\, \eta \rangle := \langle \mu \,,\, \mathrm{Ad}_g \eta \rangle \,, \qquad (4.2.5)$$

for $g \in G$, $\mu \in \mathfrak{g}^*$ and $\xi \in \mathfrak{g}$.

- Likewise, the ***coadjoint operation*** is defined by the pairing with the ad operation,

$$\mathrm{ad}^* : \mathfrak{g} \times \mathfrak{g}^* \to \mathfrak{g}^* : \quad \langle\, \mathrm{ad}^*_\xi\, \mu\,,\, \eta \,\rangle := \langle\, \mu\,,\, \mathrm{ad}_\xi\, \eta \,\rangle\,, \tag{4.2.6}$$

for $\mu \in \mathfrak{g}^*$ and $\xi, \eta \in \mathfrak{g}$.

Definition 4.2.1 (CoAdjoint action) *The map*

$$\Phi^* : G \times \mathfrak{g}^* \to \mathfrak{g}^* \quad \textit{given by} \quad (g, \mu) \mapsto \mathrm{Ad}^*_{g^{-1}}\mu \tag{4.2.7}$$

defines the ***coAdjoint action*** *of the Lie group G on its dual Lie algebra $\mathfrak{g}^*$.*

Remark 4.2.2 (Coadjoint group action with g^{-1}) Composition of coAdjoint operations with Φ^* reverses the order in the group composition law as

$$\mathrm{Ad}^*_g \mathrm{Ad}^*_h = \mathrm{Ad}^*_{hg}\,.$$

However, taking the inverse g^{-1} in Definition 4.2.1 of the coAdjoint action Φ^* restores the order and thereby allows it to represent the group composition law when acting on the dual Lie algebra, for then

$$\mathrm{Ad}^*_{g^{-1}}\mathrm{Ad}^*_{h^{-1}} = \mathrm{Ad}^*_{h^{-1}g^{-1}} = \mathrm{Ad}^*_{(gh)^{-1}}\,. \tag{4.2.8}$$

(See [MaRa1994] for further discussion of this point.) □

The following proposition will be used later in the context of Euler–Poincaré reduction.

Proposition 4.2.2 (Coadjoint motion relation) *Let $g(t)$ be a path in a Lie group G and $\mu(t)$ be a path in $\mathfrak{g}^*$. The corresponding Ad^* operation satisfies*

$$\frac{d}{dt}\mathrm{Ad}^*_{g(t)^{-1}}\mu(t) = \mathrm{Ad}^*_{g(t)^{-1}}\left[\frac{d\mu}{dt} - \mathrm{ad}^*_{\xi(t)}\mu(t)\right]\,, \tag{4.2.9}$$

where $\xi(t) = g(t)^{-1}\dot{g}(t)$.

Proof. The exercise on page 121 introduces the inverse Adjoint motion relation (4.2.3) for any fixed $\eta \in \mathfrak{g}$, repeated as

$$\frac{d}{dt}\mathrm{Ad}_{g(t)^{-1}}\eta = -\mathrm{ad}_{\xi(t)}\left(\mathrm{Ad}_{g(t)^{-1}}\eta\right).$$

Relation (4.2.3) may be proven by the following computation,

$$\begin{aligned}\left.\frac{d}{dt}\right|_{t=t_0}\mathrm{Ad}_{g(t)^{-1}}\eta &= \left.\frac{d}{dt}\right|_{t=t_0}\mathrm{Ad}_{g(t)^{-1}g(t_0)}\left(\mathrm{Ad}_{g(t_0)^{-1}}\eta\right)\\ &= -\mathrm{ad}_{\xi(t_0)}\left(\mathrm{Ad}_{g(t_0)^{-1}}\eta\right),\end{aligned}$$

in which for the last step one recalls

$$\left.\frac{d}{dt}\right|_{t=t_0} g(t)^{-1}g(t_0) = \left(-g(t_0)^{-1}\dot{g}(t_0)g(t_0)^{-1}\right)g(t_0) = -\xi(t_0).$$

Relation (4.2.3) plays a key role in demonstrating relation (4.2.9) in the theorem, as follows. Using the pairing $\langle\,\cdot\,,\,\cdot\,\rangle : \mathfrak{g}^* \times \mathfrak{g} \mapsto \mathbb{R}$ between the Lie algebra and its dual, one computes

$$\begin{aligned}\left\langle \frac{d}{dt}\mathrm{Ad}^*_{g(t)^{-1}}\mu(t),\eta\right\rangle &= \frac{d}{dt}\left\langle \mathrm{Ad}^*_{g(t)^{-1}}\mu(t),\eta\right\rangle\\ \text{by (4.2.5)} &= \frac{d}{dt}\left\langle \mu(t),\mathrm{Ad}_{g(t)^{-1}}\eta\right\rangle\\ &= \left\langle \frac{d\mu}{dt},\mathrm{Ad}_{g(t)^{-1}}\eta\right\rangle + \left\langle \mu(t),\frac{d}{dt}\mathrm{Ad}_{g(t)^{-1}}\eta\right\rangle\\ \text{by (4.2.3)} &= \left\langle \frac{d\mu}{dt},\mathrm{Ad}_{g(t)^{-1}}\eta\right\rangle + \left\langle \mu(t),-\mathrm{ad}_{\xi(t)}\left(\mathrm{Ad}_{g(t)^{-1}}\eta\right)\right\rangle\\ \text{by (4.2.6)} &= \left\langle \frac{d\mu}{dt},\mathrm{Ad}_{g(t)^{-1}}\eta\right\rangle - \left\langle \mathrm{ad}^*_{\xi(t)}\mu(t),\mathrm{Ad}_{g(t)^{-1}}\eta\right\rangle\\ \text{by (4.2.5)} &= \left\langle \mathrm{Ad}^*_{g(t)^{-1}}\frac{d\mu}{dt},\eta\right\rangle - \left\langle \mathrm{Ad}^*_{g(t)^{-1}}\mathrm{ad}^*_{\xi(t)}\mu(t),\eta\right\rangle\\ &= \left\langle \mathrm{Ad}^*_{g(t)^{-1}}\left[\frac{d\mu}{dt} - \mathrm{ad}^*_{\xi(t)}\mu(t)\right],\eta\right\rangle.\end{aligned}$$

This concludes the proof. ∎

Corollary 4.2.1 *The* ***coadjoint orbit relation***

$$\mu(t) = \mathrm{Ad}^*_{g(t)}\mu(0) \tag{4.2.10}$$

is the solution of the ***coadjoint motion equation*** *for* $\mu(t)$,

$$\frac{d\mu}{dt} - \mathrm{ad}^*_{\xi(t)}\mu(t) = 0\,. \tag{4.2.11}$$

Proof. Substituting Equation (4.2.11) into Equation (4.2.9) yields

$$\mathrm{Ad}^*_{g(t)^{-1}}\mu(t) = \mu(0)\,.$$

Operating on this equation with $\mathrm{Ad}^*_{g(t)}$ and recalling the composition rule for Ad^* from Remark 4.2.2 yields the result (4.2.10). ∎

4.3 Example: The Heisenberg Lie group

4.3.1 Definitions for the Heisenberg group

The subset of the 3×3 real matrices $SL(3,\mathbb{R})$ given by the *upper triangular matrices*

$$\left\{ H = \begin{bmatrix} 1 & a & c \\ 0 & 1 & b \\ 0 & 0 & 1 \end{bmatrix} \quad a, b, c \in \mathbb{R} \right\} \tag{4.3.1}$$

defines a noncommutative group under matrix multiplication.

The 3×3 matrix representation of this group acts on the *extended* planar vector $(x, y, 1)^T$ as

$$\begin{bmatrix} 1 & a & c \\ 0 & 1 & b \\ 0 & 0 & 1 \end{bmatrix} \begin{pmatrix} x \\ y \\ 1 \end{pmatrix} = \begin{pmatrix} x + ay + c \\ y + b \\ 1 \end{pmatrix}.$$

The group H is called the Heisenberg group and it has three parameters. To begin studying its properties, consider the matrices in H given by

$$A = \begin{bmatrix} 1 & a_1 & a_3 \\ 0 & 1 & a_2 \\ 0 & 0 & 1 \end{bmatrix}, \qquad B = \begin{bmatrix} 1 & b_1 & b_3 \\ 0 & 1 & b_2 \\ 0 & 0 & 1 \end{bmatrix}. \tag{4.3.2}$$

The matrix product gives another element of H,

$$AB = \begin{bmatrix} 1 & a_1 + b_1 & a_3 + b_3 + a_1 b_2 \\ 0 & 1 & a_2 + b_2 \\ 0 & 0 & 1 \end{bmatrix}, \tag{4.3.3}$$

and the inverses are

$$A^{-1} = \begin{bmatrix} 1 & -a_1 & a_1 a_2 - a_3 \\ 0 & 1 & -a_2 \\ 0 & 0 & 1 \end{bmatrix}, \qquad B^{-1} = \begin{bmatrix} 1 & -b_1 & b_1 b_2 - b_3 \\ 0 & 1 & -b_2 \\ 0 & 0 & 1 \end{bmatrix}. \tag{4.3.4}$$

We are dealing with a matrix (Lie) group. The ***group commutator*** is defined by

$$[A,\, B] := ABA^{-1}B^{-1} = \begin{bmatrix} 1 & 0 & a_1 b_2 - b_1 a_2 \\ 0 & 1 & 0 \\ 0 & 0 & 1 \end{bmatrix}. \tag{4.3.5}$$

Hence, the ***commutator subgroup*** $\Gamma_1(H) = [H,\, H]$ has the form

$$\Gamma_1(H) = \{[A,\, B] : A, B \in H\} \left\{ \begin{bmatrix} 1 & 0 & k \\ 0 & 1 & 0 \\ 0 & 0 & 1 \end{bmatrix}; \quad k \in \mathbb{R} \right\}. \tag{4.3.6}$$

An element C of the commutator subgroup $\Gamma_1(H)$ is of the form

$$C = \begin{bmatrix} 1 & 0 & k \\ 0 & 1 & 0 \\ 0 & 0 & 1 \end{bmatrix} \in \Gamma_1(H), \tag{4.3.7}$$

and we have the products

$$AC = \begin{bmatrix} 1 & a_1 & a_3 + k \\ 0 & 1 & a_2 \\ 0 & 0 & 1 \end{bmatrix} = CA. \tag{4.3.8}$$

Consequently, $[A, C] = AC(CA)^{-1} = AC(AC)^{-1} = I_3$. Hence, the subgroup of second commutators $\Gamma_2(H) = [\Gamma_1(H), H]$ commutes with the rest of the group, which is thus ***nilpotent of second order***.

4.3.2 Adjoint actions: AD, Ad and ad

Using the inverses in Equation (4.3.4) we compute the ***group automorphism***

$$\mathrm{AD}_B A = BAB^{-1} = \begin{bmatrix} 1 & a_1 & a_3 - a_1 b_2 + b_1 a_2 \\ 0 & 1 & a_2 \\ 0 & 0 & 1 \end{bmatrix}. \tag{4.3.9}$$

Linearising the group automorphism $\mathrm{AD}_B A$ in A at the identity yields the Ad operation,

$$\begin{aligned} \mathrm{Ad}_B \xi = B\,\xi|_{\mathrm{Id}}\,B^{-1} &= \begin{bmatrix} 1 & b_1 & b_3 \\ 0 & 1 & b_2 \\ 0 & 0 & 1 \end{bmatrix} \begin{bmatrix} 0 & \xi_1 & \xi_3 \\ 0 & 0 & \xi_2 \\ 0 & 0 & 0 \end{bmatrix} \begin{bmatrix} 1 & -b_1 & b_1 b_2 - b_3 \\ 0 & 1 & -b_2 \\ 0 & 0 & 1 \end{bmatrix} \\ &= \begin{bmatrix} 0 & \xi_1 & \xi_3 + b_1 \xi_2 - b_2 \xi_1 \\ 0 & 0 & \xi_2 \\ 0 & 0 & 0 \end{bmatrix}. \end{aligned} \tag{4.3.10}$$

This is the Ad operation of the Heisenberg group H on its Lie algebra $\mathfrak{h}(\mathbb{R}) \simeq \mathbb{R}^3$:

$$\mathrm{Ad} : H(\mathbb{R}) \times \mathfrak{h}(\mathbb{R}) \to \mathfrak{h}(\mathbb{R}). \tag{4.3.11}$$

One defines the right-invariant tangent vector,

$$\xi = \dot{A}A^{-1} = \begin{bmatrix} 0 & \dot{a}_1 & \dot{a}_3 - a_2 \dot{a}_1 \\ 0 & 0 & \dot{a}_2 \\ 0 & 0 & 0 \end{bmatrix} = \begin{bmatrix} 0 & \xi_1 & \xi_3 \\ 0 & 0 & \xi_2 \\ 0 & 0 & 0 \end{bmatrix} \in \mathfrak{h}, \tag{4.3.12}$$

and the left-invariant tangent vector,

$$\Xi = A^{-1}\dot{A} = \begin{bmatrix} 0 & \dot{a}_1 & \dot{a}_3 - a_1\dot{a}_2 \\ 0 & 0 & \dot{a}_2 \\ 0 & 0 & 0 \end{bmatrix} = \begin{bmatrix} 0 & \Xi_1 & \Xi_3 \\ 0 & 0 & \Xi_2 \\ 0 & 0 & 0 \end{bmatrix} \in \mathfrak{h}. \qquad (4.3.13)$$

Next, we linearise $\mathrm{Ad}_B\xi$ in B around the identity to find the ad operation of the Heisenberg Lie algebra $\mathfrak{h}$ on itself,

$$\mathrm{ad} : \mathfrak{h} \times \mathfrak{h} \to \mathfrak{h}\,. \qquad (4.3.14)$$

This is given explicitly by

$$\mathrm{ad}_\eta\xi = [\eta,\, \xi] := \eta\xi - \xi\eta = \begin{bmatrix} 0 & 0 & \eta_1\xi_2 - \xi_1\eta_2 \\ 0 & 0 & 0 \\ 0 & 0 & 0 \end{bmatrix}. \qquad (4.3.15)$$

Under the equivalence $\mathfrak{h} \simeq \mathbb{R}^3$ provided by

$$\begin{bmatrix} 0 & \xi_1 & \xi_3 \\ 0 & 0 & \xi_2 \\ 0 & 0 & 0 \end{bmatrix} \mapsto \begin{bmatrix} \xi_1 \\ \xi_2 \\ \xi_3 \end{bmatrix} := \boldsymbol{\xi} \qquad (4.3.16)$$

we may identify the Lie bracket with the projection onto the third component of the vector cross product:

$$[\eta,\, \xi] \mapsto \begin{bmatrix} 0 \\ 0 \\ \hat{\mathbf{3}} \cdot \boldsymbol{\eta} \times \boldsymbol{\xi} \end{bmatrix}. \qquad (4.3.17)$$

4.3.3 Coadjoint actions: Ad* and ad*

The inner product on the Heisenberg Lie algebra $\mathfrak{h} \times \mathfrak{h} \to \mathbb{R}$ is defined by the matrix trace pairing

$$\langle \eta,\, \xi \rangle = \mathrm{Tr}(\eta^T\, \xi) = \boldsymbol{\eta} \cdot \boldsymbol{\xi}\,. \qquad (4.3.18)$$

Thus, elements of the dual Lie algebra $\mathfrak{h}^*(\mathbb{R})$ may be represented as ***lower triangular matrices,***

$$\mu = \begin{bmatrix} 0 & 0 & 0 \\ \mu_1 & 0 & 0 \\ \mu_3 & \mu_2 & 0 \end{bmatrix} \in \mathfrak{h}^*(\mathbb{R}) . \tag{4.3.19}$$

The Ad* operation of the Heisenberg group $H(\mathbb{R})$ on its dual Lie algebra $\mathfrak{h}^* \simeq \mathbb{R}^3$ is defined in terms of the matrix pairing by

$$\langle \mathrm{Ad}_B^* \mu, \, \xi \rangle := \langle \mu, \, \mathrm{Ad}_B \xi \rangle . \tag{4.3.20}$$

Explicitly, one may compute

$$\begin{aligned} \langle \mu, \, \mathrm{Ad}_B \xi \rangle &= \mathrm{Tr} \left(\begin{bmatrix} 0 & 0 & 0 \\ \mu_1 & 0 & 0 \\ \mu_3 & \mu_2 & 0 \end{bmatrix} \begin{bmatrix} 0 & \xi_1 & \xi_3 + b_1\xi_2 - b_2\xi_1 \\ 0 & 0 & \xi_2 \\ 0 & 0 & 0 \end{bmatrix} \right) \\ &= \boldsymbol{\mu} \cdot \boldsymbol{\xi} + \mu_3 (b_1 \xi_2 - b_2 \xi_1) \qquad (4.3.21) \\ &= \mathrm{Tr} \left(\begin{bmatrix} 0 & 0 & 0 \\ \mu_1 - b_2\mu_3 & 0 & 0 \\ \mu_3 & \mu_2 + b_1\mu_3 & 0 \end{bmatrix} \begin{bmatrix} 0 & \xi_1 & \xi_3 \\ 0 & 0 & \xi_2 \\ 0 & 0 & 0 \end{bmatrix} \right) \\ &= \langle \mathrm{Ad}_B^* \mu, \, \xi \rangle . \qquad (4.3.22) \end{aligned}$$

Thus, we have the formula for $\mathrm{Ad}_B^* \, \mu$:

$$\mathrm{Ad}_B^* \mu = \begin{bmatrix} 0 & 0 & 0 \\ \mu_1 - b_2\mu_3 & 0 & 0 \\ \mu_3 & \mu_2 + b_1\mu_3 & 0 \end{bmatrix} . \tag{4.3.23}$$

Likewise, the ad* operation of the Heisenberg Lie algebra $\mathfrak{h}$ on its dual $\mathfrak{h}^*$ is defined in terms of the matrix pairing by

$$\langle \mathrm{ad}_\eta^* \mu, \, \xi \rangle := \langle \mu, \, \mathrm{ad}_\eta \xi \rangle \tag{4.3.24}$$

$$
\begin{aligned}
\langle \mu,\, \mathrm{ad}_\eta \xi \rangle &= \mathrm{Tr}\left(\begin{bmatrix} 0 & 0 & 0 \\ \mu_1 & 0 & 0 \\ \mu_3 & \mu_2 & 0 \end{bmatrix} \begin{bmatrix} 0 & 0 & \eta_1\xi_2 - \xi_1\eta_2 \\ 0 & 0 & 0 \\ 0 & 0 & 0 \end{bmatrix} \right) \\
&= \mu_3(\eta_1\xi_2 - \eta_2\xi_1) \qquad (4.3.25) \\
&= \mathrm{Tr}\left(\begin{bmatrix} 0 & 0 & 0 \\ -\eta_2\mu_3 & 0 & 0 \\ 0 & \eta_1\mu_3 & 0 \end{bmatrix} \begin{bmatrix} 0 & \xi_1 & \xi_3 \\ 0 & 0 & \xi_2 \\ 0 & 0 & 0 \end{bmatrix} \right) \\
&= \langle \mathrm{ad}^*_\eta \mu,\, \xi \rangle . \qquad (4.3.26)
\end{aligned}
$$

Thus, we have the formula for $\mathrm{ad}^*_\eta\,\mu$:

$$
\mathrm{ad}^*_\eta \mu = \begin{bmatrix} 0 & 0 & 0 \\ -\eta_2\mu_3 & 0 & 0 \\ 0 & \eta_1\mu_3 & 0 \end{bmatrix} . \qquad (4.3.27)
$$

4.3.4 Coadjoint motion and harmonic oscillations

According to Proposition 4.2.2, the coadjoint motion relation arises by differentiating along the coadjoint orbit. Let $A(t)$ be a path in the Heisenberg Lie group H and $\mu(t)$ be a path in $\mathfrak{h}^*$. Then we compute

$$
\frac{d}{dt}\Big(\mathrm{Ad}^*_{A(t)^{-1}}\mu(t) \Big) = \mathrm{Ad}^*_{A(t)^{-1}} \left[\frac{d\mu}{dt} - \mathrm{ad}^*_{\eta(t)}\mu(t) \right] , \qquad (4.3.28)
$$

where $\eta(t) = A(t)^{-1}\dot{A}(t)$.

With $\eta = A^{-1}\dot{A}$, Corollary 4.2.11 provides the differential equation for the coadjoint orbit,

$$
\mu(t) = \mathrm{Ad}^*_{A(t)}\mu(0) \,.
$$

The desired differential equation is the ***coadjoint motion equation***

$$
\dot{\mu} = \mathrm{ad}^*_\eta \mu \,,
$$

which may be written for the Heisenberg Lie group H as

$$\dot{\mu} = \begin{bmatrix} 0 & 0 & 0 \\ \dot{\mu}_1 & 0 & 0 \\ \dot{\mu}_3 & \dot{\mu}_2 & 0 \end{bmatrix} = \mathrm{ad}^*_\eta \mu = \begin{bmatrix} 0 & 0 & 0 \\ -\eta_2\mu_3 & 0 & 0 \\ 0 & \eta_1\mu_3 & 0 \end{bmatrix}. \quad (4.3.29)$$

That is,

$$\frac{d}{dt}(\mu_1, \mu_2, \mu_3) = (-\eta_2\mu_3, \eta_1\mu_3, 0). \quad (4.3.30)$$

Thus, the coadjoint motion equation for the Heisenberg group preserves the level sets of μ_3.

If we define the linear map $\mathfrak{h} \to \mathfrak{h}^* : (\mu_1, \mu_2) = (I_1\eta_1, I_2\eta_2)$ then the coadjoint motion equations become

$$\begin{aligned} \dot{\mu}_1 &= -\mu_3\mu_2/I_2 \,, \\ \dot{\mu}_2 &= \mu_3\mu_1/I_1 \,, \\ \dot{\mu}_3 &= 0 \,. \end{aligned} \quad (4.3.31)$$

Upon taking another time derivative, this set reduces to the equations

$$\ddot{\mu}_k = -\frac{\mu_3^2}{I_1 I_2}\mu_k \,, \quad \text{for} \quad k = 1, 2. \quad (4.3.32)$$

These are the equations for a planar isotropic harmonic oscillator on a level set of μ_3.

This calculation has proved the following.

Proposition 4.3.1 *Planar isotropic harmonic oscillations describe coadjoint orbits on the Heisenberg Lie group. The coadjoint orbits are (μ_1, μ_2) ellipses on level sets of μ_3.*

5

THE SPECIAL ORTHOGONAL GROUP $SO(3)$

Contents

5.1 Adjoint and coadjoint actions of $SO(3)$ **132**

5.1.1 Ad and ad operations for the hat map **132**

5.1.2 AD, Ad and ad actions of $SO(3)$ **133**

5.1.3 Dual Lie algebra isomorphism **135**

5.1 Adjoint and coadjoint actions of $SO(3)$

Recall that the Lie group $SO(3)$ of special orthogonal matrices is defined by

$$SO(3) := \{A \mid A \in 3 \times 3 \text{ orthogonal matrices}, \det(A) = 1\}.$$

The action of the matrix Lie group $SO(3)$ on vectors in $\mathbb{R}^3$ by left multiplication represents rotations in three dimensions. Its Lie algebra $so(3)$ comprises the 3×3 skew-symmetric matrices. Elements of $so(3)$ represent angular velocities, and elements of its dual space $so(3)^*$ under the matrix trace pairing represent angular momenta.

5.1.1 Ad and ad operations for the hat map

As shown in Theorem 2.1.1, the Lie algebra $(so(3), [\cdot, \cdot])$ with matrix commutator bracket $[\cdot\,,\,\cdot]$ maps to the Lie algebra $(\mathbb{R}^3, \times)$ with vector product $\times$, by the linear isomorphism

$$\mathbf{u} := (u^1, u^2, u^3) \in \mathbb{R}^3 \mapsto \widehat{u} := \begin{bmatrix} 0 & -u^3 & u^2 \\ u^3 & 0 & -u^1 \\ -u^2 & u^1 & 0 \end{bmatrix} \in so(3).$$

In matrix and vector components, the linear isomorphism is

$$\widehat{u}_{ij} := -\epsilon_{ijk} u^k.$$

Equivalently, this isomorphism is given by

$$\widehat{u}\mathbf{v} = \mathbf{u} \times \mathbf{v} \quad \text{for all} \quad \mathbf{u}, \mathbf{v} \in \mathbb{R}^3.$$

This is the hat map $\widehat{\ }\,:\ (so(3), [\cdot, \cdot]) \to (\mathbb{R}^3, \times)$ defined earlier in (2.1.10) and (2.1.11) using

$$\widehat{u} = \mathbf{u} \cdot \widehat{\boldsymbol{J}} = u^a \widehat{J}_a,$$

which holds for the $so(3)$ basis set (2.1.4) of skew-symmetric 3×3 matrices $\widehat{J}_a$, with $a = 1, 2, 3$.

Exercise. Verify the following formulas for $\mathbf{u}, \mathbf{v}, \mathbf{w} \in \mathbb{R}^3$:

$$
\begin{aligned}
(\mathbf{u} \times \mathbf{v})\widehat{\ } &= \widehat{u}\,\widehat{v} - \widehat{v}\,\widehat{u} =: [\widehat{u}, \widehat{v}] \,, \\
[\widehat{u}, \widehat{v}]\,\mathbf{w} &= (\mathbf{u} \times \mathbf{v}) \times \mathbf{w} \,, \\
((\mathbf{u} \times \mathbf{v}) \times \mathbf{w})\widehat{\ } &= [[\widehat{u}, \widehat{v}]\,, \widehat{w}] \,, \\
\mathbf{u} \cdot \mathbf{v} &= -\frac{1}{2}\,\mathrm{trace}(\widehat{u}\,\widehat{v}) \\
&=: \big\langle \,\widehat{u}\,,\, \widehat{v}\,\big\rangle \,,
\end{aligned}
$$

in which the dot product of vectors is also the natural pairing of 3×3 skew-symmetric matrices.

★

Exercise. (Jacobi identity under the hat map) Verify that the Jacobi identity for the cross product of vectors in $\mathbb{R}^3$ is equivalent to the Jacobi identity for the commutator product of 3×3 skew matrices by proving the following identity satisfied by the hat map,

$$
\begin{aligned}
&((\mathbf{u} \times \mathbf{v}) \times \mathbf{w} + (\mathbf{v} \times \mathbf{w}) \times \mathbf{u} + (\mathbf{w} \times \mathbf{u}) \times \mathbf{v})\widehat{\ } \\
&= 0 = [[\widehat{u}, \widehat{v}]\,, \widehat{w}] + [[\widehat{v}, \widehat{w}]\,, \widehat{u}] + [[\widehat{w}, \widehat{u}]\,, \widehat{v}] \,.
\end{aligned}
$$

★

5.1.2 AD, Ad and ad actions of $SO(3)$

- AD action of $SO(3)$ on itself: The AD action for $SO(3)$ is conjugation by matrix multiplication

$$I_A(B) = ABA^{-1} \,.$$

- Ad action of $SO(3)$ on its Lie algebra $so(3)$: The corresponding adjoint action of $SO(3)$ on $so(3)$ may be obtained as follows.

Differentiating $B(t)$ at $B(0) = Id$ gives

$$\mathrm{Ad}_A\,\widehat{v} = \frac{d}{dt}\Big|_{t=0} AB(t)A^{-1} = A\widehat{v}A^{-1}\,, \quad \text{with} \quad \widehat{v} = B'(0)\,.$$

One calculates the pairing with a vector $\mathbf{w} \in \mathbb{R}^3$ as

$$\mathrm{Ad}_A\widehat{v}(\mathbf{w}) = A\widehat{v}(A^{-1}\mathbf{w}) = A(\mathbf{v} \times A^{-1}\mathbf{w}) = A\mathbf{v} \times \mathbf{w} = (A\mathbf{v})\widehat{\ }\mathbf{w}\,,$$

where we have used the relation

$$A(\mathbf{u} \times \mathbf{v}) = A\mathbf{u} \times A\mathbf{v}\,,$$

which holds for any $\mathbf{u}, \mathbf{v} \in \mathbb{R}^3$ and $A \in SO(3)$. Consequently,

$$\mathrm{Ad}_A\widehat{v} = (A\mathbf{v})\widehat{\ }.$$

Identifying $so(3) \simeq \mathbb{R}^3$ then gives

$$\boxed{\mathrm{Ad}_A\mathbf{v} = A\mathbf{v}.}$$

So (speaking prose all our lives) the adjoint (Ad) action of the Lie group $SO(3)$ on its Lie algebra $so(3)$ may be identified with multiplication of a matrix in $SO(3)$ times a vector in $\mathbb{R}^3$.

- ad action of $so(3)$ on itself: Differentiating again gives the ad action of the Lie algebra $so(3)$ on itself:

$$\boxed{[\,\widehat{u}, \widehat{v}\,] = \mathrm{ad}_{\widehat{u}}\,\widehat{v} = \frac{d}{dt}\Big|_{t=0}\left(e^{t\widehat{u}}\mathbf{v}\right)\widehat{\ } = (\widehat{u}\mathbf{v})\widehat{\ } = (\mathbf{u} \times \mathbf{v})\widehat{\ }\,.}$$

So the ad action of the Lie algebra $so(3)$ on itself is by the matrix commutator of skew-symmetric matrices, which the hat map (isomorphism) identifies with the vector cross product.

- Infinitesimal generator: Likewise, the ***infinitesimal generator*** corresponding to $\mathbf{u} \in \mathbb{R}^3$ has the expression

$$\mathbf{u}_{\mathbb{R}^3}(\mathbf{x}) := \frac{d}{dt}\Big|_{t=0} e^{t\widehat{u}}\mathbf{x} = \widehat{u}\,\mathbf{x} = \mathbf{u} \times \mathbf{x}\,.$$

5.1.3 Dual Lie algebra isomorphism $\breve{}$: $so(3)^* \to \mathbb{R}^3$

Proposition 5.1.1 (Coadjoint actions) *The dual $so(3)^*$ is identified with $\mathbb{R}^3$ by the isomorphism*

$$\mathbf{\Pi} \in \mathbb{R}^3 \mapsto \breve{\Pi} \in so(3)^* : \quad \left\langle \breve{\Pi}\,,\,\widehat{u} \right\rangle := \mathbf{\Pi} \cdot \mathbf{u} \quad \textit{for any} \quad \mathbf{u} \in \mathbb{R}^3\,.$$

In terms of this isomorphism, the coAdjoint action of $SO(3)$ on $so(3)^*$ is given by

$$\mathrm{Ad}^*_{A^{-1}}\,\breve{\Pi} = (A\mathbf{\Pi})\breve{}\ , \tag{5.1.1}$$

and the coadjoint action of $so(3)$ on $so(3)^*$ is given by

$$\mathrm{ad}^*_{\widehat{u}}\,\breve{\Pi} = (\mathbf{\Pi} \times \mathbf{u})\breve{}\,. \tag{5.1.2}$$

Proof.

- Computing the coAdjoint Ad^* action of $SO(3)$ on $so(3)^*$: One computes from the definition,

$$\begin{aligned}\left\langle \mathrm{Ad}^*_{A^{-1}}\,\breve{\Pi}\,,\,\widehat{u}\right\rangle &= \left\langle \breve{\Pi}\,,\,\mathrm{Ad}_{A^{-1}}\,\widehat{u}\right\rangle = \left\langle \breve{\Pi}\,,\,(A^{-1}\mathbf{u})\widehat{}\,\right\rangle = \mathbf{\Pi} \cdot A^T\mathbf{u}\\ &= A\mathbf{\Pi} \cdot \mathbf{u} = \left\langle (A\mathbf{\Pi})\breve{}\ ,\,\widehat{u}\right\rangle .\end{aligned}$$

  That is, the coAdjoint action of $SO(3)$ on $so(3)^*$ has the expression in (5.1.1),

$$\mathrm{Ad}^*_{A^{-1}}\,\breve{\Pi} = (A\mathbf{\Pi})\breve{}\ .$$

- Computing the ad^* action of $so(3)$ on its dual $so(3)^*$: Let $\mathbf{u}, \mathbf{v} \in \mathbb{R}^3$ and note that

$$\begin{aligned}\left\langle \mathrm{ad}^*_{\widehat{u}}\,\breve{\Pi}, \widehat{v}\right\rangle &= \left\langle \breve{\Pi}, [\widehat{u}, \widehat{v}]\right\rangle = \left\langle \breve{\Pi}, (\mathbf{u} \times \mathbf{v})\widehat{}\,\right\rangle = \mathbf{\Pi} \cdot (\mathbf{u} \times \mathbf{v})\\ &= (\mathbf{\Pi} \times \mathbf{u}) \cdot \mathbf{v} = \left\langle (\mathbf{\Pi} \times \mathbf{u})\breve{}, \widehat{v}\right\rangle,\end{aligned}$$

  which shows that $\mathrm{ad}^*_{\widehat{u}}\,\breve{\Pi} = (\mathbf{\Pi} \times \mathbf{u})\breve{}$, thereby proving (5.1.2).

■

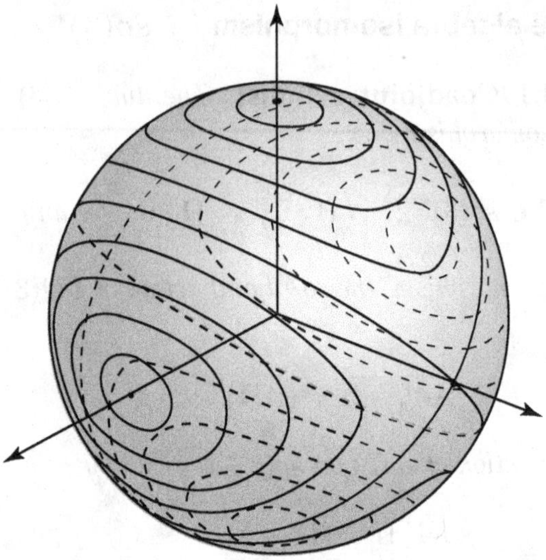

Figure 5.1. A coAdjoint orbit of the action of $SO(3)$ on $so(3)^*$ is a sphere of radius $|\mathbf{\Pi}|$. The curves on this coAdjoint orbit are its intersections with the level sets of kinetic energy of a rigid body.

Remark 5.1.1

- The coAdjoint orbit

$$\mathcal{O} = \{A\mathbf{\Pi} \mid A \in SO(3)\} \subset \mathbb{R}^3$$

of $SO(3)$ through $\mathbf{\Pi} \in \mathbb{R}^3$ is a sphere S^2 of radius $|\mathbf{\Pi}|$ centred at the origin (Figure 5.1).

- The set $\{\mathbf{\Pi} \times \mathbf{u} \mid \mathbf{u} \in \mathbb{R}^3\} = T_{\mathbf{\Pi}}\mathcal{O}$ is the plane perpendicular to $\mathbf{\Pi}$, i.e., the tangent space to the sphere S^2 of radius $|\mathbf{\Pi}|$ centred at the origin.

□

Exercise. What are the analogues of the hat map

$$\widehat{\ }: (so(3), \mathrm{ad}) \to (\mathbb{R}^3, \times),$$

and its dual

$$\breve{\ }: (so(3)^*, \mathrm{ad}^*) \to (\mathbb{R}^3, \times),$$

- for the three-dimensional Lie algebras $sp(2, \mathbb{R})$, $so(2,1)$, $su(1,1)$ and $sl(2, \mathbb{R})$?
- for the six-dimensional Lie algebra $so(4)$? ★

Exercise. Compute formula (4.2.9) from Proposition 4.2.2 for the matrix Lie group $SO(3)$. ★

Answer. Let $A(t)$ be a path in the matrix Lie group $SO(3)$ and $\breve{\Pi}(t)$ be a path in $\mathfrak{g}^*$. Then compute

$$\frac{d}{dt}\mathrm{Ad}^*_{A(t)^{-1}}\breve{\Pi}(t) = \mathrm{Ad}^*_{A(t)^{-1}}\left[\frac{d\breve{\Pi}}{dt} - \mathrm{ad}^*_{\widehat{\Omega}}\breve{\Pi}(t)\right], \qquad (5.1.3)$$

where $\widehat{\Omega}(t) = A(t)^{-1}\dot{A}(t) \in so(3)$ is the body angular velocity.

$$\left\langle \frac{d}{dt}\mathrm{Ad}^*_{A(t)^{-1}}\,\breve{\Pi}(t)\,,\,\widehat{u}\right\rangle = \frac{d}{dt}\left\langle \breve{\Pi}(t)\,,\,\mathrm{Ad}_{A(t)^{-1}}\,\widehat{u}\right\rangle$$

$$= \frac{d}{dt}\left\langle \breve{\Pi}(t)\,,\,(A(t)^{-1}\mathbf{u})\widehat{\ }\right\rangle$$

$$= \frac{d}{dt}\Big(\mathbf{\Pi}(t)\cdot A(t)^{-1}\mathbf{u}\Big)$$

$$
\begin{aligned}
&= \Big(\dot{\mathbf{\Pi}} \cdot A(t)^{-1}\mathbf{u}\Big) - \Big(\mathbf{\Pi} \cdot A(t)^{-1}\dot{A}A(t)^{-1}\mathbf{u}\Big) \\
&= \Big(\dot{\mathbf{\Pi}} \cdot A(t)^{-1}\mathbf{u}\Big) - \Big(\mathbf{\Pi} \cdot \widehat{\Omega}A(t)^{-1}\mathbf{u}\Big) \\
&= \Big(\dot{\mathbf{\Pi}} \cdot A(t)^{-1}\mathbf{u}\Big) - \Big(\mathbf{\Pi} \cdot \mathbf{\Omega}\times A(t)^{-1}\mathbf{u}\Big) \\
&= \Big(\big(\dot{\mathbf{\Pi}} + \mathbf{\Omega} \times \mathbf{\Pi}\big) \cdot A(t)^{-1}\mathbf{u}\Big) \\
&= \Big(A(t)\big(\dot{\mathbf{\Pi}} + \mathbf{\Omega} \times \mathbf{\Pi}\big)\Big) \cdot \mathbf{u} \\
&= \left\langle \mathrm{Ad}^*_{A(t)^{-1}} \left[\frac{d\breve{\Pi}}{dt} - \mathrm{ad}^*_{\widehat{\Omega}}\breve{\Pi}(t)\right], \widehat{u} \right\rangle .
\end{aligned}
$$

▲

Remark 5.1.2 (Spatial angular momentum conservation) This computation provides a geometrical proof that the equation of motion for the ***body angular momentum*** $\mathbf{\Pi}$,

$$
\frac{d\mathbf{\Pi}}{dt} + \mathbf{\Omega} \times \mathbf{\Pi} = 0 ,
$$

implies conservation of the ***spatial angular momentum*** $\boldsymbol{\pi}$ given by

$$
\boldsymbol{\pi}(t) = A(t)\mathbf{\Pi}(t) . \tag{5.1.4}
$$

Namely, the quantity $\mathbf{\Pi}(t) \cdot A(t)^{-1}\mathbf{u}$ is conserved for any fixed vector $\mathbf{u} \in \mathbb{R}^3$, and

$$
\mathbf{\Pi}(t) \cdot A(t)^{-1}\mathbf{u} = A(t)\mathbf{\Pi}(t) \cdot \mathbf{u} = \boldsymbol{\pi}(t) \cdot \mathbf{u} .
$$

This conservation law makes perfect sense in the light of Lemma 2.1.4 for the rigid body, which summarises the two equivalent sets of equations in the spatial and body frames as

$$
\frac{d\boldsymbol{\pi}}{dt} = 0 \quad \text{and} \quad \frac{d\mathbf{\Pi}}{dt} + \big(\mathbb{I}^{-1}\mathbf{\Pi}\big) \times \mathbf{\Pi} = 0 . \tag{5.1.5}
$$

Since $d\boldsymbol{\pi}/dt = 0$, we may invert the relation (5.1.4) as

$$\boldsymbol{\Pi}(t) = A(t)^{-1}\boldsymbol{\pi} = A(t)^{-1}\boldsymbol{\Pi}(0) \tag{5.1.6}$$

and interpret $\boldsymbol{\pi}$ as the initial value of $\boldsymbol{\Pi}(t)$ by setting $A(0)^{-1} = Id$. This interpretation is also consistent with the representation of rigid-body motion as a time-dependent curve $A(t)$ in the Lie group $SO(3)$. □

6 ADJOINT AND COADJOINT SEMIDIRECT-PRODUCT GROUP ACTIONS

Contents

6.1 Special Euclidean group $SE(3)$ — 142
6.2 Adjoint operations for $SE(3)$ — 144
6.3 Adjoint actions of $SE(3)$'s Lie algebra — 148
    6.3.1 The ad action of $se(3)$ on itself — 148
    6.3.2 The ad* action of $se(3)$ on its dual $se(3)^*$ — 149
    6.3.3 Left versus right — 151
6.4 Special Euclidean group $SE(2)$ — 153
6.5 Semidirect-product group $SL(2,\mathbb{R})Ⓢ\mathbb{R}^2$ — 156
    6.5.1 Definitions for $SL(2,\mathbb{R})Ⓢ\mathbb{R}^2$ — 156
    6.5.2 AD, Ad, and ad actions — 158
    6.5.3 Ad* and ad* actions — 160
    6.5.4 Coadjoint motion relation — 162
6.6 Galilean group — 164
    6.6.1 Definitions for $G(3)$ — 164
    6.6.2 AD, Ad, and ad actions of $G(3)$ — 165
6.7 Iterated semidirect products — 167

6.1 Special Euclidean group $SE(3)$

As a set, the special Euclidean group in three dimensions $SE(3)$ is the Cartesian product $S = SO(3) \times \mathbb{R}^3$. The group $SE(3)$ acts on $x \in \mathbb{R}^3$ by rotations and translations: $x \to Rx + v$, with $R \in SO(3)$ and $v \in \mathbb{R}^3$. This group action may be represented by multiplication from the *left* of 4×4 block matrices of the form

$$E(R,v)\begin{bmatrix} x \\ 1 \end{bmatrix} = \begin{pmatrix} R & v \\ 0 & 1 \end{pmatrix}\begin{bmatrix} x \\ 1 \end{bmatrix} = \begin{bmatrix} Rx+v \\ 1 \end{bmatrix}.$$

Group multiplication in $SE(3)$ may also be represented by 4×4 ***matrix multiplication,*** as

$$\begin{aligned} E(\tilde{R},\tilde{v})E(R,v) &= \begin{pmatrix} \tilde{R} & \tilde{v} \\ 0 & 1 \end{pmatrix}\begin{pmatrix} R & v \\ 0 & 1 \end{pmatrix} \\ &= \begin{pmatrix} \tilde{R}R & \tilde{R}v+\tilde{v} \\ 0 & 1 \end{pmatrix} = E(\tilde{R}R, \tilde{R}v+\tilde{v}) . \end{aligned}$$

One may abbreviate this multiplication by merely writing the top row as

$$(\tilde{R}\,,\,\tilde{v})(R\,,\,v) = (\tilde{R}R\,,\,\tilde{R}v+\tilde{v}) .$$

The inverse of a group element $(R,v)^{-1}$ is naturally identified with the matrix inverse as

$$E(R,v)^{-1} = \begin{pmatrix} R & v \\ 0 & 1 \end{pmatrix}^{-1} = \begin{pmatrix} R^{-1} & -R^{-1}v \\ 0 & 1 \end{pmatrix},$$

so that

$$E(R,v)^{-1}E(R,v) = \begin{pmatrix} R^{-1} & -R^{-1}v \\ 0 & 1 \end{pmatrix}\begin{pmatrix} R & v \\ 0 & 1 \end{pmatrix} = \begin{pmatrix} 1 & 0 \\ 0 & 1 \end{pmatrix},$$

or in the abbreviated top-row notation

$$(R\,,\,v)^{-1} = (R^{-1}\,,\,-R^{-1}v)\,.$$

Remark 6.1.1 More generally, these formulas for $SE(3)$ are identified as the left action of a ***semidirect-product Lie group,***

$$SE(3) \simeq SO(3)\,\circledS\,\mathbb{R}^3\,.$$

Semidirect-product Lie groups were defined in Chapter 1. For more details, see, e.g., [MaRa1994]. Semidirect-product group multiplication is defined as follows for the case treated here, in which the normal subgroup is a vector space. □

Definition 6.1.1 (Semidirect-product group action) *Suppose a Lie group G acts from the* left *by linear maps on a vector space V. (This will also induce a left action of G on the dual space V^*.) The* ***semidirect-product group*** $S = G\circledS V$ *is the Cartesian product of sets $S = G \times V$ whose group multiplication is defined by*

$$(g_1, v_1)(g_2, v_2) = (g_1 g_2, v_1 + g_1 v_2), \tag{6.1.1}$$

where the action of $g \in G$ on $v \in V$ is denoted simply as gv (on the left). The identity element is $(e, 0)$ where e is the identity in G. The inverse of an element is given by

$$(g, v)^{-1} = (g^{-1}, -g^{-1}v). \tag{6.1.2}$$

Remark 6.1.2 These formulas show that $SE(3)$ is a semidirect-product Lie group. □

6.2 Adjoint operations for $SE(3)$

AD operation

The AD operation $\mathrm{AD} : SE(3) \times SE(3) \mapsto SE(3)$ is conveniently expressed in the top-row notation as

$$\begin{aligned}\mathrm{AD}_{(R,v)}(\tilde{R},\tilde{v}) &= (R,v)(\tilde{R},\tilde{v})(R,v)^{-1}\\ &= (R,v)(\tilde{R},\tilde{v})(R^{-1},-R^{-1}v)\\ &= (R,v)(\tilde{R}R^{-1},\tilde{v}-\tilde{R}R^{-1}v)\\ &= (R\tilde{R}R^{-1},v+R\tilde{v}-R\tilde{R}R^{-1}v)\,. \qquad (6.2.1)\end{aligned}$$

Remark 6.2.1 This formula has its counterpart for general semidirect products in which the normal subgroup is a vector space,

$$\begin{aligned}\mathrm{AD}_{(g_1,v_1)}(g_2,v_2) &= (g_1,v_1)(g_2,v_2)(g_1,v_1)^{-1}\\ &= (g_1g_2g_1^{-1},v_1+g_1v_2-g_1g_2g_1^{-1}v_1)\,.\end{aligned}$$

□

Ad operation

Taking time derivatives of quantities adorned with the tilde $(\tilde{\cdot})$ in formula (6.2.1) for $\mathrm{AD}_{(R,v)}(\tilde{R}(t),\tilde{v}(t))$ and evaluating at the identity $t=0$ yields

$$\mathrm{Ad}_{(R,v)}(\dot{\tilde{R}}(0),\dot{\tilde{v}}(0)) = (\mathrm{Ad}_R\dot{\tilde{R}}(0),\,-\mathrm{Ad}_R\dot{\tilde{R}}(0)v+R\dot{\tilde{v}}(0))\,.$$

Setting $\dot{\tilde{R}}(0)=\tilde{\xi}$ and $\dot{\tilde{v}}(0)=\tilde{\alpha}$ defines the Ad action of $SE(3)$ on its

Lie algebra with elements $(\tilde{\xi},\tilde{\alpha}) \in se(3)$ as $\mathrm{Ad} : SE(3) \times se(3) \to se(3)$,

$$\begin{aligned}\mathrm{Ad}_{(R\,,v)}(\tilde{\xi}\,,\,\tilde{\alpha}) &= (\mathrm{Ad}_R\tilde{\xi}\,,\,-\,\mathrm{Ad}_R\tilde{\xi}v + R\tilde{\alpha}) \\ &= (R\tilde{\xi}R^{-1}\,,\,-\,R\tilde{\xi}R^{-1}v + R\tilde{\alpha})\,. \qquad (6.2.2)\end{aligned}$$

Remark 6.2.2 In vector form under the hat map this becomes

$$\mathrm{Ad}_{(R\,,\mathbf{v})}(\tilde{\boldsymbol{\xi}}\,,\,\tilde{\boldsymbol{\alpha}}) = (R\tilde{\boldsymbol{\xi}}\,,\,-\,R\tilde{\boldsymbol{\xi}} \times \mathbf{v} + R\tilde{\boldsymbol{\alpha}})\,. \qquad (6.2.3)$$

□

Remark 6.2.3 The Ad operation for left actions of a semidirect-product group S is given by

$$\mathrm{Ad}_{(g\,,v)}(\xi\,,\,\alpha) = (g\xi g^{-1}\,,\,g\alpha - \,g\xi g^{-1}v)\,.$$

By Equation (6.1.2) one then finds the Adjoint action of the inverse

$$\mathrm{Ad}_{(g\,,v)^{-1}}(\xi\,,\,\alpha) = \mathrm{Ad}_{(g^{-1}\,,\,-g^{-1}v)}(\xi\,,\,\alpha) = (g^{-1}\xi g\,,\,g^{-1}\alpha + \,g^{-1}\xi v)\,,$$

where the left action of the Lie algebra $\mathfrak{g}$ on V is denoted by concatenation, as in ξv. □

Ad* operation

The pairing $\langle\,\cdot\,,\cdot\,\rangle\,:\,se(3)^* \times se(3) \mapsto \mathbb{R}$ is obtained by identifying $SE(3) \simeq SO(3) \times \mathbb{R}^3$ and taking the sum,

$$\langle\,(\mu,\beta)\,,\,(\xi\,,\,\alpha)\,\rangle \equiv \frac{1}{2}\mathrm{tr}(\mu^T\xi\,) + \boldsymbol{\beta}\cdot\boldsymbol{\alpha}\,, \qquad (6.2.4)$$

with $\mu \in so(3)^*$, $\xi \in so(3)$ represented as skew-symmetric 3×3 matrices and $\boldsymbol{\beta}$, $\boldsymbol{\alpha} \in \mathbb{R}^3$ represented in usual vector notation. Thus, one computes the Ad* operation as

$$\begin{aligned}\langle\,\mathrm{Ad}^*_{(R\,,v)^{-1}}(\mu,\beta)\,,\,(\xi\,,\,\alpha)\,\rangle &= \langle\,(\mu,\beta)\,,\,\mathrm{Ad}_{(R\,,v)^{-1}}(\xi\,,\,\alpha)\,\rangle \\ &= \langle\,(\mu,\beta)\,,\,(R^{-1}\xi R\,,\,R^{-1}\alpha + \,R^{-1}\xi v)\,\rangle \\ &= \langle\,\mu\,,\,R^{-1}\xi R\rangle + \langle\,\beta\,,\,R^{-1}\alpha + \,R^{-1}\xi v\,\rangle\end{aligned}$$

$$
\begin{aligned}
&= \langle R\mu R^{-1}, \xi \rangle + \langle R\beta, \alpha + \xi v \rangle \\
&= \langle R\mu R^{-1}, \xi \rangle + \langle R\beta, \alpha \rangle \\
&\quad + \langle \mathrm{skew}(v \otimes R\beta), \xi \rangle \\
&= \langle R\mu R^{-1} + \mathrm{skew}(v \otimes R\beta), \xi \rangle \\
&\quad + \langle R\beta, \alpha \rangle ,
\end{aligned}
$$

where $\mathrm{skew}(v \otimes R\beta)$ is the skew-symmetric part of $v \otimes R\beta$, which arises upon taking the trace of the product $v \otimes R\beta$ with the skew-symmetric 3×3 matrix ξ. One also uses the induced pairing

$$
\langle \beta, R^{-1}\alpha \rangle = \langle R\beta, \alpha \rangle ,
$$

since $\beta^T R^{-1}\alpha = (R\beta)^T \alpha$.

Remark 6.2.4 This computation expresses the Ad^* operation $\mathrm{Ad}^* : SE(3) \times se(3)^* \mapsto se(3)^*$ in its $SO(3)$ and $\mathbb{R}^3$ components explicitly as

$$
\mathrm{Ad}^*_{(R,v)^{-1}}(\mu, \beta) = \Big(R\mu R^{-1} + \mathrm{skew}(v \otimes R\beta), R\beta \Big) . \tag{6.2.5}
$$

In vector form under the dual of the hat map the previous formula becomes

$$
\mathrm{Ad}^*_{(R,\mathbf{v})^{-1}}(\boldsymbol{\mu}, \boldsymbol{\beta}) = (R\boldsymbol{\mu} + \mathbf{v} \times R\boldsymbol{\beta}, R\boldsymbol{\beta}) . \tag{6.2.6}
$$

□

Remark 6.2.5 (Semidirect-product Ad, Ad* actions) Upon denoting the various group and algebra actions by concatenation from the left, the Adjoint and coAdjoint actions for semidirect products may be expressed in slightly simpler form as (see, e.g., [MaRa1994])

$$
(g, v)(\xi, \alpha) = (g\xi, g\alpha - (g\xi)v) \tag{6.2.7}
$$

and

$$(g, v)(\mu, \beta) = (g\mu + v \diamond (g\beta), g\beta), \tag{6.2.8}$$

where $(g, v) \in S = G \times V$, $(\xi, \alpha) \in \mathfrak{s} = \mathfrak{g} \times V$, $(\mu, \beta) \in \mathfrak{s}^* = \mathfrak{g}^* \times V^*$, $g\xi = \mathrm{Ad}_g \xi$, $g\mu = \mathrm{Ad}^*_{g^{-1}}\mu$, $g\beta$ denotes the induced *left* action of g on β (the *left* action of G on V induces a *left* action of G on V^* – the inverse of the transpose of the action on V).

□

Definition 6.2.1 (The diamond operation $\diamond$) *The diamond operation $\diamond$ that appears in Equation (6.2.8) is defined by*

$$\langle\, v \diamond (g\beta)\,, \xi \,\rangle = -\,\langle\, (g\beta) \diamond v\,, \xi \,\rangle = \langle\, (g\beta)\,, \xi v \,\rangle\,. \tag{6.2.9}$$

That is, the diamond operation minus the dual of the (left) Lie algebra action. In the present case for $SE(3)$ one has

$$v \diamond (R\beta) = \mathrm{skew}(v \otimes R\beta) = (v \times R\beta)\widehat{\ },$$

by the hat map, as in Equation (6.2.6).

Exercise. (Coadjoint action $\mathrm{Ad}^*_{g^{-1}}$ of semidirect-product groups) Show that the map $\mathrm{Ad}^*_{(g,v)^{-1}}$ preserves the action

$$(g_1, v_1)(g_2, v_2) = (g_1 g_2, v_1 + g_1 v_2)$$

of a semidirect-product group by computing the composition,

$$\mathrm{Ad}^*_{(g_1,v_1)^{-1}} \mathrm{Ad}^*_{(g_2,v_1)^{-1}} = \mathrm{Ad}^*_{(g_1 g_2, v_1 + g_1 v_2)^{-1}}\,.$$

★

6.3 Adjoint actions of $SE(3)$'s Lie algebra

6.3.1 The ad action of $se(3)$ on itself

To express the operation $\mathrm{ad}^* : se(3)^* \times se(3) \to se(3)^*$, one begins by computing its corresponding ad operation, $\mathrm{ad} : se(3) \times se(3) \to se(3)$. This is done by taking time derivatives of *un*adorned quantities of $\mathrm{Ad}_{(R(t)\,,\,v(t))}(\tilde{\xi}\,,\,\tilde{\alpha})$ in Equation (6.2.2) evaluated at the identity to find

$$
\begin{aligned}
&\mathrm{ad}_{(\dot{R}(0)\,,\,\dot{v}(0))}(\tilde{\xi}\,,\,\tilde{\alpha})\\
&\quad= \Big(\dot{R}\tilde{\xi}R^{-1} - R\tilde{\xi}R^{-1}\dot{R}R^{-1}\,,\\
&\qquad\quad - \dot{R}\tilde{\xi}R^{-1}v + R\tilde{\xi}R^{-1}\dot{R}R^{-1}v - R\tilde{\xi}R^{-1}\dot{v} + \dot{R}\tilde{\alpha}\Big)\Big|_{\mathrm{Id}}\,.
\end{aligned}
$$

As before, one sets $\dot{R}(0) = \xi$, $\dot{v}(0) = \alpha$, $R(0) = \mathrm{Id}$ and $v(0) = 0$. In this notation, the ad operation $\mathrm{ad}_{(\xi\,,\,\alpha)}$ for the right-invariant Lie algebra action of $se(3)$ may thus be rewritten as

$$
\begin{aligned}
\mathrm{ad}_{(\xi\,,\,\alpha)}(\tilde{\xi}\,,\,\tilde{\alpha}) &= (\xi\tilde{\xi} - \tilde{\xi}\xi\,,\,-(\xi\tilde{\xi} - \tilde{\xi}\xi)v - \tilde{\xi}(\xi v + \alpha) + \xi\tilde{\alpha})\Big|_{\mathrm{Id}}\\
&= ([\xi\,,\,\tilde{\xi}]\,,\,-\xi\tilde{\xi}v + \xi\tilde{\alpha} - \tilde{\xi}\alpha)\Big|_{\mathrm{Id}}\\
&= (\mathrm{ad}_\xi\,\tilde{\xi}\,,\,\xi\tilde{\alpha} - \tilde{\xi}\alpha)\,,
\end{aligned}
$$

where the last step uses $v(0) = 0$. The result is just the matrix commutator,

$$
\mathrm{ad}_{(\xi\,,\,\alpha)}(\tilde{\xi}\,,\,\tilde{\alpha}) = \left[\begin{pmatrix} \xi & \alpha \\ 0 & 0 \end{pmatrix}, \begin{pmatrix} \tilde{\xi} & \tilde{\alpha} \\ 0 & 0 \end{pmatrix}\right] = \begin{pmatrix} [\xi\,,\,\tilde{\xi}] & \xi\tilde{\alpha} - \tilde{\xi}\alpha \\ 0 & 0 \end{pmatrix}.
$$

Remark 6.3.1 (The semidirect-product Lie bracket) The (left) Lie algebra of the semidirect-product Lie group S is the semidirect-product Lie algebra, $\mathfrak{s} = \mathfrak{g} \circledS V$, whose Lie bracket is expressed as

$$[(\xi_1, v_1), (\xi_2, v_2)] = ([\xi_1, \xi_2], \xi_1 v_2 - \xi_2 v_1)\,, \qquad (6.3.1)$$

where the induced action of $\mathfrak{g}$ on V is denoted by concatenation, as in $\xi_1 v_2$. □

Remark 6.3.2 In vector notation, using the hat map $\widehat{\ }: \mathbb{R}^3 \mapsto \mathfrak{so}(3)$ given by $(\widehat{v})_{ij} = -\epsilon_{ijk} v_k$ so that $\widehat{v}\,\mathbf{w} = \mathbf{v} \times \mathbf{w}$ and $[\widehat{v}\,,\, \widehat{w}] = (\mathbf{v} \times \mathbf{w})\widehat{\ }$, one has $[\xi\,,\, \tilde{\xi}] = (\boldsymbol{\xi} \times \tilde{\boldsymbol{\xi}})\widehat{\ }$and finds the correspondence (isomorphism)

$$\begin{aligned} \mathrm{ad}_{(\xi\,,\alpha)}(\tilde{\xi}\,,\,\tilde{\alpha}) &= [(\xi, \alpha)\,,\,(\tilde{\xi}, \tilde{\alpha})\,] \\ &= ([\xi\,,\,\tilde{\xi}\,]\,,\,\xi\tilde{\alpha} - \tilde{\xi}\alpha) \\ &= \Big((\boldsymbol{\xi} \times \tilde{\boldsymbol{\xi}})\widehat{\ },\,(\boldsymbol{\xi} \times \tilde{\boldsymbol{\alpha}} - \tilde{\boldsymbol{\xi}} \times \boldsymbol{\alpha})\Big)\,. \end{aligned}$$

This expression will be useful in interpreting the ad and ad* actions as motion on $\mathbb{R}^3$. □

6.3.2 The ad* action of $se(3)$ on its dual $se(3)^*$

One computes the ad* action of $se(3)$ on its dual $se(3)^*$ by using the pairing,

$$\begin{aligned} \langle \mathrm{ad}^*_{(\xi\,,\alpha)}(\mu\,,\,\beta)\,,\,(\tilde{\xi}\,,\,\tilde{\alpha})\rangle &= \langle(\mu\,,\,\beta)\,,\,\mathrm{ad}_{(\xi\,,\alpha)}(\tilde{\xi}\,,\,\tilde{\alpha})\rangle \\ &= \langle(\mu\,,\,\beta)\,,\,(\mathrm{ad}_{\xi}\,\tilde{\xi}\,,\,\xi\tilde{\alpha} - \tilde{\xi}\alpha)\rangle \qquad (6.3.2) \\ &= \langle\mu\,,\,\mathrm{ad}_{\xi}\,\tilde{\xi}\rangle + \langle\beta\,,\,\xi\tilde{\alpha}\rangle - \langle\beta\,,\,\tilde{\xi}\alpha\rangle \end{aligned}$$

$$= \langle \mathrm{ad}^*_\xi \mu\,,\, \tilde{\xi} \rangle + \langle -\xi\beta\,,\, \tilde{\alpha} \rangle + \langle \beta \diamond \alpha\,,\, \tilde{\xi} \rangle$$

$$= \langle (\mathrm{ad}^*_\xi \mu + \beta \diamond \alpha\,,\, -\xi\beta)\,,\, (\tilde{\xi}\,,\, \tilde{\alpha}) \rangle.$$

Remark 6.3.3 Again the diamond operation

$$\diamond : \mathbb{R}^3 \times \mathbb{R}^3 \mapsto so(3)^* \simeq \mathbb{R}^3$$

arises, as defined in Equation (6.2.9) by the dual Lie algebra actions,

$$\langle \beta \diamond \alpha\,,\, \tilde{\xi} \rangle \;=\; -\langle \beta\,,\, \tilde{\xi}\alpha \rangle\,.$$

In vector notation, this becomes

$$-\langle \beta\,,\, \tilde{\xi}\alpha \rangle \;=\; -\boldsymbol{\beta} \cdot \tilde{\boldsymbol{\xi}} \times \boldsymbol{\alpha} = \boldsymbol{\beta} \times \boldsymbol{\alpha} \cdot \tilde{\boldsymbol{\xi}} = \langle \beta \diamond \alpha\,,\, \tilde{\xi} \rangle\,.$$

Thus, the diamond operation for $se(3)$ is simply the cross product of vectors in $\mathbb{R}^3$. □

Under the hat map, the pairing $\langle\,\cdot\,,\,\cdot\,\rangle \,:\, se(3)^* \times se(3) \to \mathbb{R}$ in (6.2.4) transforms into the dot product of vectors in $\mathbb{R}^3$,

$$\langle (\mu\,,\, \beta)\,,\, (\xi\,,\, \alpha) \rangle = \boldsymbol{\mu} \cdot \boldsymbol{\xi} + \boldsymbol{\beta} \cdot \boldsymbol{\alpha}\,.$$

Thus, the ad^* action of $se(3)$ may be expressed in terms of vector operations, as

$$\begin{aligned}
&\langle \mathrm{ad}^*_{(\xi\,,\,\alpha)}(\mu\,,\, \beta)\,,\, (\tilde{\xi}\,,\, \tilde{\alpha}) \rangle \\
&\quad = \langle (\mu\,,\, \beta)\,,\, \mathrm{ad}_{(\xi\,,\,\alpha)}(\tilde{\xi}\,,\, \tilde{\alpha}) \rangle \\
&\quad = \boldsymbol{\mu} \cdot (\boldsymbol{\xi} \times \tilde{\boldsymbol{\xi}}) + \boldsymbol{\beta} \cdot (\boldsymbol{\xi} \times \tilde{\boldsymbol{\alpha}} - \tilde{\boldsymbol{\xi}} \times \boldsymbol{\alpha}) \\
&\quad = (\boldsymbol{\mu} \times \boldsymbol{\xi} - \boldsymbol{\alpha} \times \boldsymbol{\beta}) \cdot \tilde{\boldsymbol{\xi}} - \boldsymbol{\xi} \times \boldsymbol{\beta} \cdot \tilde{\boldsymbol{\alpha}}\,. \qquad (6.3.3)
\end{aligned}$$

Exercise. Find the conservation laws for the equation of coadjoint motion

$$\frac{d}{dt}(\boldsymbol{\mu}\,,\,\boldsymbol{\beta}) = \mathrm{ad}^*_{(\boldsymbol{\xi}\,,\,\boldsymbol{\alpha})}(\boldsymbol{\mu}\,,\,\boldsymbol{\beta})$$

on $se(3)^*$, the dual of the special Euclidean Lie algebra in three dimensions, when $(\boldsymbol{\mu}\,,\,\boldsymbol{\beta})$ are linearly related to $(\boldsymbol{\xi}\,,\,\boldsymbol{\alpha})$ by $(\boldsymbol{\mu}\,,\,\boldsymbol{\beta}) = (\mathbf{I}\boldsymbol{\xi}\,,\,\mathbf{K}\boldsymbol{\alpha})$, for symmetric matrices $(\mathbf{I}, \mathbf{K})$. ★

Summary. The adjoint and coadjoint actions for $SE(3) \simeq SO(3) \circledS \mathbb{R}^3$ are

$$\begin{aligned}
\mathrm{AD}_{(R\,,\,v)}(\tilde{R}\,,\,\tilde{v}) &= (R\tilde{R}R^{-1}\,,\,v + R\tilde{v} - R\tilde{R}R^{-1}v),\\
\mathrm{Ad}_{(R\,,\,v)^{-1}}(\xi\,,\,\alpha) &= (R^{-1}\xi R,\, R^{-1}\alpha + R^{-1}\xi v),\\
\mathrm{ad}_{(\xi\,,\,\alpha)}(\tilde{\xi}\,,\,\tilde{\alpha}) &= \Big((\boldsymbol{\xi} \times \tilde{\boldsymbol{\xi}})\hat{},\, (\boldsymbol{\xi} \times \tilde{\boldsymbol{\alpha}} - \tilde{\boldsymbol{\xi}} \times \boldsymbol{\alpha})\Big),\\
\mathrm{Ad}^*_{(R\,,\,v)^{-1}}(\boldsymbol{\mu}\,,\,\boldsymbol{\beta}) &= (R\boldsymbol{\mu} + \boldsymbol{v} \times R\boldsymbol{\beta},\, R\boldsymbol{\beta}),\\
\mathrm{ad}^*_{(\xi\,,\,\alpha)}(\mu\,,\,\beta) &= \Big(\boldsymbol{\mu} \times \boldsymbol{\xi} - \boldsymbol{\alpha} \times \boldsymbol{\beta},\, -\boldsymbol{\xi} \times \boldsymbol{\beta}\Big).
\end{aligned}$$

6.3.3 Left versus right

When working with various models of continuum mechanics and plasmas it is convenient to work with *right* representations of G on the vector space V (as in, for example, [HoMaRa1998]). We shall denote the semidirect product by the same symbol, $S = G \circledS V$, the action of G on V being denoted by vg. The formulas change under these conventions as follows. Group multiplication (the analogue of (6.1.1)) is given by

$$(g_1, v_1)(g_2, v_2) = (g_1 g_2, v_2 + v_1 g_2), \tag{6.3.4}$$

and the Lie algebra bracket on $\mathfrak{s} = \mathfrak{g} Ⓢ V$ (the analogue of (6.3.1)) has the expression

$$[(\xi_1, v_1), (\xi_2, v_2)] = ([\xi_1, \xi_2], v_1\xi_2 - v_2\xi_1), \qquad (6.3.5)$$

where we denote the induced action of $\mathfrak{g}$ on V by concatenation, as in $v_1\xi_2$. The adjoint and coadjoint actions have the formulas (analogues of (6.2.7) and (6.2.8))

$$(g, v)(\xi, u) = (g\xi, (u + v\xi)g^{-1}), \qquad (6.3.6)$$

$$(g, v)(\mu, a) = (g\mu + (vg^{-1}) \diamond (ag^{-1}), ag^{-1}), \qquad (6.3.7)$$

where, as usual, $g\xi = \mathrm{Ad}_g\xi$, $g\mu = \mathrm{Ad}^*_{g^{-1}}\mu$, ag denotes the inverse of the dual isomorphism defined by $g \in G$ (so that $g \mapsto ag$ is a *right* action). Note that the adjoint and coadjoint actions are *left* actions. In this case, the $\mathfrak{g}$-actions on $\mathfrak{g}^*$ and V^* are defined as before to be minus the dual map given by the $\mathfrak{g}$-actions on $\mathfrak{g}$ and V and are denoted, respectively, by $\xi\mu$ (because it is a left action) and $a\xi$ (because it is a right action).

Left-invariant tangent vectors

The left-invariant tangent vectors to (R, v) at the identity $(\mathrm{Id}, 0)$ are given by

$$(\xi, \alpha) = (R, v)^{-1}(\dot{R}, \dot{v}), \qquad (6.3.8)$$

or, in matrix form,

$$\begin{pmatrix} \xi & \alpha \\ 0 & 0 \end{pmatrix} = \begin{pmatrix} R^{-1} & -R^{-1}v \\ 0 & 1 \end{pmatrix} \begin{pmatrix} \dot{R} & \dot{v} \\ 0 & 0 \end{pmatrix}$$

$$= \begin{pmatrix} R^{-1}\dot{R} & R^{-1}\dot{v} \\ 0 & 0 \end{pmatrix} = \begin{pmatrix} \xi & R^{-1}\dot{v} \\ 0 & 0 \end{pmatrix}.$$

This gives the ***reconstruction formula*** for left-invariant tangent vectors,

$$(\dot{R}, \dot{v}) = (R, v)(\xi, \alpha) = (R\xi, R\alpha), \qquad (6.3.9)$$

which may also be expressed in matrix form as

$$\begin{pmatrix} \dot{R} & \dot{v} \\ 0 & 0 \end{pmatrix} = \begin{pmatrix} R & v \\ 0 & 1 \end{pmatrix} \begin{pmatrix} \xi & \alpha \\ 0 & 0 \end{pmatrix} = \begin{pmatrix} R\xi & R\alpha \\ 0 & 0 \end{pmatrix}.$$

Exercise. (Right-invariant tangent vectors) The right-invariant tangent vectors to (R, v) at the identity $(\mathrm{Id}, 0)$ are given by

$$(\xi, \alpha) = (\dot{R}, \dot{v})(R, v)^{-1} = (\dot{R}R^{-1}, -\dot{R}R^{-1}v + \dot{v}). \tag{6.3.10}$$

Show that the reconstruction formula for right-invariant tangent vectors is given by

$$(\dot{R}, \dot{v}) = (\xi, \alpha)(R, v) = (\xi R, \xi v + \alpha). \tag{6.3.11}$$

★

Remark 6.3.4 (Right- vs left-invariant reconstructions) The reconstruction formulas for right-invariant (6.3.11) and left-invariant (6.3.9) tangent vectors in $SE(3)$ are completely different. In particular, right-invariant reconstruction involves translations, while left-invariant reconstruction does not. □

6.4 Special Euclidean group $SE(2)$

The special Euclidean group of the plane $SE(2) \simeq SO(2) \circledS \mathbb{R}^2$ has coordinates

$$(R_\theta, v) = \begin{pmatrix} R_\theta & v \\ 0 & 1 \end{pmatrix},$$

where $v \in \mathbb{R}^2$ is a vector in the plane and R_θ is the rotation matrix

$$R_\theta = \begin{pmatrix} \cos\theta & -\sin\theta \\ \sin\theta & \cos\theta \end{pmatrix}.$$

Exercise. Calculate the inverse of (R_θ, v) and show that the Lie algebra $se(2)$ of $SE(2)$ consists of the 3×3 block matrices of the form

$$\begin{pmatrix} -\xi\mathbb{J} & \alpha \\ 0 & 0 \end{pmatrix}, \quad \text{where} \quad \mathbb{J} = \begin{pmatrix} 0 & 1 \\ -1 & 0 \end{pmatrix}.$$

(The skew-symmetric 2×2 matrix $\mathbb{J} = -\mathbb{J}^T = -\mathbb{J}^{-1}$ represents rotation by $-\pi/2$.) ★

Exercise. Identify the Lie algebra $se(2)$ with $\mathbb{R}^3$ via the isomorphism

$$\begin{pmatrix} -\xi\mathbb{J} & \alpha \\ 0 & 0 \end{pmatrix} \in se(2) \mapsto (\xi, \alpha) \in \mathbb{R}^3,$$

and compute the expression for the Lie algebra bracket as

$$\begin{aligned} \left[(\xi, \alpha_1, \alpha_2), (\tilde{\xi}, \tilde{\alpha}_1, \tilde{\alpha}_2)\right] &= (0, -\xi\tilde{\alpha}_2 + \tilde{\xi}\alpha_2, \xi\tilde{\alpha}_1 - \tilde{\xi}\alpha_1) \\ &= (0, -\xi\mathbb{J}\tilde{\alpha} + \tilde{\xi}\mathbb{J}\alpha), \end{aligned}$$

where $\alpha = (\alpha_1, \alpha_2)$ and $\tilde{\alpha} = (\tilde{\alpha}_1, \tilde{\alpha}_2)$. ★

Exercise. Check using $R_\theta\mathbb{J} = \mathbb{J}R_\theta$ that the adjoint action for $SE(2)$ of

$$(R_\theta, v) = \begin{pmatrix} R_\theta & v \\ 0 & 1 \end{pmatrix} \quad \text{on} \quad (\xi, \alpha) = \begin{pmatrix} -\xi\mathbb{J} & \alpha \\ 0 & 0 \end{pmatrix}$$

is given by

$$(R_\theta, v)(\xi, \alpha)(R_\theta, v)^{-1} = \begin{pmatrix} -\xi\mathbb{J} & \xi\mathbb{J}v + R_\theta\alpha \\ 0 & 0 \end{pmatrix},$$

or in coordinates

$$\mathrm{Ad}_{(R_\theta,v)}(\xi,\alpha) = (\xi,\, \xi\mathbb{J}v + R_\theta\alpha)\,. \tag{6.4.1}$$

★

The pairing $\langle\,\cdot\,,\,\cdot\,\rangle : se(2)^* \times se(2) \mapsto \mathbb{R}$ is obtained by identifying $SE(2) \simeq SO(2) \times \mathbb{R}^2$ and taking the sum,

$$\langle\, (\mu,\beta)\,,\, (\xi\,,\, \alpha)\,\rangle \equiv \mu\xi + \beta\cdot\alpha\,, \tag{6.4.2}$$

with $\mu \in so(2)^*$, $\xi \in so(2)$ represented as skew-symmetric 2×2 matrices and β, $\alpha \in \mathbb{R}^2$ represented as planar vectors. Elements of the dual Lie algebra $se(2)^*$ may be written as block matrices of the form

$$(\mu,\beta) = \begin{pmatrix} \frac{\mu}{2}\mathbb{J} & 0 \\ \beta & 0 \end{pmatrix},$$

since

$$\mathrm{tr}\left[\begin{pmatrix} \frac{\mu}{2}\mathbb{J} & 0 \\ \beta & 0 \end{pmatrix}\begin{pmatrix} -\xi\mathbb{J} & \alpha \\ 0 & 0 \end{pmatrix}\right] = \mu\xi + \beta\cdot\alpha\,,$$

via the nondegenerate pairing provided by the trace of the matrix product. Thus, we may identify the dual Lie algebra $se(2)^*$ with $\mathbb{R}^3$ via the isomorphism

$$\begin{pmatrix} \frac{\mu}{2}\mathbb{J} & 0 \\ \beta & 0 \end{pmatrix} \in se(2)^* \mapsto (\mu,\beta) \in \mathbb{R}^3\,,$$

so that in these coordinates the pairing (6.4.2) between $se(2)^*$ and $se(2)$ becomes the usual dot product in $\mathbb{R}^3$.

Exercise. Check that the coadjoint action of $SE(2)$ on $se(2)^*$ is given by

$$\mathrm{Ad}^*_{(R_\theta,v)^{-1}}(\mu,\beta) = (\mu - R_\theta\beta\cdot\mathbb{J}v,\; R_\theta\beta)\,. \tag{6.4.3}$$

★

Exercise. Show that the coadjoint orbits for $SE(2)$ are the cylinders $T^*S^1_\alpha = \{(\mu, \beta) : |\beta| = \text{constant}\}$, together with points on the μ-axis. ★

Exercise. What are the Casimirs for $SE(2)$? ★

6.5 Semidirect-product group $SL(2,\mathbb{R})Ⓢ\mathbb{R}^2$

As a further example, we compute the adjoint and coadjoint actions for the semidirect-product group $SL(2,\mathbb{R})Ⓢ\mathbb{R}^2$. The change to $SL(2,\mathbb{R})Ⓢ\mathbb{R}^2$ from $SE(2) \cong SO(2)Ⓢ\mathbb{R}^2$ incorporates area-preserving dilations and both left and right $SO(2)$ rotations into $SE(2)$, as well as the translations $\mathbb{R}^2$. These additional degrees of freedom may be recognised in the polar decomposition of $R \in SL(2,\mathbb{R})$ into $R = O_1 S O_2$, in which S is a 2×2 diagonal matrix of unit determinant, and O_1 and O_2 are the left and right $SO(2)$ rotations, respectively. After defining the matrix representations of the Lie group and its Lie algebra, we derive its AD, Ad and ad actions. We then define the dual Lie algebra and derive its Ad* and ad* actions. Finally, we compute its coadjoint motion equations.

6.5.1 Definitions for $SL(2,\mathbb{R})Ⓢ\mathbb{R}^2$

We consider the semidirect-product Lie group $G = SL(2,\mathbb{R})Ⓢ\mathbb{R}^2$ in which $SL(2,\mathbb{R})$ acts on $\mathbb{R}^2$ by matrix multiplication from the left. The group composition rule is

$$(\tilde{R}, \tilde{v})(R, v) = (\tilde{R}R, \tilde{R}v + \tilde{v}) , \tag{6.5.1}$$

which can be represented by multiplication of 3×3 matrices. That is, the action of G on $\mathbb{R}^3$ has a matrix representation, given by

$$(R, v) \mapsto \begin{pmatrix} R & v \\ 0 & 1 \end{pmatrix}, \tag{6.5.2}$$

for $R \in SL(2,\mathbb{R})$, so that $\det R = 1$ and $v \in \mathbb{R}^2$. The matrix multiplication

$$\begin{pmatrix} \tilde{R} & \tilde{v} \\ 0 & 1 \end{pmatrix}\begin{pmatrix} R & v \\ 0 & 1 \end{pmatrix} = \begin{pmatrix} \tilde{R}R & \tilde{R}v + \tilde{v} \\ 0 & 1 \end{pmatrix} \tag{6.5.3}$$

agrees with (6.5.1). The inverse is given by

$$(\tilde{R}, \tilde{v})^{-1} = (\tilde{R}^{-1}, -\tilde{R}^{-1}\tilde{v}) \tag{6.5.4}$$

and the identity element is $(\mathbb{1}, 0)$, where $\mathbb{1}$ is the 2×2 identity element of $SL(2,\mathbb{R})$.

The 3×3 matrix representation of this group in (6.5.2) acts on the *extended* vector $(r, 1)^T$ as

$$\begin{pmatrix} R & v \\ 0 & 1 \end{pmatrix}\begin{pmatrix} r \\ 1 \end{pmatrix} = \begin{pmatrix} Rr + v \\ 1 \end{pmatrix}.$$

The Lie group $G = SL(2,\mathbb{R})\circledS\mathbb{R}^2$ has five parameters. These may be identified by considering their action on an ellipse whose centre is initially at the origin of coordinates in the plane. In the polar decomposition of the matrix $R = O_2 S O_1 \in SL(2,\mathbb{R})$, the orthogonal matrix $O_1 \in SO(2)$ rotates the planar reference coordinates about the centre at the origin into the principal axes of the fixed ellipse. The diagonal matrix S with $\det S = 1$ then stretches the principal axes while preserving the area of the ellipse. Next, the orthogonal matrix $O_2 \in SO(2)$ rigidly rotates the rescaled ellipse about its centre. And finally, the vector $v \in \mathbb{R}^2$ translates the centre of the rescaled rotated ellipse to a new location in the plane.

6.5.2 AD, Ad, and ad actions

A matrix calculation represents the AD action $G \times G \to G$ as

$$\begin{aligned} \mathrm{AD}_{(\tilde{R},\tilde{v})}(R,v) &= (\tilde{R},\tilde{v})(R,v)(\tilde{R}^{-1},-\tilde{R}^{-1}\tilde{v}) \\ &= (\mathrm{AD}_{\tilde{R}}\, R, -\mathrm{AD}_{\tilde{R}}\, R\tilde{v} + \tilde{R}v + \tilde{v}) \\ &= \begin{pmatrix} \tilde{R}R\tilde{R}^{-1} & -\tilde{R}R\tilde{R}^{-1}\tilde{v} + \tilde{R}v + \tilde{v} \\ 0 & 1 \end{pmatrix}. \end{aligned}$$

Next, the Ad action $G \times \mathfrak{g} \to \mathfrak{g}$ may be computed. By taking derivatives of the matrix representation (6.5.2) for the Lie group $G = SL(2,\mathbb{R})Ⓢ\mathbb{R}^2$ at the identity, one defines the basis

$$X = \begin{pmatrix} 1 & 0 \\ 0 & -1 \end{pmatrix}, \quad Y = \begin{pmatrix} 0 & 1 \\ 1 & 0 \end{pmatrix}, \quad Z = \begin{pmatrix} 0 & 1 \\ -1 & 0 \end{pmatrix},$$
$$H_1 = \begin{pmatrix} 1 \\ 0 \end{pmatrix}, \quad H_2 = \begin{pmatrix} 0 \\ 1 \end{pmatrix}, \tag{6.5.5}$$

for the matrix representation of the Lie algebra $\mathfrak{g} = sl(2,\mathbb{R})\oplus\mathbb{R}^2$. The traceless property of the 2×2 $sl(2,\mathbb{R})$ matrices follows from taking the derivative at the identity of the unit determinant condition for $SL(2,\mathbb{R})$. A matrix representation of the Lie algebra

$$\xi = (A,h) \in \mathfrak{g} = sl(2,\mathbb{R}) \oplus \mathbb{R}^2$$

is given in this basis by $A = xX + yY + zZ$ and $h = h_1H_1 + h_2H_2$, with $(x,y,z) \in \mathbb{R}^3$ and $(h_1,h_2) \in \mathbb{R}^2$, so that

$$A = \begin{pmatrix} x & y+z \\ y-z & -x \end{pmatrix} \in sl(2,\mathbb{R}) \quad \text{and} \quad h = \begin{pmatrix} h_1 \\ h_2 \end{pmatrix}. \tag{6.5.6}$$

Consequently, one finds the Ad action in matrix form,

$$\begin{aligned} \mathrm{Ad}_{(\tilde{R},\tilde{v})}(A,h) &= \frac{d}{dt}\bigg|_{t=0} \mathrm{AD}_{(\tilde{R},\tilde{v})}(e^{tA}, th) \\ &= \begin{pmatrix} \tilde{R}A\tilde{R}^{-1} & -\tilde{R}A\tilde{R}^{-1}\tilde{v} + \tilde{R}h \\ 0 & 0 \end{pmatrix}. \end{aligned} \tag{6.5.7}$$

The matrix Lie group representation (6.5.2) implies the following 3×3 matrix Lie algebra representation, denoted by M,

$$(A,h) \mapsto M(x,y,z,h_1,h_2) = \begin{pmatrix} x & y+z & h_1 \\ y-z & -x & h_2 \\ 0 & 0 & 0 \end{pmatrix}. \tag{6.5.8}$$

This may be abbreviated as

$$(A,h) \mapsto M(A,h) = \begin{pmatrix} A & h \\ 0 & 0 \end{pmatrix}. \tag{6.5.9}$$

In this matrix representation, one obtains

$$\mathrm{Ad}_{(\tilde{R},\tilde{v})}(A,h) = \begin{pmatrix} \tilde{R} & \tilde{v} \\ 0 & 1 \end{pmatrix} \begin{pmatrix} A & h \\ 0 & 0 \end{pmatrix} \begin{pmatrix} \tilde{R}^{-1} & -\tilde{R}^{-1}\tilde{v} \\ 0 & 1 \end{pmatrix}, \tag{6.5.10}$$

which recovers the result in formula (6.5.7).

The ad action may now be computed by a matrix commutation,

$$\begin{aligned} \mathrm{ad}_{(\tilde{A},\tilde{h})}(A,h) &= \left[\begin{pmatrix} \tilde{A} & \tilde{h} \\ 0 & 0 \end{pmatrix}, \begin{pmatrix} A & h \\ 0 & 0 \end{pmatrix} \right] \\ &= \begin{pmatrix} [\tilde{A},A] & \tilde{A}h - A\tilde{h} \\ 0 & 0 \end{pmatrix} \\ &=: \left([\tilde{A},A], \tilde{A}h - A\tilde{h} \right). \end{aligned} \tag{6.5.11}$$

This is the standard form of the ad action of a semidirect-product Lie algebra.

6.5.3 Ad* and ad* actions

A 3×3 matrix representation of the Lie algebra $\mathfrak{g} = sl(2,\mathbb{R}) \oplus \mathbb{R}^2$ is defined by the basis

$$X = \begin{pmatrix} 1 & 0 & 0 \\ 0 & -1 & 0 \\ 0 & 0 & 0 \end{pmatrix}, \quad Y = \begin{pmatrix} 0 & 1 & 0 \\ 1 & 0 & 0 \\ 0 & 0 & 0 \end{pmatrix}, \quad Z = \begin{pmatrix} 0 & 1 & 0 \\ -1 & 0 & 0 \\ 0 & 0 & 0 \end{pmatrix},$$
$$H_1 = \begin{pmatrix} 0 & 0 & 1 \\ 0 & 0 & 0 \\ 0 & 0 & 0 \end{pmatrix}, \quad H_2 = \begin{pmatrix} 0 & 0 & 0 \\ 0 & 0 & 1 \\ 0 & 0 & 0 \end{pmatrix}. \tag{6.5.12}$$

The corresponding basis elements in $\mathfrak{g}^*$ dual to these are their transposes, denoted by $X^*, Y^*, Z^*, H_1^*, H_2^*$. An element of $\mathfrak{g}^*$ can be written as a row component vector with respect to this basis. Namely,

$$aX^* + bY^* + cZ^* + k_1 H_1^* + k_2 H_2^* =: \begin{pmatrix} a & b & c & k_1 & k_2 \end{pmatrix}. \tag{6.5.13}$$

The matrix representation of the dual Lie algebra $\mathfrak{g}^*$ is defined via the following map,

$$aX^* + bY^* + cZ^* + k_1 H_1^* + k_2 H_2^* \mapsto M^*(a,b,c,k_1,k_2), \tag{6.5.14}$$

with the traceless matrix representation

$$M^*(a,b,c,k_1,k_2) = \begin{pmatrix} a & b-c & 0 \\ b+c & -a & 0 \\ k_1 & k_2 & 0 \end{pmatrix}. \tag{6.5.15}$$

To compute the Ad* action, denote

$$D = \begin{pmatrix} a & b-c \\ b+c & -a \end{pmatrix}, \quad k = \begin{pmatrix} k_1 & k_2 \end{pmatrix}. \tag{6.5.16}$$

Then consider $(D,k) \in \mathfrak{g}^*$ defined as the element $aX^* + \cdots + k_2 H_2^*$, which in the matrix representation (6.5.15) corresponds to

$$(D,k) \mapsto \begin{pmatrix} D & 0 \\ k & 0 \end{pmatrix}. \tag{6.5.17}$$

Let A and h be as in (6.5.6). Then, upon denoting the map $\mathfrak{g}^*\times\mathfrak{g}\to\mathbb{R}$ as the pairing $\langle\,\cdot\,,\,\cdot\,\rangle_{\mathfrak{g}^*\times\mathfrak{g}}$, one finds

$$\begin{aligned}\left\langle \mathrm{Ad}^*_{(R,v)}(D,k),(A,h)\right\rangle_{\mathfrak{g}\times\mathfrak{g}^*} &= \langle (D,k), \mathrm{Ad}_{(R,v)}(A,h)\rangle \\ &= \frac{1}{2}\mathrm{Tr}\left(\begin{pmatrix} D & 0 \\ k & 0\end{pmatrix} g \begin{pmatrix} A & h \\ 0 & 0\end{pmatrix} g^{-1}\right) \\ &= \frac{1}{2}\mathrm{Tr}\left(g^{-1}\begin{pmatrix} D & 0 \\ k & 0\end{pmatrix} g \begin{pmatrix} A & h \\ 0 & 0\end{pmatrix}\right)\end{aligned}$$

where, as in (6.5.2),

$$g = \begin{pmatrix} R & v \\ 0 & 1\end{pmatrix}. \tag{6.5.18}$$

A side calculation yields

$$g^{-1}\begin{pmatrix} D & 0 \\ k & 0\end{pmatrix} g = \begin{pmatrix} R^{-1}(D-vk)R & * \\ kR & * \end{pmatrix}, \tag{6.5.19}$$

in which the entries at positions $*$ are immaterial, since taking the trace with the last factor in the pairing above will eliminate them. One may also make the upper left part of this matrix traceless, by subtracting the product of its trace times half the 2×2 identity, $\mathbb{1}$. This is allowed, since the trace of this product with the traceless matrix A will not contribute in the pairing.

Hence, one arrives at the matrix representation of the Ad^* action,

$$\mathrm{Ad}^*_{(R,v)}(D,k) = \begin{pmatrix} R^{-1}(D-vk)R + \frac{1}{2}(\mathbf{v}\cdot\mathbf{k})\mathbb{1} & 0 \\ kR & 0\end{pmatrix}, \tag{6.5.20}$$

in which $\mathrm{Tr}(R^{-1}DR)=0$, so the trace of the 2×2 upper left submatrix vanishes. The corresponding result for $(R,v)^{-1}$ is

$$\mathrm{Ad}^*_{(R,v)^{-1}}(D,k) = \begin{pmatrix} RDR^{-1} + (vk)R^{-1} - \frac{1}{2}\mathrm{Tr}((vk)R^{-1}) & 0 \\ kR^{-1} & 0\end{pmatrix}.$$

The next step is to compute ad* from

$$\left\langle \mathrm{ad}^*_{(A,h)}(D,k),(\tilde{A},\tilde{h})\right\rangle = \left\langle (D,k),\mathrm{ad}_{(A,h)}(\tilde{A},\tilde{h})\right\rangle \tag{6.5.21}$$
$$= \frac{1}{2}\mathrm{Tr}\left(\begin{pmatrix} D & 0 \\ k & 0 \end{pmatrix}\left[\begin{pmatrix} A & h \\ 0 & 0 \end{pmatrix},\begin{pmatrix} \tilde{A} & \tilde{h} \\ 0 & 0 \end{pmatrix}\right]\right).$$

This leads to

$$\mathrm{ad}^*_{(A,h)}\begin{pmatrix} D & 0 \\ k & 0 \end{pmatrix} = \begin{pmatrix} [D,A]-hk+\frac{1}{2}(\mathbf{h}\cdot\mathbf{k})\mathbb{1} & 0 \\ kA & 0 \end{pmatrix}. \tag{6.5.22}$$

The commutator $[D,\ A]$ in this equation may be computed in the 2×2 matrix basis (X,Y,Z) for $sl(2,\mathbb{R})$, since in that basis we have $A = xX + yY + zZ$ and $D = aX + bY - cZ$, the latter because $(X^T,Y^T,Z^T) = (X,Y,-Z)$.

Summary. The adjoint and coadjoint actions for $SL(2,\mathbb{R})Ⓢ\mathbb{R}^2$ are

$$\begin{aligned}
\mathrm{AD}_{(\tilde{R},\tilde{v})}(R,v) &= \left(\tilde{R}R\tilde{R}^{-1},\ -\tilde{R}R\tilde{R}^{-1}\tilde{v}+\tilde{R}v+\tilde{v}\right),\\
\mathrm{Ad}_{(\tilde{R},\tilde{v})}(A,h) &= \left(\tilde{R}A\tilde{R}^{-1},\ -\tilde{R}A\tilde{R}^{-1}\tilde{v}+\tilde{R}h\right),\\
\mathrm{ad}_{(\tilde{A},\tilde{h})}(A,h) &= \left([\tilde{A},A],\tilde{A}h-A\tilde{h}\right),\\
\mathrm{Ad}^*_{(R,v)}(D,k) &= \left(R^{-1}(D-vk)R+\tfrac{1}{2}(\mathbf{v}\cdot\mathbf{k})\mathbb{1},\ kR\right),\\
\mathrm{ad}^*_{(A,h)}(D,k) &= \left([D,A]-hk+\tfrac{1}{2}(\mathbf{h}\cdot\mathbf{k})\mathbb{1},\ kA\right).
\end{aligned}$$

6.5.4 Coadjoint motion relation

According to Proposition 4.2.2, the coadjoint motion relation arises from differentiating the coadjoint orbit relation. Let $g(t) = (R(t),v(t))$ be a path in the Lie group $G = SL(2,\mathbb{R})Ⓢ\mathbb{R}^2$ and let $\mu(t) = (D(t),l(t))$ be a path in the dual Lie algebra $\mathfrak{g}^*$. Then Proposition 4.2.2 supplies the coadjoint motion relation

$$\frac{d}{dt}\left(\mathrm{Ad}^*_{g(t)^{-1}}\mu\right) = \mathrm{Ad}^*_{g(t)^{-1}}\left[\frac{d\mu}{dt} - \mathrm{ad}^*_{\xi(t)}\mu(t)\right], \tag{6.5.23}$$

where $\xi(t) = g(t)^{-1}\dot{g}(t) = (A(t),h(t))$.

Corollary 4.2.11 provides the differential equation for the coadjoint orbit relation

$$\mu(t) = \mathrm{Ad}^*_{g(t)}\mu(0)\,.$$

The desired differential equation is the ***coadjoint motion equation***

$$\frac{d\mu}{dt} = \mathrm{ad}^*_{\xi(t)}\mu(t)\,.$$

This equation is expressed for the Lie group $G = SL(2,\mathbb{R})\circledS\mathbb{R}^2$ as

$$\frac{d}{dt}(D,\,k) = \Big(\big[D,A\big] - hk + \frac{1}{2}(\mathbf{h}\cdot\mathbf{k})\mathbb{1},\; kA\Big)\,, \tag{6.5.24}$$

with $(D,k) \in \mathfrak{g}^*$ defined in (6.5.17) and $(A,h) \in \mathfrak{g}$ defined in (6.5.6). When (A,h) is known as a smooth invertible function of (D,k), e.g., $(A,h) = (\partial H/\partial D, \partial H/\partial k)$ for a Hamiltonian $H(D,k)$, then this equation becomes a Lie–Poisson Hamiltonian system for the dynamics of a rotating, stretching, circulating and translating ellipse, whose area is preserved by the motion. See [Ho1991] for the corresponding discussion of coadjoint motion on $GL(2,\mathbb{R})$.

Exercise. Recompute the ad action and ad* action for the semidirect-product group $SL(2,\mathbb{R})\circledS\mathbb{R}^2$ after introducing the polar decomposition $R = O_1 S O_2$, in which S is a diagonal matrix and O_1 and O_2 are $SO(2)$ rotations. Identify two types of centrifugal terms in the coadjoint orbit equations. ★

Exercise. Compare the ad and ad* actions for the semidirect-product groups $SL(2,\mathbb{R})\circledS\mathbb{R}^2$ and $Sp(2)\circledS\mathbb{R}^2$. ★

Exercise. Compute the ad action and ad* action for the Galilean group $SE(3)\circledS\mathbb{R}^4$. ★

Exercise. Write the semidirect-product action $SU(n)Ⓢ H_n$ of n-dimensional unitary transformations on the vector space of $n \times n$ Hermitian matrices. Compute its ad action and ad* action.

How is this related to $O(n)ⓈS_n$, the semidirect-product group of n-dimensional orthogonal transformations $O(n)$ acting on the vector space of $n \times n$ symmetric matrices S_n?

★

6.6 Galilean group

6.6.1 Definitions for $G(3)$

As discussed in Chapter 1, the Galiliean group in three dimensions $G(3)$ has ten parameters,

$$(O \in SO(3)\,,\ \mathbf{r}_0 \in \mathbb{R}^3\,,\ \mathbf{v}_0 \in \mathbb{R}^3\,,\ t_0 \in \mathbb{R}\,).$$

The Galilean group is a semidirect-product Lie group, which may be written as

$$G(3) = SE(3)\,Ⓢ\,\mathbb{R}^4 = \Big(SO(3)\,Ⓢ\,\mathbb{R}^3\Big)Ⓢ\,\mathbb{R}^4\,. \qquad (6.6.1)$$

That is, the subgroup of Euclidean motions, which comprises rotations and Galilean velocity boosts $(O, \mathbf{v}_0) \in SE(3)$, acts homogeneously on the subgroups of space and time translations $(\mathbf{r}_0, t_0) \in \mathbb{R}^4$ which commute with each other.

Matrix representation of $G(3)$

The formula for group composition $G(3) \times G(3) \to G(3)$ may be represented by matrix multiplication as in Equation (1.5.3),

$$\begin{aligned} \tilde{g}g &= \begin{pmatrix} \widetilde{O} & \tilde{\mathbf{v}} & \tilde{\mathbf{r}} \\ \mathbf{0} & 1 & \tilde{t} \\ \mathbf{0} & 0 & 1 \end{pmatrix} \begin{pmatrix} O & \mathbf{v} & \mathbf{r} \\ \mathbf{0} & 1 & t \\ \mathbf{0} & 0 & 1 \end{pmatrix} \\ &= \begin{pmatrix} \widetilde{O}O & \widetilde{O}\mathbf{v} + \tilde{\mathbf{v}} & \widetilde{O}\mathbf{r} + \tilde{\mathbf{v}}t + \tilde{\mathbf{r}} \\ \mathbf{0} & 1 & \tilde{t} + t \\ \mathbf{0} & 0 & 1 \end{pmatrix}. \end{aligned} \tag{6.6.2}$$

This may also be expressed succinctly in row notation as

$$\begin{aligned} \tilde{g}g &= (\widetilde{O}, \tilde{\mathbf{v}}, \tilde{\mathbf{r}}, \tilde{t})(O, \mathbf{v}, \mathbf{r}, t) \\ &= (\widetilde{O}O,\ \widetilde{O}\mathbf{v} + \tilde{\mathbf{v}},\ \widetilde{O}\mathbf{r} + \tilde{\mathbf{v}}t + \tilde{\mathbf{r}},\ \tilde{t} + t). \end{aligned} \tag{6.6.3}$$

The inverse operation is given in matrix form as

$$\begin{aligned} \tilde{g}^{-1} &= \begin{pmatrix} \widetilde{O}^{-1} & -\widetilde{O}^{-1}\tilde{\mathbf{v}} & -\widetilde{O}^{-1}(\tilde{\mathbf{r}} - \tilde{t}\tilde{\mathbf{v}}) \\ \mathbf{0} & 1 & -\tilde{t} \\ \mathbf{0} & 0 & 1 \end{pmatrix} \\ &= (\widetilde{O}^{-1},\ -\widetilde{O}^{-1}\tilde{\mathbf{v}},\ -\widetilde{O}^{-1}(\tilde{\mathbf{r}} - \tilde{t}\tilde{\mathbf{v}}),\ -\tilde{t}). \end{aligned} \tag{6.6.4}$$

6.6.2 AD, Ad, and ad actions of $G(3)$

A matrix calculation represents the AD action $G(3) \times G(3) \to G(3)$ as

$$\begin{aligned} \mathrm{AD}_{\tilde{g}}\, g &= \mathrm{AD}_{(\widetilde{O}, \tilde{\mathbf{v}}, \tilde{\mathbf{r}}, \tilde{t})}(O, \mathbf{v}, \mathbf{r}, t) \\ &= (\widetilde{O}, \tilde{\mathbf{v}}, \tilde{\mathbf{r}}, \tilde{t})(O, \mathbf{v}, \mathbf{r}, t)(\widetilde{O}^{-1}, -\widetilde{O}^{-1}\tilde{\mathbf{v}}, -\widetilde{O}^{-1}(\tilde{\mathbf{r}} - \tilde{t}\,\tilde{\mathbf{v}}), -\tilde{t}) \\ &= \Big(\mathrm{AD}_{\widetilde{O}}\, O,\ -(\mathrm{AD}_{\widetilde{O}}\, O)\tilde{\mathbf{v}} + \widetilde{O}\mathbf{v} + \tilde{\mathbf{v}}, \\ &\qquad -(\mathrm{AD}_{\widetilde{O}}\, O)(\mathbf{r} - \mathbf{v}t) - (\widetilde{O}\mathbf{v} + \tilde{\mathbf{v}})\tilde{t} + \widetilde{O}\mathbf{r} + \tilde{\mathbf{v}}t + \tilde{\mathbf{r}},\ t \Big). \end{aligned}$$

Next, the Ad action $G(3) \times \mathfrak{g}(3) \to \mathfrak{g}(3)$ may be computed. By taking derivatives of the matrix representation (1.5.3) for the Galilean group $G(3)$ at the identity, one finds

$$\Xi := g^{-1}(s)g'(s)|_{s=0} = \begin{pmatrix} \widehat{\Xi} & \mathbf{v} & \mathbf{r} - \mathbf{v}t \\ 0 & 0 & t \\ 0 & 0 & 0 \end{pmatrix} =: \left(\widehat{\Xi},\, \mathbf{v},\, \mathbf{r} - \mathbf{v}t,\, t\right),$$

in terms of the 3×3 skew-symmetric matrix $\widehat{\Xi} = O^{-1}(s)O'(s)|_{s=0}$ and the Galilean shift parameters $\mathbf{v}$, $\mathbf{r}$, t. Consequently, one finds the Ad action in matrix form,

$$\mathrm{Ad}_{\tilde{g}}\Xi = \mathrm{Ad}_{(\widetilde{O},\tilde{\mathbf{v}},\tilde{\mathbf{r}},\tilde{t})}(\widehat{\Xi}, \mathbf{v}, \mathbf{r} - \mathbf{v}t, t) \tag{6.6.5}$$

$$= \begin{pmatrix} \mathrm{Ad}_{\widetilde{O}}\widehat{\Xi} & -(\mathrm{Ad}_{\widetilde{O}}\widehat{\Xi})\tilde{\mathbf{v}} + \widetilde{O}\mathbf{v} & -(\mathrm{Ad}_{\widetilde{O}}\widehat{\Xi})(\tilde{\mathbf{r}} - \tilde{\mathbf{v}}\tilde{t}) - \widetilde{O}\mathbf{v}\tilde{t} + \widetilde{O}(\mathbf{r} - \mathbf{v}t) + \tilde{\mathbf{v}}t \\ 0 & 0 & t \\ 0 & 0 & 0 \end{pmatrix}.$$

Taking the derivative of the tilde variables and evaluating at the identity recovers the commutation relation asserted in Section 1.7,

$$\begin{aligned} \mathrm{ad}_{\tilde{\Xi}}\Xi &= \left[(\widehat{\widetilde{\Xi}},\, \tilde{\mathbf{v}},\, \tilde{\mathbf{r}} - \tilde{\mathbf{v}}\tilde{t},\, \tilde{t}),\, (\widehat{\Xi},\, \mathbf{v},\, \mathbf{r} - \mathbf{v}t,\, t)\right] \\ &= \left(\left[\widehat{\widetilde{\Xi}}, \widehat{\Xi}\right],\, \widehat{\widetilde{\Xi}}\mathbf{v} - \widehat{\Xi}\tilde{\mathbf{v}},\, \widehat{\widetilde{\Xi}}(\mathbf{r}, \mathbf{v}, t) - \widehat{\Xi}(\tilde{\mathbf{r}}, \tilde{\mathbf{v}}, \tilde{t}),\, 0\right), \end{aligned}$$

where

$$\begin{aligned} &\widehat{\widetilde{\Xi}}(\mathbf{r}, \mathbf{v}, t) - \widehat{\Xi}(\tilde{\mathbf{r}}, \tilde{\mathbf{v}}, \tilde{t}) \\ &:= \left(\widehat{\widetilde{\Xi}}(\mathbf{r} - \mathbf{v}t) + \tilde{\mathbf{v}}t\right) - \left(\widehat{\Xi}(\tilde{\mathbf{r}} - \tilde{\mathbf{v}}\tilde{t}) + \mathbf{v}\tilde{t}\right). \end{aligned}$$

The Galilean group's Ad^* and ad^* actions may now be obtained by using the matrix transpose pairing to define the dual Lie algebra.

Exercise. Compute the coadjoint actions of $G(3)$, the Galilean group in three dimensions. ★

6.7 Iterated semidirect products

ad action

Let G be a Lie group with Lie algebra $\mathfrak{g}$. Consider the Lie group obtained by iterating the semidirect-product action of G on itself n times,

$$G_1 \circledS (\ldots (G_{n-2} \circledS (G_{n-1} \circledS G_n)) \ldots). \tag{6.7.1}$$

Its Lie algebra elements are denoted as

$$(u_1, u_2, \ldots, u_n) \in \mathfrak{g}^{\times n} := \mathfrak{g}_1 \times (\ldots (\mathfrak{g}_{n-2} \times (\mathfrak{g}_{n-1} \times \mathfrak{g}_n)) \ldots).$$

These possess the iterated semidirect product Lie algebra action

$$(u_1, u_2, \ldots, u_n) \in \mathfrak{g}^{\circledS n} := \mathfrak{g}_1 \circledS (\ldots (\mathfrak{g}_{n-2} \circledS (\mathfrak{g}_{n-1} \circledS \mathfrak{g}_n)) \ldots)$$

given by

$$\begin{aligned} &\mathrm{ad}_{(u_1, u_2, \ldots, u_n)}(v_1, v_2, \ldots, v_n) \\ &= \Big(\mathrm{ad}_{u_1} v_1,\ \mathrm{ad}_{u_1+u_2} v_2 + \mathrm{ad}_{u_2} v_1,\ \mathrm{ad}_{u_1+u_2+u_3} v_3 + \mathrm{ad}_{u_3}(v_1 + v_2), \\ &\quad \ldots, \sum_{k=1}^{n} \mathrm{ad}_{u_k} v_n + \sum_{k=1}^{n-1} \mathrm{ad}_{u_n} v_k \Big) \end{aligned} \tag{6.7.2}$$

in which the level $n = 1$ and $n = 2$ formulas are already familiar.

Exercise. Prove formula (6.7.2) by induction, in which the m-th step is given by

$$\ldots, \mathrm{ad}_{u_m} v_m + \mathrm{ad}_{u_m} \left(\sum_{k=1}^{m-1} v_k \right) - \mathrm{ad}_{v_m} \left(\sum_{k=1}^{m-1} u_k \right), \ldots$$

This is how infinitesimal transformations compose under iteration of semidirect-product Lie algebra action. ★

ad* action

One calculates ad* for the iterated semidirect product by using (6.7.2) in the pairing $\langle\,\cdot\,,\,\cdot\,\rangle : (\mathfrak{g}^{\circledS n})^* \times \mathfrak{g}^{\circledS n} \to \mathbb{R}$

$$\begin{aligned}&\big\langle (\mu_1, \mu_2, \dots, \mu_n),\, \mathrm{ad}_{(u_1,u_2,\dots,u_n)}(v_1, v_2, \dots, v_n)\big\rangle \\ &\quad = \big\langle \mathrm{ad}^*{}_{(u_1,u_2,\dots,u_n)}(\mu_1, \mu_2, \dots, \mu_n),\, (v_1, v_2, \dots, v_n)\big\rangle\end{aligned}$$

to find the coadjoint action

$$\begin{aligned}&\mathrm{ad}^*{}_{(u_1,u_2,\dots,u_n)}(\mu_1, \mu_2, \dots, \mu_n) \\ &= \left(\sum_{k=1}^{n} \mathrm{ad}^*_{u_k}\mu_k\,, \dots, \sum_{k=1}^{m} \mathrm{ad}^*_{u_k}\mu_m + \sum_{k=m+1}^{n} \mathrm{ad}^*_{u_k}\mu_k\,, \dots, \sum_{k=1}^{n} \mathrm{ad}^*_{u_k}\mu_n \right).\end{aligned} \tag{6.7.3}$$

Exercise. Derive the system of Euler–Poincaré equations for coadjoint motion on $(\mathfrak{g}^{\circledS n})^*$.

Show that its Lie–Poisson structure may be diagonalised by taking a linear combination of the variables. ★

Exercise. Use the hat map to write the system of Euler–Poincaré equations for coadjoint motion on $(\mathfrak{so}(3)^{\circledS n})^*$ in $\mathbb{R}^3$ vector form.

Also write its diagonal Lie–Poisson structure. ★

7

EULER–POINCARÉ AND LIE–POISSON EQUATIONS ON $SE(3)$

Contents

7.1 Euler–Poincaré equations for left-invariant Lagrangians under $SE(3)$ **170**

7.1.1 Legendre transform from $se(3)$ to $se(3)^*$ **172**

7.1.2 Lie–Poisson bracket on $se(3)^*$ **172**

7.1.3 Coadjoint motion on $se(3)^*$ **173**

7.2 Kirchhoff equations on $se(3)^*$ **176**

7.2.1 Looks can be deceiving: The heavy top **178**

7.1 Euler–Poincaré equations for left-invariant Lagrangians under $SE(3)$

The matrix Euler–Poincaré equation for left-invariant Lagrangians was introduced in Proposition 2.4.3, in the context of Manakov's [Man1976] commutator formulation of the Euler rigid-body equations. As we shall see, Euler–Poincaré evolution is naturally expressed as coadjoint motion.

The variational derivatives of a given left-invariant Lagrangian $\ell(\xi,\,\alpha) : se(3) \to \mathbb{R}$ are expressed as

$$(\mu\,,\,\beta) = \Big(\frac{\delta\ell}{\delta\xi}\,,\,\frac{\delta\ell}{\delta\alpha}\Big) \in se(3)^*. \tag{7.1.1}$$

The commutator form of the matrix Euler–Poincaré Equation (2.4.27) is

$$\frac{dM}{dt} = [\,M\,,\,\Omega\,] \quad \text{with} \quad M = \frac{\delta l}{\delta\Omega}\,, \tag{7.1.2}$$

for any Lagrangian $l(\Omega)$, where $\Omega = g^{-1}\dot{g} \in \mathfrak{g}$ and $\mathfrak{g}$ is the left-invariant matrix Lie algebra of any matrix Lie group G. Recall that $M = \delta l/\delta\Omega \in \mathfrak{g}^*$, where the dual $\mathfrak{g}^*$ is defined in terms of the matrix trace pairing. Hence, we may rewrite the Euler–Poincaré Equation (7.1.2) in terms of the left action of the Lie algebra on its dual. This happens by the ad* action; so the Euler–Poincaré equation becomes

$$\frac{dM}{dt} = \mathrm{ad}^*_\Omega\, M \quad \text{with} \quad M = \frac{\delta l}{\delta\Omega}\,. \tag{7.1.3}$$

Proposition 7.1.1 *The corresponding Euler–Poincaré equation for the special Euclidean group in three dimensions, $SE(3)$, describes coadjoint motion on $se(3)^*$. Namely,*

$$\begin{aligned}\Big\langle \Big(\frac{d\mu}{dt}\,,\,\frac{d\beta}{dt}\Big)\,,\,(\tilde\xi\,,\,\tilde\alpha)\Big\rangle &= \langle \mathrm{ad}^*_{(\xi\,,\alpha)}(\mu\,,\,\beta)\,,\,(\tilde\xi\,,\,\tilde\alpha)\rangle \qquad (7.1.4)\\ &= \langle(\mathrm{ad}^*_\xi\,\mu + \beta\diamond\alpha\,,\,-\xi\beta)\,,\,(\tilde\xi\,,\,\tilde\alpha)\rangle\,.\end{aligned}$$

Proof. The proof follows immediately from the definition of the Euler–Poincaré equation for $SE(3)$. ∎

Definition 7.1.1 (Euler–Poincaré equation for $SE(3)$) *The Euler–Poincaré equation for the special Euclidean group in three dimensions, $SE(3)$, is*

$$\left(\frac{d\mu}{dt}, \frac{d\beta}{dt}\right) = \mathrm{ad}^*_{(\xi\,,\,\alpha)}(\mu\,,\,\beta)\,. \tag{7.1.5}$$

In the $so(3)^*$ and $\mathbb{R}^3$ components of $se(3)^*$ this is

$$\frac{d\mu}{dt} = \mathrm{ad}^*_\xi\,\mu + \beta\diamond\alpha \quad \textit{and} \quad \frac{d\beta}{dt} = -\,\xi\beta\,, \tag{7.1.6}$$

or in terms of variational quantities,

$$\frac{d}{dt}\frac{\delta l}{\delta\xi} = \mathrm{ad}^*_\xi\,\frac{\delta l}{\delta\xi} + \frac{\delta l}{\delta\alpha}\diamond\alpha \quad \textit{and} \quad \frac{d}{dt}\frac{\delta l}{\delta\alpha} = -\,\xi\,\frac{\delta l}{\delta\alpha}\,. \tag{7.1.7}$$

Remark 7.1.1 (Vector Euler–Poincaré equation for $SE(3)$) In vector form, the EP equations become, cf. the vector form of ad^* for $se(3)$ in (6.3.3),

$$\dot{\boldsymbol{\mu}} = \boldsymbol{\mu}\times\boldsymbol{\xi} - \boldsymbol{\alpha}\times\boldsymbol{\beta} \quad \text{and} \quad \dot{\boldsymbol{\beta}} = -\,\boldsymbol{\xi}\times\boldsymbol{\beta}\,. \tag{7.1.8}$$

That is, in terms of vector variational quantities,

$$\frac{d}{dt}\frac{\delta l}{\delta\boldsymbol{\xi}} = \frac{\delta l}{\delta\boldsymbol{\xi}}\times\boldsymbol{\xi} - \boldsymbol{\alpha}\times\frac{\delta l}{\delta\boldsymbol{\alpha}} \quad \text{and} \quad \frac{d}{dt}\frac{\delta l}{\delta\boldsymbol{\alpha}} = -\,\boldsymbol{\xi}\times\frac{\delta l}{\delta\boldsymbol{\alpha}}\,. \tag{7.1.9}$$

These vector EP equations on $se(3)^*$ are readily seen to conserve the quantities $C_1 = \boldsymbol{\mu}\cdot\boldsymbol{\beta}$ and $C_2 = |\boldsymbol{\beta}|^2$, corresponding to a reference direction $\boldsymbol{\beta}(0)$ and the projection of the angular momentum vector $\boldsymbol{\mu}$ in this direction at any time. The quantities C_1 and C_2 will turn out to be ***Casimirs*** for the heavy top. Level surfaces of the quantities C_1 and C_2 define the coadjoint orbits on which the motion takes place for *any* choice of Lagrangian. □

Exercise. What are the corresponding vector EP equations on $se(2)^*$? ★

7.1.1 Legendre transform from $se(3)$ to $se(3)^*$

We Legendre-transform the reduced Lagrangian $\ell(\boldsymbol{\xi}, \boldsymbol{\alpha}) : se(3) \to \mathbb{R}$ to the Hamiltonian

$$h(\boldsymbol{\mu}, \boldsymbol{\beta}) = \boldsymbol{\mu} \cdot \boldsymbol{\xi} + \boldsymbol{\beta} \cdot \boldsymbol{\alpha} - \ell(\boldsymbol{\xi}, \boldsymbol{\alpha}),$$

whose variations are given by

$$\begin{aligned} \delta h(\boldsymbol{\mu}, \boldsymbol{\beta}) &= \frac{\partial h}{\partial \boldsymbol{\mu}} \cdot \delta \boldsymbol{\mu} + \frac{\partial h}{\partial \boldsymbol{\beta}} \cdot \delta \boldsymbol{\beta} \\ &= \boldsymbol{\xi} \cdot \delta \boldsymbol{\mu} + \boldsymbol{\alpha} \cdot \delta \boldsymbol{\beta} + \left(\boldsymbol{\mu} - \frac{\partial \ell}{\partial \boldsymbol{\xi}} \right) \cdot \delta \boldsymbol{\xi} + \left(\boldsymbol{\beta} - \frac{\partial \ell}{\partial \boldsymbol{\alpha}} \right) \cdot \delta \boldsymbol{\alpha} . \end{aligned}$$

Consequently, the Hamiltonian has derivatives $\partial h / \partial \boldsymbol{\mu} = \boldsymbol{\xi}$ and $\partial h / \partial \boldsymbol{\beta} = \boldsymbol{\alpha}$.

Exercise. What is the Legendre transform from $se(2)$ to $se(2)^*$? ★

7.1.2 Lie–Poisson bracket on $se(3)^*$

Rearranging the time derivative of a smooth function f yields

$$\begin{aligned} \frac{df}{dt}(\boldsymbol{\mu}, \boldsymbol{\beta}) &= \frac{\partial f}{\partial \boldsymbol{\mu}} \cdot \dot{\boldsymbol{\mu}} + \frac{\partial f}{\partial \boldsymbol{\beta}} \cdot \dot{\boldsymbol{\beta}} \\ &= \frac{\partial f}{\partial \boldsymbol{\mu}} \cdot \left(\boldsymbol{\mu} \times \frac{\partial h}{\partial \boldsymbol{\mu}} + \boldsymbol{\beta} \times \frac{\partial h}{\partial \boldsymbol{\beta}} \right) + \frac{\partial f}{\partial \boldsymbol{\beta}} \cdot \boldsymbol{\beta} \times \frac{\partial h}{\partial \boldsymbol{\mu}} \end{aligned}$$

$$
\begin{aligned}
&= -\boldsymbol{\mu}\cdot\frac{\partial f}{\partial\boldsymbol{\mu}}\times\frac{\partial h}{\partial\boldsymbol{\mu}} - \boldsymbol{\beta}\cdot\left(\frac{\partial f}{\partial\boldsymbol{\beta}}\times\frac{\partial h}{\partial\boldsymbol{\mu}} - \frac{\partial h}{\partial\boldsymbol{\beta}}\times\frac{\partial f}{\partial\boldsymbol{\mu}}\right) \\
&=: \{f\,,h\}\,.
\end{aligned}
$$

This is the ***Lie–Poisson bracket*** defined on $se(3)^*$ and expressed in terms of vectors $(\boldsymbol{\mu},\,\boldsymbol{\beta})\in\mathbb{R}^3\times\mathbb{R}^3$. By construction, it returns the equations of motion in Hamiltonian form. The Lie–Poisson bracket may be written in matrix form as

$$
\begin{aligned}
\{f\,,h\} &= \partial_{(\boldsymbol{\mu},\,\boldsymbol{\beta})}f^T\,\mathbb{J}\,\partial_{(\boldsymbol{\mu},\,\boldsymbol{\beta})}h \\
&= \begin{bmatrix}\partial f/\partial\boldsymbol{\mu}\\ \partial f/\partial\boldsymbol{\beta}\end{bmatrix}^T \begin{bmatrix}\boldsymbol{\mu}\times & \boldsymbol{\beta}\times\\ \boldsymbol{\beta}\times & 0\end{bmatrix} \begin{bmatrix}\partial h/\partial\boldsymbol{\mu}\\ \partial h/\partial\boldsymbol{\beta}\end{bmatrix}.
\end{aligned}
\tag{7.1.10}
$$

The Hamiltonian matrix $\mathbb{J}$ has null eigenvectors

$$
[0,\boldsymbol{\beta}]^T = \partial_{(\boldsymbol{\mu},\,\boldsymbol{\beta})}|\boldsymbol{\beta}|^2 \quad\text{and}\quad [\boldsymbol{\beta},\,\boldsymbol{\mu}]^T = \partial_{(\boldsymbol{\mu},\,\boldsymbol{\beta})}(\boldsymbol{\beta}\cdot\boldsymbol{\mu})\,.
$$

Consequently the ***distinguished functions*** $C_1 = |\boldsymbol{\beta}|^2$ and $C_2 = \boldsymbol{\mu}\cdot\boldsymbol{\beta}$ whose derivatives are the null eigenvectors of the Hamiltonian matrix $\mathbb{J}$ will ***Poisson-commute*** with any smooth function $f(\boldsymbol{\mu},\,\boldsymbol{\beta})$. Such distinguished functions are called the ***Casimirs*** of the Lie–Poisson bracket in (7.1.10).

Exercise. What is the Lie–Poisson Hamiltonian formulation on $se(2)^*$? ★

7.1.3 Coadjoint motion on $se(3)^*$

Theorem 7.1.1 (Conservation of $se(3)^*$ momentum) *The Euler–Poincaré Equation (7.1.5) for $SE(3)$ conserves the momentum $\mathrm{Ad}^*_{(R(t)\,,\,v(t))^{-1}}$ $(\mu(t),\beta(t))$. That is,*

$$
\frac{d}{dt}\Big(\mathrm{Ad}^*_{(R(t)\,,\,v(t))^{-1}}\big(\mu(t),\beta(t)\big)\Big) = 0\,. \tag{7.1.11}
$$

Proof. By Equation (4.2.9) in Proposition 4.2.2 one may rewrite the Euler–Poincaré Equation (7.1.5) for $SE(3)$ equivalently as

$$\frac{d}{dt}\Big(\mathrm{Ad}^*_{(R(t)\,,\,v(t))^{-1}}\big(\mu(t),\beta(t)\big)\Big) \tag{7.1.12}$$
$$= \mathrm{Ad}^*_{(R(t)\,,\,v(t))^{-1}}\left[\left(\frac{d\mu}{dt},\frac{d\beta}{dt}\right) - \mathrm{ad}^*_{(\xi\,,\,\alpha)}\big(\mu(t)\,,\,\beta(t)\big)\right] = 0\,,$$

where the left-invariant tangent vectors

$$(\xi(t),\alpha(t)) = (R^{-1}\dot{R}(t), R^{-1}\dot{v}(t))$$

to (R, v) at the identity are given in Equation (6.3.8) and the operation $\mathrm{Ad}^*_{(R(t)\,,\,v(t))^{-1}}$ is given for $se(3)^*$ explicitly in formula (6.2.5). ∎

Remark 7.1.2 (First integral) The evolution Equation (7.1.11) conserves the first integral

$$\mathrm{Ad}^*_{(R(t)\,,\,v(t))^{-1}}\big(\mu(t),\beta(t)\big) = \big(\mu(0),\beta(0)\big)\,. \tag{7.1.13}$$

This is the analogue for coadjoint motion on $se(3)^*$ of spatial angular momentum conservation for $so(3)^*$. By Equation (6.2.5) the independently conserved $so(3)^*$ and $\mathbb{R}^3$ components of the $se(3)^*$ momentum are

$$\Big(R(t)\mu(t)R(t)^{-1} + \mathrm{skew}(v(t)\otimes R(t)\beta(t))\,, R(t)\beta(t)\Big)$$
$$= \big(\mu(0),\beta(0)\big)\,. \tag{7.1.14}$$

This formula is particularly convenient, because $\beta(t) = R(t)^{-1}\beta(0)$ implies that $\beta(t)$ simply states how the spatial vector $\beta(0)$ looks in the rotating frame. □

Corollary 7.1.1 *The solution of the EP Equation (7.1.5) for $SE(3)$ evolves by coadjoint motion on the dual Lie algebra $se(3)^*$.*

Proof. The first integral of the evolution Equation (7.1.11) yields

$$(\mu(t)\,,\,\beta(t)) = \mathrm{Ad}^*_{(R(t)\,,\,v(t))}(\mu(0)\,,\,\beta(0))\,. \tag{7.1.15}$$

This expresses the solution as coadjoint motion on $se(3)^*$. ∎

Remark 7.1.3 Equation (7.1.12) for conservation of $se(3)^*$ momentum is written in vector notation as

$$
\begin{aligned}
(\boldsymbol{\mu}(0)\,,\,\boldsymbol{\beta}(0)) &= \mathrm{Ad}^*_{(R(t)\,,\,\mathbf{v}(t))^{-1}}(\boldsymbol{\mu}(t)\,,\,\boldsymbol{\beta}(t)) \qquad (7.1.16)\\
&= \Big(R(t)\boldsymbol{\mu}(t) + \mathbf{v}(t)\times R(t)\boldsymbol{\beta}(t)\,,\, R(t)\boldsymbol{\beta}(t)\Big),
\end{aligned}
$$

which may be rearranged into the evolution operator,

$$
\begin{aligned}
(\boldsymbol{\mu}(t)\,,\,\boldsymbol{\beta}(t)) &= \mathrm{Ad}^*_{(R(t)\,,\,\mathbf{v}(t))}(\boldsymbol{\mu}(0)\,,\,\boldsymbol{\beta}(0)) \qquad (7.1.17)\\
&= \Big(R^{-1}(t)\big(\boldsymbol{\mu}(0) - \mathbf{v}(t)\times\boldsymbol{\beta}(0)\big)\,,\, R^{-1}(t)\boldsymbol{\beta}(0)\Big).
\end{aligned}
$$

This is the vector form of coadjoint motion on $se(3)^*$. It is also the solution of the Hamiltonian system of equations defined by the Lie–Poisson bracket (7.1.10). □

Exercise. Verify directly that this solution satisfies the vector EP equations on $se(3)^*$ in (7.1.8) and conserves the quantities $C_1 = \boldsymbol{\mu}\cdot\boldsymbol{\beta}$ and $C_2 = |\boldsymbol{\beta}|^2$. ★

Exercise. Compute the coadjoint solution of the vector EP equations on $se(2)^*$ that corresponds to (7.1.17) on $se(3)^*$. What quantities are conserved in this case? ★

Remark 7.1.4 Completing the solution requires the group parameters $R(t)$ and $v(t)$ to be reconstructed from $\xi(t)$ and $\alpha(t)$ according to $\dot{R} = R\xi$ and $\dot{v} = R\alpha$ in Equation (6.3.9). In vector notation, the latter equations become $\dot{R} = R\boldsymbol{\xi}$ and $\dot{\mathbf{v}} = R\boldsymbol{\alpha}$, where $\mathbf{v}$ is linear displacement of a moving body, $\boldsymbol{\alpha}$ is its linear velocity and $\boldsymbol{\xi}$ is its angular velocity. □

Theorem 7.1.2 (Kelvin–Noether) *The Euler–Poincaré Equation (7.1.5) for* $SE(3)$ *preserves the natural pairing*

$$\frac{d}{dt}\Big\langle (\mu(t),\beta(t))\,,\,\mathrm{Ad}_{(R(t)\,,\,v(t))^{-1}}(\eta,\gamma)\Big\rangle = 0\,, \tag{7.1.18}$$

for any fixed $(\eta,\gamma)\in\mathfrak{se}(3)$.

Proof. Verifying directly,

$$\begin{aligned}\frac{d}{dt}\Big\langle (\mu(t),\beta(t))\,,\,\mathrm{Ad}_{(R(t)\,,\,v(t))^{-1}}\eta\Big\rangle \\ = \Big\langle \frac{d}{dt}\Big(\mathrm{Ad}^*_{(R(t)\,,\,v(t))^{-1}}(\mu(t),\beta(t))\Big)\,,\,\eta\Big\rangle = 0\,,\end{aligned} \tag{7.1.19}$$

by Equation (7.1.12). ■

Remark 7.1.5 Theorem 7.1.2 is associated with Kelvin and Noether because it arises from symmetry (Noether) and it happens to coincide with the Kelvin circulation theorem in the Euler–Poincaré formulation of ideal fluid motion [HoMaRa1998]. □

7.2 Kirchhoff equations on $se(3)^*$

Suppose the Lagrangian is chosen to be the sum of the kinetic energies of rotational and translational motion of an ellipsoidal underwater vehicle with coincident centres of gravity and buoyancy. In this case, the Lagrangian is given by the sum of the rotational and translational kinetic energies as [HoJeLe1998]

$$l(\xi,\alpha) = \frac{1}{2}\langle \xi\,,\,\mathbb{I}\xi\,\rangle + \frac{1}{2}\langle \alpha\,,\,M\alpha\rangle = \frac{1}{2}\boldsymbol{\xi}\cdot\mathbb{I}\boldsymbol{\xi} + \frac{1}{2}\boldsymbol{\alpha}\cdot\mathbb{M}\boldsymbol{\alpha}\,. \tag{7.2.1}$$

The two 3×3 symmetric matrices (metrics) represent the moment of inertia ($\mathbb{I}$) and the body mass matrix ($\mathbb{M}$) of the ellipsoidal underwater vehicle. The corresponding angular and linear momenta are given as the pair of vectors

$$(\boldsymbol{\mu}, \boldsymbol{\beta}) = \left(\frac{\delta \ell}{\delta \boldsymbol{\xi}}, \frac{\delta \ell}{\delta \boldsymbol{\alpha}}\right) = (\mathbb{I}\boldsymbol{\xi}, \mathbb{M}\boldsymbol{\alpha}).$$

Consequently, the evolution of an ellipsoidal underwater vehicle is governed by a vector Euler–Poincaré equation on $se(3)^*$ of the form (7.1.8),

$$\begin{aligned} (\mathbb{I}\boldsymbol{\xi})^{\cdot} &= -\boldsymbol{\xi} \times \mathbb{I}\boldsymbol{\xi} - \boldsymbol{\alpha} \times \mathbb{M}\boldsymbol{\alpha}, \\ (\mathbb{M}\boldsymbol{\alpha})^{\cdot} &= -\boldsymbol{\xi} \times \mathbb{M}\boldsymbol{\alpha}. \end{aligned} \tag{7.2.2}$$

Remark 7.2.1 Of these vectors, $\boldsymbol{\beta} = \mathbb{M}\boldsymbol{\alpha}$ is linear momentum and $\boldsymbol{\mu} = \mathbb{M}\boldsymbol{\xi}$ is angular momentum. The Lagrangian $l(\xi, \alpha)$ in (7.2.1) is the kinetic energy for the motion of an ellipsoidal underwater vehicle. □

The Hamiltonian for the ellipsoidal underwater vehicle is the quadratic form

$$h(\boldsymbol{\mu}, \boldsymbol{\beta}) = \frac{1}{2}\boldsymbol{\mu} \cdot \mathbb{I}^{-1}\boldsymbol{\mu} + \frac{1}{2}\boldsymbol{\beta} \cdot \mathbb{M}^{-1}\boldsymbol{\beta}. \tag{7.2.3}$$

The Lie–Poisson equations corresponding to (7.1.8) are

$$\begin{bmatrix} \dot{\boldsymbol{\mu}} \\ \dot{\boldsymbol{\beta}} \end{bmatrix} = \begin{bmatrix} \boldsymbol{\mu}\times & \boldsymbol{\beta}\times \\ \boldsymbol{\beta}\times & 0 \end{bmatrix} \begin{bmatrix} \partial h/\partial \boldsymbol{\mu} \\ \partial h/\partial \boldsymbol{\beta} \end{bmatrix} = \begin{bmatrix} \boldsymbol{\mu}\times & \boldsymbol{\beta}\times \\ \boldsymbol{\beta}\times & 0 \end{bmatrix} \begin{bmatrix} \mathbb{I}^{-1}\boldsymbol{\mu} \\ \mathbb{M}^{-1}\boldsymbol{\beta} \end{bmatrix}. \tag{7.2.4}$$

These are the Lie–Poisson equations for geodesic motion on $se(3)^*$ with respect to the metric given by the Hamiltonian in (7.2.5).

Remark 7.2.2 The stability of the equilibrium solutions of (7.2.2) for ellipsoidal underwater vehicles may be investigated within the present Euler–Poincaré framework. See [HoJeLe1998] and [GaMi1995]. □

Exercise. Compute the Lie–Poisson equations for geodesic motion on $se(2)^*$ with respect to a quadratic Hamiltonian metric. ★

7.2.1 Looks can be deceiving: The heavy top

The Hamiltonian for the heavy top is the sum of its kinetic and potential energies,

$$h(\boldsymbol{\mu}, \boldsymbol{\beta}) = \underbrace{\frac{1}{2}\boldsymbol{\mu} \cdot \mathbb{I}^{-1}\boldsymbol{\mu}}_{\text{kinetic}} + \underbrace{mg\boldsymbol{\chi} \cdot \boldsymbol{\beta}}_{\text{potential}}, \tag{7.2.5}$$

in which $\boldsymbol{\mu}$ is the body angular momentum, $\boldsymbol{\chi}$ is the distance in the body from its point of support to its centre of mass, mg is its weight and $\boldsymbol{\beta} = O^{-1}(t)\hat{\mathbf{z}}$ is the vertical direction, as seen from the body. The derivatives of this Hamiltonian are

$$\frac{\partial h}{\partial \boldsymbol{\mu}} = \boldsymbol{\xi} \quad \text{and} \quad \frac{\partial h}{\partial \boldsymbol{\beta}} = mg\boldsymbol{\chi}.$$

The correct equations of motion for the heavy-top emerge from this Hamiltonian and the Lie–Poisson bracket on $se(3)^*$:

$$\begin{aligned} \dot{\boldsymbol{\mu}} = \{\boldsymbol{\mu}, h\} &= \boldsymbol{\mu} \times \frac{\partial h}{\partial \boldsymbol{\mu}} + \boldsymbol{\beta} \times \frac{\partial h}{\partial \boldsymbol{\beta}} \\ &= \boldsymbol{\mu} \times \boldsymbol{\xi} + \boldsymbol{\beta} \times mg\boldsymbol{\chi}, \qquad (7.2.6) \\ \dot{\boldsymbol{\beta}} = \{\boldsymbol{\beta}, h\} &= \boldsymbol{\beta} \times \frac{\partial h}{\partial \boldsymbol{\mu}} = \boldsymbol{\beta} \times \boldsymbol{\xi}. \qquad (7.2.7) \end{aligned}$$

It may seem surprising that the heavy-top dynamics would emerge from the $SE(3)$ Lie–Poisson bracket, because the heavy top has no linear velocity or linear momentum. There is a story behind how this happened, which will be discussed in the next chapter.

Exercise. Does the Hamiltonian in (7.2.5) follow from the Legendre transformation of a Lagrangian $\ell(\boldsymbol{\xi}, \boldsymbol{\alpha})$? If so, prove it. If not, what goes wrong? ★

Exercise. What are the equations corresponding to (7.2.6) and (7.2.7) for $se(2)^*$? To what physical system do these equations correspond? ★

Exercise. Is the dynamical system governing $(M, \mathbb{S})$ in Equations (2.5.25) and (2.5.26) Hamiltonian? Prove it.

How is that system analogous to a heavy top? ★

Exercise. Write the Lie–Poisson Hamiltonian formulations for motion on the dual of each of the Lie algebras for

- the Heisenberg Lie group; and
- the semidirect-product Lie group $SL(2,\mathbb{R})\circledS\mathbb{R}^2$. ★

8 HEAVY-TOP EQUATIONS

Contents

8.1	Introduction and definitions	182
8.2	Heavy-top action principle	183
8.3	Lie–Poisson brackets	184
8.3.1	Lie–Poisson brackets and momentum maps	185
8.3.2	Lie–Poisson brackets for the heavy top	186
8.4	Clebsch action principle	187
8.5	Kaluza–Klein construction	188

8.1 Introduction and definitions

A top is a rigid body of mass m rotating with a fixed point of support in a constant gravitational field of acceleration $-g\hat{\mathbf{z}}$ pointing vertically downward. The orientation of the body relative to the vertical axis $\hat{\mathbf{z}}$ is defined by the unit vector $\mathbf{\Gamma} = \mathbf{R}^{-1}(t)\hat{\mathbf{z}}$ for a curve $\mathbf{R}(t) \in SO(3)$. According to its definition, the unit vector $\mathbf{\Gamma}$ represents the motion of the vertical direction as seen from the rotating body. Consequently, it satisfies the auxiliary motion equation,

$$\dot{\mathbf{\Gamma}} = -\mathbf{R}^{-1}\dot{\mathbf{R}}(t)\mathbf{\Gamma} = -\widehat{\mathbf{\Omega}}(t)\mathbf{\Gamma} = \mathbf{\Gamma} \times \mathbf{\Omega}\,. \tag{8.1.1}$$

Here the rotation matrix $\mathbf{R}(t) \in SO(3)$, the skew matrix $\widehat{\mathbf{\Omega}} = \mathbf{R}^{-1}\dot{\mathbf{R}} \in so(3)$ and the body angular frequency vector $\mathbf{\Omega} \in \mathbb{R}^3$ are related by the hat map, $\mathbf{\Omega} = \big(\mathbf{R}^{-1}\dot{\mathbf{R}}\big)\widehat{\ }$, where

$$\text{hat map,}\quad \widehat{\ }: (so(3), [\cdot,\cdot]) \to (\mathbb{R}^3, \times)\,,$$

with $\widehat{\mathbf{\Omega}}\mathbf{v} = \mathbf{\Omega} \times \mathbf{v}$ for any $\mathbf{v} \in \mathbb{R}^3$.

The motion of a top is determined from Euler's equations in vector form,

$$\mathbb{I}\dot{\mathbf{\Omega}} = \mathbb{I}\mathbf{\Omega} \times \mathbf{\Omega} + mg\mathbf{\Gamma} \times \boldsymbol{\chi}\,, \tag{8.1.2}$$

$$\dot{\mathbf{\Gamma}} = \mathbf{\Gamma} \times \mathbf{\Omega}\,, \tag{8.1.3}$$

where $\mathbf{\Omega}$, $\mathbf{\Gamma}$, $\boldsymbol{\chi} \in \mathbb{R}^3$ are vectors in the rotating body frame. Here

- $\mathbf{\Omega} = (\Omega_1, \Omega_2, \Omega_3)$ is the body angular velocity vector.
- $\mathbb{I} = \mathrm{diag}(I_1, I_2, I_3)$ is the moment of inertia tensor, diagonalised in the body principal axes.
- $\mathbf{\Gamma} = R^{-1}(t)\hat{\mathbf{z}}$ represents the motion of the unit vector along the vertical axis, as seen from the body.
- $\boldsymbol{\chi}$ is the constant vector in the body from the point of support to the body's centre of mass.
- m is the total mass of the body and g is the constant acceleration of gravity.

8.2 Heavy-top action principle

Proposition 8.2.1 *The heavy-top motion equation (8.1.2) is equivalent to the* ***heavy-top action principle*** $\delta S_{\rm red}=0$ *for a* ***reduced action,***

$$S_{\rm red}=\int_a^b l(\boldsymbol{\Omega},\boldsymbol{\Gamma})\,dt=\int_a^b \frac{1}{2}\Big\langle\,\mathbb{I}\boldsymbol{\Omega}\,,\,\boldsymbol{\Omega}\Big\rangle-\Big\langle mg\boldsymbol{\chi}\,,\,\boldsymbol{\Gamma}\Big\rangle\,dt\,, \qquad (8.2.1)$$

where variations of vectors $\boldsymbol{\Omega}$ *and* $\boldsymbol{\Gamma}$ *are restricted to be of the form*

$$\delta\boldsymbol{\Omega}=\dot{\boldsymbol{\Sigma}}+\boldsymbol{\Omega}\times\boldsymbol{\Sigma} \quad \textit{and} \quad \delta\boldsymbol{\Gamma}=\boldsymbol{\Gamma}\times\boldsymbol{\Sigma}\,, \qquad (8.2.2)$$

arising from variations of the corresponding definitions $\widehat{\boldsymbol{\Omega}}=\mathbf{R}^{-1}\dot{\mathbf{R}}$ *and* $\boldsymbol{\Gamma}=\mathbf{R}^{-1}(t)\hat{\mathbf{z}}$ *in which* $\widehat{\boldsymbol{\Sigma}}(t)=\mathbf{R}^{-1}\delta\mathbf{R}$ *is a curve in* $\mathbb{R}^3$ *that vanishes at the endpoints in time.*

Proof. Since $\mathbb{I}$ is symmetric and $\boldsymbol{\chi}$ is constant, one finds the variation,

$$\begin{aligned}
\delta\int_a^b l(\boldsymbol{\Omega},\boldsymbol{\Gamma})\,dt &= \int_a^b \Big\langle\,\mathbb{I}\boldsymbol{\Omega}\,,\,\delta\boldsymbol{\Omega}\Big\rangle-\Big\langle mg\boldsymbol{\chi}\,,\,\delta\boldsymbol{\Gamma}\Big\rangle\,dt\\
&= \int_a^b \Big\langle\,\mathbb{I}\boldsymbol{\Omega}\,,\,\dot{\boldsymbol{\Sigma}}+\boldsymbol{\Omega}\times\boldsymbol{\Sigma}\Big\rangle-\Big\langle mg\boldsymbol{\chi}\,,\,\boldsymbol{\Gamma}\times\boldsymbol{\Sigma}\Big\rangle\,dt\\
=\int_a^b \Big\langle-\frac{d}{dt}\mathbb{I}\boldsymbol{\Omega}\,,\,\boldsymbol{\Sigma}\Big\rangle &+\Big\langle\,\mathbb{I}\boldsymbol{\Omega}\,,\,\boldsymbol{\Omega}\times\boldsymbol{\Sigma}\Big\rangle-\Big\langle mg\boldsymbol{\chi}\,,\,\boldsymbol{\Gamma}\times\boldsymbol{\Sigma}\Big\rangle dt\\
&= \int_a^b \Big\langle-\frac{d}{dt}\mathbb{I}\boldsymbol{\Omega}+\mathbb{I}\boldsymbol{\Omega}\times\boldsymbol{\Omega}+mg\,\boldsymbol{\Gamma}\times\boldsymbol{\chi}\,,\,\boldsymbol{\Sigma}\Big\rangle\,dt,
\end{aligned}$$

upon integrating by parts and using the endpoint conditions, $\boldsymbol{\Sigma}(b)=\boldsymbol{\Sigma}(a)=0$. Since $\boldsymbol{\Sigma}$ is otherwise arbitrary, (8.2.1) is equivalent to

$$-\frac{d}{dt}\mathbb{I}\boldsymbol{\Omega}+\mathbb{I}\boldsymbol{\Omega}\times\boldsymbol{\Omega}+mg\,\boldsymbol{\Gamma}\times\boldsymbol{\chi}=0\,,$$

which is Euler's motion equation for the heavy top (8.1.2). This motion equation is completed by the auxiliary equation $\dot{\boldsymbol{\Gamma}}=\boldsymbol{\Gamma}\times\boldsymbol{\Omega}$ in (8.1.3) arising from the definition of $\boldsymbol{\Gamma}$. ∎

The Legendre transformation for $l(\mathbf{\Omega}, \mathbf{\Gamma})$ gives the body angular momentum

$$\mathbf{\Pi} = \frac{\partial l}{\partial \mathbf{\Omega}} = \mathbb{I}\mathbf{\Omega}\,.$$

The well-known energy Hamiltonian for the heavy top then emerges as

$$h(\mathbf{\Pi}, \mathbf{\Gamma}) = \mathbf{\Pi} \cdot \mathbf{\Omega} - l(\mathbf{\Omega}, \mathbf{\Gamma}) = \frac{1}{2}\langle \mathbf{\Pi}\,,\, \mathbb{I}^{-1}\mathbf{\Pi} \rangle + \langle mg\,\boldsymbol{\chi}\,,\, \mathbf{\Gamma}\,\rangle\,, \quad (8.2.3)$$

which is the sum of the kinetic and potential energies of the top.

The Lie–Poisson equations

Let $f, h : \mathfrak{g}^* \to \mathbb{R}$ be real-valued functions on the dual space $\mathfrak{g}^*$. Denoting elements of $\mathfrak{g}^*$ by μ, the functional derivative of f at μ is defined as the unique element $\delta f/\delta\mu$ of $\mathfrak{g}$ defined by

$$\lim_{\varepsilon \to 0} \frac{1}{\varepsilon}[f(\mu + \varepsilon\delta\mu) - f(\mu)] = \left\langle \delta\mu, \frac{\delta f}{\delta\mu} \right\rangle, \quad (8.2.4)$$

for all $\delta\mu \in \mathfrak{g}^*$, where $\langle \cdot\,,\, \cdot \rangle$ denotes the pairing between $\mathfrak{g}^*$ and $\mathfrak{g}$.

8.3 Lie–Poisson brackets

Definition 8.3.1 (Lie–Poisson equations) *The* $(\pm)$ ***Lie–Poisson brackets*** *are defined by*

$$\{f, h\}_{\pm}(\mu) = \pm\left\langle \mu, \left[\frac{\delta f}{\delta\mu}, \frac{\delta h}{\delta\mu}\right] \right\rangle = \mp\left\langle \mu, \mathrm{ad}_{\delta h/\delta\mu}\frac{\delta f}{\delta\mu} \right\rangle. \quad (8.3.1)$$

The corresponding ***Lie–Poisson equations****, determined by* $\dot{f} = \{f, h\}$, *read*

$$\dot{\mu} = \{\mu, h\} = \mp\,\mathrm{ad}^*_{\delta h/\delta\mu}\,\mu\,, \quad (8.3.2)$$

where one defines the ad* *operation in terms of the pairing* $\langle \cdot\,, \cdot \rangle$, *by*

$$\{f, h\} = \left\langle \mu, \operatorname{ad}_{\delta h/\delta\mu} \frac{\delta f}{\delta\mu} \right\rangle = \left\langle \operatorname{ad}^*_{\delta h/\delta\mu}\, \mu, \frac{\delta f}{\delta\mu} \right\rangle .$$

Remark 8.3.1 The Lie–Poisson setting of mechanics is a special case of the general theory of systems on Poisson manifolds, for which there is now extensive theoretical development. (See [MaRa1994] for a start on this literature.) □

8.3.1 Lie–Poisson brackets and momentum maps

An important feature of the rigid-body bracket carries over to general Lie algebras. Namely, *Lie–Poisson brackets on* $\mathfrak{g}^*$ *arise from canonical brackets on the cotangent bundle (phase space)* T^*G associated with a Lie group G which has $\mathfrak{g}$ as its associated Lie algebra. Thus, the process by which the Lie–Poisson brackets arise is the momentum map

$$T^*G \mapsto \mathfrak{g}^* .$$

For example, a rigid body is free to rotate about its centre of mass and G is the (proper) rotation group $SO(3)$. The choice of T^*G as the primitive phase space is made according to the classical procedures of mechanics described earlier. For the description using Lagrangian mechanics, one forms the velocity phase space TG. The Hamiltonian description on T^*G is then obtained by standard procedures: Legendre transforms, etc.

The passage from T^*G to the space of $\mathbf{\Pi}$'s (body angular momentum space) is determined by *left* translation on the group. This mapping is an example of a ***momentum map***; that is, a mapping whose components are the "Noether quantities" associated with a symmetry group. That the map from T^*G to $\mathfrak{g}^*$ is a Poisson map *is a general fact about momentum maps*. The Hamiltonian point of view of all this is a standard subject reviewed in Chapter 11.

Remark 8.3.2 (Lie–Poisson description of the heavy top) As it turns out, the underlying Lie algebra for the Lie–Poisson description of the heavy top consists of the Lie algebra $se(3, \mathbb{R})$ of infinitesimal Euclidean motions in $\mathbb{R}^3$. This is a bit surprising, because heavy-top motion itself does *not* actually arise through spatial translations by the Euclidean group; in fact, the body has a fixed point! Instead, the Lie algebra $se(3, \mathbb{R})$ arises for another reason associated with the breaking of the $SO(3)$ isotropy by the presence of the gravitational field. This symmetry breaking introduces a semidirect-product Lie–Poisson structure which happens to coincide with the dual of the Lie algebra $se(3, \mathbb{R})$ in the case of the heavy top. □

8.3.2 Lie–Poisson brackets for the heavy top

The Lie algebra of the special Euclidean group in three dimensions is $se(3) = \mathbb{R}^3 \times \mathbb{R}^3$ with the Lie bracket

$$[(\boldsymbol{\xi}, \mathbf{u}), (\boldsymbol{\eta}, \mathbf{v})] = (\boldsymbol{\xi} \times \boldsymbol{\eta}, \boldsymbol{\xi} \times \mathbf{v} - \boldsymbol{\eta} \times \mathbf{u}) . \tag{8.3.3}$$

We identify the dual space with pairs $(\boldsymbol{\Pi}, \boldsymbol{\Gamma})$; the corresponding $(-)$ Lie–Poisson bracket called the ***heavy-top bracket*** is

$$\{f\,,\,h\}(\boldsymbol{\Pi}, \boldsymbol{\Gamma}) = -\,\boldsymbol{\Pi} \cdot \frac{\partial f}{\partial \boldsymbol{\Pi}} \times \frac{\partial h}{\partial \boldsymbol{\Pi}} - \boldsymbol{\Gamma} \cdot \left(\frac{\partial f}{\partial \boldsymbol{\Pi}} \times \frac{\partial h}{\partial \boldsymbol{\Gamma}} - \frac{\partial h}{\partial \boldsymbol{\Pi}} \times \frac{\partial f}{\partial \boldsymbol{\Gamma}} \right) .$$

This Lie–Poisson bracket and the Hamiltonian (8.2.3) recover Equations (8.1.2) and (8.1.3) for the heavy top, as

$$\begin{aligned} \dot{\boldsymbol{\Pi}} = \{\boldsymbol{\Pi}\,,\,h\} &= \boldsymbol{\Pi} \times \frac{\partial h}{\partial \boldsymbol{\Pi}} + \boldsymbol{\Gamma} \times \frac{\partial h}{\partial \boldsymbol{\Gamma}} \\ &= \boldsymbol{\Pi} \times \mathbb{I}^{-1}\boldsymbol{\Pi} + \boldsymbol{\Gamma} \times mg\boldsymbol{\chi} , \\ \dot{\boldsymbol{\Gamma}} = \{\boldsymbol{\Gamma}\,,\,h\} &= \boldsymbol{\Gamma} \times \frac{\partial h}{\partial \boldsymbol{\Pi}} = \boldsymbol{\Gamma} \times \mathbb{I}^{-1}\boldsymbol{\Pi} . \end{aligned}$$

Remark 8.3.3 (Semidirect products and symmetry breaking) The Lie algebra of the Euclidean group has a structure which is a special case of what is called a ***semidirect product***. Here, it is the semidirect-product action $so(3) \circledS \mathbb{R}^3$ of the Lie algebra of rotations $so(3)$ acting on the infinitesimal translations $\mathbb{R}^3$, which happens to coincide with $se(3, \mathbb{R})$.

In general, the Lie bracket for semidirect-product action $\mathfrak{g} \circledS V$ of a Lie algebra $\mathfrak{g}$ on a vector space V is given by, cf. Equation (6.3.1),

$$\Big[(X,a),(\overline{X},\overline{a})\Big] = \Big([X,\overline{X}],\overline{X}(a) - X(\overline{a})\Big),$$

in which $X,\overline{X} \in \mathfrak{g}$ and $a,\overline{a} \in V$. Here, the action of the Lie algebra on the vector space is denoted, e.g., $X(\overline{a})$. Usually, this action would be the Lie derivative. □

Lie–Poisson brackets defined on the dual spaces of semidirect-product Lie algebras tend to occur under rather general circumstances when the symmetry in T^*G is broken, e.g., reduced to an isotropy subgroup of a set of parameters. In particular, there are similarities in structure between the Poisson bracket for compressible flow and that for the heavy top. In the latter case, the vertical direction of gravity breaks the isotropy of $\mathbb{R}^3$ from $SO(3)$ to $SO(2)$. The general theory for semidirect products is reviewed in a variety of places, including [MaRaWe1984a, MaRaWe1984b].

Many interesting examples of Lie–Poisson brackets on semidirect products exist for fluid dynamics. These semidirect-product Lie–Poisson Hamiltonian theories range from simple fluids, to charged fluid plasmas, to magnetised fluids, to multiphase fluids, to super fluids, to Yang–Mills fluids, relativistic or not, and to liquid crystals. Many of these theories are discussed from the Euler–Poincaré viewpoint in [HoMaRa1998] and [Ho2002].

8.4 Clebsch action principle

Proposition 8.4.1 (Clebsch heavy-top action principle) *The heavy-top Equations (8.1.2) and (8.1.3) follow from a Clebsch constrained action principle, $\delta S = 0$, with*

$$S = \int_a^b \frac{1}{2}\Big\langle \mathbb{I}\boldsymbol{\Omega}\,,\,\boldsymbol{\Omega}\Big\rangle - \Big\langle mg\boldsymbol{\chi}\,,\,\boldsymbol{\Gamma}\Big\rangle + \Big\langle \boldsymbol{\Xi}\,,\,\dot{\boldsymbol{\Gamma}} + \boldsymbol{\Omega}\times\boldsymbol{\Gamma}\Big\rangle\, dt\,. \tag{8.4.1}$$

Remark 8.4.1 The last term in this action is the ***Clebsch constraint*** for the auxiliary equation satisfied by the unit vector $\boldsymbol{\Gamma}$. From its definition $\mathbf{\Gamma} = \mathbf{R}^{-1}(t)\hat{\mathbf{z}}$ and the definition of the body angular velocity $\mathbf{\Omega} = \mathbf{R}^{-1}(t)\dot{\mathbf{R}}$, this unit vector must satisfy

$$\dot{\mathbf{\Gamma}} = -\,\mathbf{R}^{-1}\dot{\mathbf{R}}(t)\mathbf{\Gamma} = -\,\widehat{\Omega}(t)\mathbf{\Gamma} = -\,\mathbf{\Omega}\times\mathbf{\Gamma}\,.$$

(The third equality invokes the hat map.) According to the Clebsch construction, the Lagrange multiplier Ξ enforcing the auxiliary Equation (8.4.1) will become the momentum canonically conjugate to the auxiliary variable $\mathbf{\Gamma}$. □

Proof. The stationary variations of the constrained action (8.4.1) yield the following three ***Clebsch relations***, cf. Equations (2.5.22) for the rigid body,

$$\begin{aligned} \delta\mathbf{\Omega}: &\quad \mathbb{I}\mathbf{\Omega} + \mathbf{\Gamma}\times\mathbf{\Xi} = 0\,,\\ \delta\mathbf{\Xi}: &\quad \dot{\mathbf{\Gamma}} + \mathbf{\Omega}\times\mathbf{\Gamma} = 0\,,\\ \delta\mathbf{\Gamma}: &\quad \dot{\mathbf{\Xi}} + \mathbf{\Omega}\times\mathbf{\Xi} + mg\,\boldsymbol{\chi} = 0\,. \end{aligned}$$

As we shall see in Chapter 11, the first Clebsch relation defines the momentum map $T^*\mathbb{R}^3 \to so(3)^*$ for the body angular momentum $\mathbb{I}\mathbf{\Omega}$. From the other two Clebsch relations, the equation of motion for the body angular momentum may be computed as

$$\begin{aligned} \mathbb{I}\dot{\mathbf{\Omega}} &= -\,\dot{\mathbf{\Gamma}}\times\mathbf{\Xi} - \mathbf{\Gamma}\times\dot{\mathbf{\Xi}}\\ &= (\mathbf{\Omega}\times\mathbf{\Gamma})\times\mathbf{\Xi} + \mathbf{\Gamma}\times(\mathbf{\Omega}\times\mathbf{\Xi} + mg\,\boldsymbol{\chi})\\ &= \mathbf{\Omega}\times(\mathbf{\Gamma}\times\mathbf{\Xi}) + \mathbf{\Gamma}\times mg\,\boldsymbol{\chi}\\ &= -\,\mathbf{\Omega}\times(\mathbb{I}\mathbf{\Omega}) + mg\,\mathbf{\Gamma}\times\boldsymbol{\chi}\,, \end{aligned}$$

which recovers Euler's motion Equation (8.1.2) for the heavy top. ■

8.5 Kaluza–Klein construction

The Lagrangian in the heavy-top action principle (8.2.1) may be transformed into quadratic form. This is accomplished by suspending the system in a higher-dimensional space via the ***Kaluza–Klein construction***. This construction proceeds for the heavy top as a slight modification of the well-known Kaluza–Klein construction for a charged particle in a prescribed magnetic field.

Let Q_{KK} be the manifold $SO(3) \times \mathbb{R}^3$ with variables $(\mathbf{R}, \mathbf{q})$. On Q_{KK} introduce the ***Kaluza–Klein Lagrangian***

$$L_{KK} : TQ_{KK} \simeq TSO(3) \times T\mathbb{R}^3 \mapsto \mathbb{R},$$

as

$$\begin{aligned} L_{KK}(\mathbf{R}, \mathbf{q}, \dot{\mathbf{R}}, \dot{\mathbf{q}}; \hat{\mathbf{z}}) &= L_{KK}(\boldsymbol{\Omega}, \boldsymbol{\Gamma}, \mathbf{q}, \dot{\mathbf{q}}) \\ &= \frac{1}{2}\langle \mathbb{I}\boldsymbol{\Omega}\,,\, \boldsymbol{\Omega}\rangle + \frac{1}{2}|\boldsymbol{\Gamma} + \dot{\mathbf{q}}|^2\,, \end{aligned} \tag{8.5.1}$$

with $\boldsymbol{\Omega} = \big(\mathbf{R}^{-1}\dot{\mathbf{R}}\big)\widehat{\ }$ and $\boldsymbol{\Gamma} = \mathbf{R}^{-1}\hat{\mathbf{z}}$. The Lagrangian L_{KK} is positive-definite in $(\boldsymbol{\Omega}, \boldsymbol{\Gamma}, \dot{\mathbf{q}})$; so it may be regarded as a kinetic energy which defines a metric, the ***Kaluza–Klein metric*** on TQ_{KK}.

The Legendre transformation for L_{KK} gives the momenta

$$\boldsymbol{\Pi} = \mathbb{I}\boldsymbol{\Omega} \qquad \text{and} \qquad \mathbf{p} = \boldsymbol{\Gamma} + \dot{\mathbf{q}}\,. \tag{8.5.2}$$

Since L_{KK} does not depend on $\mathbf{q}$, the Euler–Lagrange equation

$$\frac{d}{dt}\frac{\partial L_{KK}}{\partial \dot{\mathbf{q}}} = \frac{\partial L_{KK}}{\partial \mathbf{q}} = 0$$

shows that $\mathbf{p} = \partial L_{KK}/\partial\dot{\mathbf{q}}$ is conserved. The ***constant vector*** $\mathbf{p}$ is now identified as the vector in the body,

$$\mathbf{p} = \boldsymbol{\Gamma} + \dot{\mathbf{q}} = -\,mg\,\boldsymbol{\chi}\,.$$

After this identification, the heavy-top action principle in Proposition 8.2.1 with the Kaluza–Klein Lagrangian returns Euler's motion equation for the heavy top (8.1.2).

The Hamiltonian H_{KK} associated with L_{KK} by the Legendre transformation (8.5.2) is

$$\begin{aligned} H_{KK}(\boldsymbol{\Pi}, \boldsymbol{\Gamma}, \mathbf{q}, \mathbf{p}) &= \boldsymbol{\Pi}\cdot\boldsymbol{\Omega} + \mathbf{p}\cdot\dot{\mathbf{q}} - L_{KK}(\boldsymbol{\Omega}, \boldsymbol{\Gamma}, \mathbf{q}, \dot{\mathbf{q}}) \\ &= \frac{1}{2}\boldsymbol{\Pi}\cdot\mathbb{I}^{-1}\boldsymbol{\Pi} - \mathbf{p}\cdot\boldsymbol{\Gamma} + \frac{1}{2}|\mathbf{p}|^2 \\ &= \frac{1}{2}\boldsymbol{\Pi}\cdot\mathbb{I}^{-1}\boldsymbol{\Pi} + \frac{1}{2}|\mathbf{p} - \boldsymbol{\Gamma}|^2 - \frac{1}{2}|\boldsymbol{\Gamma}|^2\,. \end{aligned}$$

Recall that $\mathbf{\Gamma}$ is a unit vector. On the constant level set $|\mathbf{\Gamma}|^2 = 1$, the Kaluza–Klein Hamiltonian H_{KK} is a positive quadratic function, shifted by a constant. Likewise, on the constant level set $\mathbf{p} = -mg\boldsymbol{\chi}$, the Kaluza–Klein Hamiltonian H_{KK} is a function of only the variables $(\mathbf{\Pi}, \mathbf{\Gamma})$ and is equal to the Hamiltonian (8.2.3) for the heavy top up to an additive constant. As a result we have the following.

Proposition 8.5.1 *The Lie–Poisson equations for the Kaluza–Klein Hamiltonian H_{KK} recover Euler's equations for the heavy top, (8.1.2) and (8.1.3).*

Proof. The Lie–Poisson bracket may be written in matrix form explicitly as

$$\{f, h\} = \begin{bmatrix} \partial f/\partial \mathbf{\Pi} \\ \partial f/\partial \mathbf{\Gamma} \\ \partial f/\partial \mathbf{q} \\ \partial f/\partial \mathbf{p} \end{bmatrix}^T \begin{bmatrix} \mathbf{\Pi}\times & \mathbf{\Gamma}\times & 0 & 0 \\ \mathbf{\Gamma}\times & 0 & 0 & 0 \\ 0 & 0 & 0 & Id \\ 0 & 0 & -Id & 0 \end{bmatrix} \begin{bmatrix} \partial h/\partial \mathbf{\Pi} \\ \partial h/\partial \mathbf{\Gamma} \\ \partial h/\partial \mathbf{q} \\ \partial h/\partial \mathbf{p} \end{bmatrix}. \tag{8.5.3}$$

Consequently, one obtains the following Hamiltonian equations for $h = H_{KK}(\mathbf{\Pi}, \mathbf{\Gamma}, \mathbf{q}, \mathbf{p})$,

$$\begin{bmatrix} \dot{\mathbf{\Pi}} \\ \dot{\mathbf{\Gamma}} \\ \dot{\mathbf{q}} \\ \dot{\mathbf{p}} \end{bmatrix} = \begin{bmatrix} \mathbf{\Pi}\times & \mathbf{\Gamma}\times & 0 & 0 \\ \mathbf{\Gamma}\times & 0 & 0 & 0 \\ 0 & 0 & 0 & Id \\ 0 & 0 & -Id & 0 \end{bmatrix} \begin{bmatrix} \mathbf{\Omega} \\ -\mathbf{p} \\ 0 \\ \mathbf{p} - \mathbf{\Gamma} \end{bmatrix}. \tag{8.5.4}$$

These recover the heavy-top Equations (8.1.2) and (8.1.3) upon evaluating $\mathbf{p} = -mg\boldsymbol{\chi}$. ■

Exercise. In an attempt to mimic Manakov's beautiful idea for showing the integrability of the rigid body on $SO(n)$, one might imagine writing the three-dimensional heavy-top Equations (8.1.2) and (8.1.3) by inserting a spectral parameter λ as

$$\frac{d}{dt}(\mathbf{\Gamma} + \lambda\mathbf{\Pi} + \lambda^2\boldsymbol{J}) = (\mathbf{\Gamma} + \lambda\mathbf{\Pi} + \lambda^2\boldsymbol{J}) \times (\mathbf{\Omega} + \lambda\boldsymbol{K}),$$

with constant vectors $\boldsymbol{J}$ and $\boldsymbol{K}$ in $\mathbb{R}^3$. Does this formulation provide enough constants of motion to show the integrability of the heavy-top equations for some values of χ and $\mathbb{I}$? If so, which types of tops may be shown to be integrable this way? ★

Answer. The polynomial equation above implies the following relations, at various powers of λ:

$$
\begin{aligned}
\lambda^3 &: \ \boldsymbol{J} \times \boldsymbol{K} = \mathbf{0} \implies \boldsymbol{J} \parallel \boldsymbol{K}, \implies \boldsymbol{J} = \alpha \boldsymbol{K}, \ \alpha = const. \\
\lambda^2 &: \ \dot{\boldsymbol{J}} = \mathbf{0} = \boldsymbol{\Pi} \times \boldsymbol{K} + \boldsymbol{J} \times \boldsymbol{\Omega}, \implies (\mathbb{I}\boldsymbol{\Omega} - \alpha\boldsymbol{\Omega}) \times \boldsymbol{K} = \mathbf{0}. \\
\lambda^1 &: \ \dot{\boldsymbol{\Pi}} = \boldsymbol{\Pi} \times \boldsymbol{\Omega} + \boldsymbol{\Gamma} \times \boldsymbol{K}, \implies \boldsymbol{K} = mg\boldsymbol{\chi}. \\
\lambda^0 &: \ \dot{\boldsymbol{\Gamma}} = \boldsymbol{\Gamma} \times \boldsymbol{\Omega}.
\end{aligned}
$$

These relationships hold, provided the moment of inertia $\mathbb{I}$ is either proportional to the identity (Euler top), or has two equal entries that make it cylindrically symmetric about the vector χ (Lagrange top).

This system conserves each of the coefficients of the powers of λ in $|\boldsymbol{\Gamma} + \lambda\boldsymbol{\Pi} + \lambda^2\boldsymbol{J}|^2$. That is, besides the kinematic constant $|\boldsymbol{J}|^2$, it conserves

$$
|\boldsymbol{\Gamma}|^2, \quad \boldsymbol{\Gamma}\cdot\boldsymbol{\Pi}, \quad \frac{1}{2\alpha}|\boldsymbol{\Pi}|^2 + mg\boldsymbol{\Gamma}\cdot\boldsymbol{\chi}, \quad \boldsymbol{\Pi}\cdot\boldsymbol{\chi}.
$$

The first two are the Casimirs of the Lie–Poisson bracket in (8.3.4), the third is the Hamiltonian and the last is the χ-component of the angular momentum, which is conserved when the moment of inertia $\mathbb{I}$ is cylindrically symmetric about the vector χ. This symmetry holds for the Euler top and the Lagrange top, which are indeed known to be integrable. For in-depth discussions of this approach to heavy-top dynamics, see [Ra1982, RaVM1982]. ▲

Exercise. Manakov's approach for the heavy top in the vector notation of the previous exercise suggests a similar application to the $n \times n$ matrix commutator equation

$$\frac{d}{dt}(\Gamma + \lambda\Pi + \lambda^2 J) = [\Gamma + \lambda\Pi + \lambda^2 J, \Omega + \lambda K]$$

with skew-symmetric $(\Gamma, \Pi, \Omega, J, K)$ with constant (J, K). Determine whether this approach could be used to extend Manakov's treatment of the rigid body in n dimensions to the n-dimensional versions of the Euler top and the Lagrange top. ★

Exercise. Extend the Manakov approach even further by computing the system of $n \times n$ matrix equations for

$$\begin{aligned}\frac{d}{dt}&(\Gamma + \lambda M + \lambda^2 N + \lambda^3 J)\\ &= [\Gamma + \lambda M + \lambda^2 N + \lambda^3 J, \Omega + \lambda\omega + \lambda^2 K].\end{aligned}$$

Is this extended matrix system Hamiltonian? If so, what is its Lie–Poisson bracket? ★

9 THE EULER–POINCARÉ THEOREM

Contents

9.1	Action principles on Lie algebras	194
9.2	Hamilton–Pontryagin principle	198
9.3	Clebsch approach to Euler–Poincaré	199
9.3.1	Defining the Lie derivative	201
9.3.2	Clebsch Euler–Poincaré principle	202
9.4	Lie–Poisson Hamiltonian formulation	206
9.4.1	Cotangent-lift momentum maps	207

9.1 Action principles on Lie algebras

Hamilton's principle for stationary action was explained earlier for deriving Euler's equations for rigid-body rotations in either their vector or quaternion forms. In the notation for the AD, Ad and ad actions of Lie groups and Lie algebras, Hamilton's principle (that the equations of motion arise from stationarity of the action) for Lagrangians defined on Lie algebras may be expressed as follows. This is the Euler–Poincaré theorem [Po1901].

Theorem 9.1.1 (Euler–Poincaré theorem) *Stationarity*

$$\delta S(\xi) = \delta \int_a^b l(\xi)\, dt = 0 \tag{9.1.1}$$

of an action

$$S(\xi) = \int_a^b l(\xi)\, dt\,,$$

whose Lagrangian is defined on the (left-invariant) Lie algebra $\mathfrak{g}$ of a Lie group G by $l(\xi) : \mathfrak{g} \mapsto \mathbb{R}$, yields the ***Euler–Poincaré equation*** *on $\mathfrak{g}^*$,*

$$\frac{d}{dt}\frac{\delta l}{\delta \xi} = \mathrm{ad}^*_\xi \frac{\delta l}{\delta \xi}\,, \tag{9.1.2}$$

for variations of the left-invariant Lie algebra element

$$\xi = g^{-1}\dot{g}(t) \in \mathfrak{g}$$

that are restricted to the form

$$\delta\xi = \dot{\eta} + \mathrm{ad}_\xi\, \eta\,, \tag{9.1.3}$$

in which $\eta(t) \in \mathfrak{g}$ is a curve in the Lie algebra $\mathfrak{g}$ that vanishes at the endpoints in time.

Exercise. What is the solution to the Euler–Poincaré Equation (9.1.2) in terms of $\mathrm{Ad}^*_{g(t)}$?

Hint: Take a look at the earlier equation (4.2.10). ★

Remark 9.1.1 The earlier forms (4.1.21) and (4.1.23) of the variational formula for vectors and quaternions are now seen to apply more generally. Namely, such variations are defined for any Lie algebra. □

Proof. A direct computation proves Theorem 9.1.1. Later, we will explain the source of the constraint (9.1.3) on the form of the variations on the Lie algebra. One verifies the statement of the theorem by computing with a nondegenerate pairing $\langle\,\cdot\,,\,\cdot\,\rangle : \mathfrak{g}^* \times \mathfrak{g} \to \mathbb{R}$,

$$
\begin{aligned}
0 = \delta \int_a^b l(\xi)\, dt &= \int_a^b \Big\langle \frac{\delta l}{\delta \xi}\,,\, \delta\xi \Big\rangle\, dt \\
&= \int_a^b \Big\langle \frac{\delta l}{\delta \xi}\,,\, \dot{\eta} + \mathrm{ad}_\xi\, \eta \Big\rangle\, dt \\
&= \int_a^b \Big\langle - \frac{d}{dt}\frac{\delta l}{\delta \xi} + \mathrm{ad}^*_\xi \frac{\delta l}{\delta \xi}\,,\, \eta \Big\rangle dt + \Big\langle \frac{\delta l}{\delta \xi}\,,\, \eta \Big\rangle \Big|_a^b\,,
\end{aligned}
$$

upon integrating by parts. The last term vanishes, by the endpoint conditions, $\eta(b) = \eta(a) = 0$.

Since $\eta(t) \in \mathfrak{g}$ is otherwise arbitrary, (9.1.1) is equivalent to

$$
-\frac{d}{dt}\frac{\delta l}{\delta \xi} + \mathrm{ad}^*_\xi \frac{\delta l}{\delta \xi} = 0\,,
$$

which recovers the Euler–Poincaré Equation (9.1.2) in the statement of the theorem. ■

Corollary 9.1.1 (Noether's theorem for Euler–Poincaré) *If η is an infinitesimal symmetry of the Lagrangian, then $\langle \frac{\delta l}{\delta \xi}\,,\, \eta\rangle$ is its associated constant of the Euler–Poincaré motion.*

Proof. Consider the endpoint terms $\langle \frac{\delta l}{\delta \xi}, \eta \rangle|_a^b$ arising in the variation δS in (9.1.1) and note that this implies for any time $t \in [a, b]$ that

$$\left\langle \frac{\delta l}{\delta \xi(t)}, \eta(t) \right\rangle = \text{constant},$$

when the Euler–Poincaré Equations (9.1.2) are satisfied. ■

Corollary 9.1.2 (Interpretation of Noether's theorem) *Noether's theorem for the Euler–Poincaré stationary principle may be interpreted as conservation of the* spatial momentum *quantity*

$$\left(\mathrm{Ad}^*_{g^{-1}(t)} \frac{\delta l}{\delta \xi(t)} \right) = \text{constant},$$

as a consequence of the Euler–Poincaré Equation (9.1.2).

Proof. Invoke left-invariance of the Lagrangian $l(\xi)$ under $g \to h_\epsilon g$ with $h_\epsilon \in G$. For this symmetry transformation, one has $\delta g = \zeta g$ with $\zeta = \frac{d}{d\epsilon}\big|_{\epsilon=0} h_\epsilon$, so that

$$\eta = g^{-1}\delta g = \mathrm{Ad}_{g^{-1}}\zeta \in \mathfrak{g}\,.$$

In particular, along a curve $\eta(t)$ we have

$$\eta(t) = \mathrm{Ad}_{g^{-1}(t)}\eta(0) \quad \text{on setting} \quad \zeta = \eta(0),$$

at any initial time $t = 0$ (assuming of course that $[0, t] \in [a, b]$). Consequently,

$$\left\langle \frac{\delta l}{\delta \xi(t)}, \eta(t) \right\rangle = \left\langle \frac{\delta l}{\delta \xi(0)}, \eta(0) \right\rangle = \left\langle \frac{\delta l}{\delta \xi(t)}, \mathrm{Ad}_{g^{-1}(t)}\eta(0) \right\rangle.$$

For the nondegenerate pairing $\langle \cdot, \cdot \rangle$, this means that

$$\frac{\delta l}{\delta \xi(0)} = \left(\mathrm{Ad}^*_{g^{-1}(t)} \frac{\delta l}{\delta \xi(t)} \right) = \text{constant}.$$

The constancy of this quantity under the Euler–Poincaré dynamics in (9.1.2) is verified, upon taking the time derivative and using the coadjoint motion relation (4.2.9) in Proposition 4.2.2. ■

Remark 9.1.2 The form of the variation in (9.1.3) arises directly by

(i) computing the variations of the left-invariant Lie algebra element $\xi = g^{-1}\dot{g} \in \mathfrak{g}$ induced by taking variations δg in the group;

(ii) taking the time derivative of the variation $\eta = g^{-1}g' \in \mathfrak{g}$; and

(iii) using the equality of cross derivatives ($g^{\cdot\,\prime} = d^2g/dtds = g^{\prime\,\cdot}$).

Namely, one computes, cf. Proposition (2.4.1) for the rigid body,

$$\begin{aligned} \xi' &= (g^{-1}\dot{g})' = -\,g^{-1}g'g^{-1}\dot{g} + g^{-1}g^{\cdot\,\prime} = -\,\eta\xi + g^{-1}g^{\cdot\,\prime}\,, \\ \dot{\eta} &= (g^{-1}g')^{\cdot} = -\,g^{-1}\dot{g}g^{-1}g' + g^{-1}g^{\prime\,\cdot} = -\,\xi\eta + g^{-1}g^{\prime\,\cdot}\,. \end{aligned}$$

On taking the difference, the terms with cross derivatives cancel and one finds the variational formula (9.1.3),

$$\xi' - \dot{\eta} = [\xi\,,\,\eta] \quad \text{with} \quad [\xi\,,\,\eta] := \xi\,\eta - \eta\,\xi = \mathrm{ad}_\xi\,\eta\,. \qquad (9.1.4)$$

Thus, the same formal calculations as for vectors and quaternions also apply to Hamilton's principle on (matrix) Lie algebras. □

Example 9.1.1 (Euler–Poincaré equation for $SE(3)$) *The Euler–Poincaré Equation (9.1.2) for $SE(3)$ is equivalent to*

$$\left(\frac{d}{dt}\frac{\delta l}{\delta \xi}\,,\,\frac{d}{dt}\frac{\delta l}{\delta \alpha}\right) = \mathrm{ad}^*_{(\xi\,,\,\alpha)}\left(\frac{\delta l}{\delta \xi}\,,\,\frac{\delta l}{\delta \alpha}\right). \qquad (9.1.5)$$

This formula recovers the Euler–Poincaré Equation (7.1.7) for $SE(3)$ upon using the definition of the ad^* *operation for $se(3)$ in Equation (6.3.2).*

Remark 9.1.3 Corollary 9.1.2 is again the Kelvin–Noether Theorem 7.1.2, seen earlier for $SE(3)$ and now proven for the Euler–Poincaré equations on an arbitrary Lie group. □

9.2 Hamilton–Pontryagin principle

Formula (9.1.4) for the variation of the vector $\xi = g^{-1}\dot{g} \in \mathfrak{g}$ may be imposed as a constraint in Hamilton's principle and thereby provide an immediate derivation of the Euler–Poincaré Equation (9.1.2). This constraint is incorporated into the following theorem [BoMa2009].

Theorem 9.2.1 (Hamilton–Pontryagin principle) *The Euler–Poincaré equation*

$$\frac{d}{dt}\frac{\delta l}{\delta \xi} = \mathrm{ad}^*_\xi \frac{\delta l}{\delta \xi} \tag{9.2.1}$$

on the dual Lie algebra $\mathfrak{g}^*$ *is equivalent to the following implicit variational principle,*

$$\delta S(\xi, g, \dot{g}) = \delta \int_a^b l(\xi, g, \dot{g})\, dt = 0, \tag{9.2.2}$$

for a constrained action

$$\begin{aligned} S(\xi, g, \dot{g}) &= \int_a^b l(\xi, g, \dot{g})\, dt \\ &= \int_a^b \Big[l(\xi) + \langle \mu \,,\, (g^{-1}\dot{g} - \xi) \rangle \Big]\, dt\,. \end{aligned} \tag{9.2.3}$$

Proof. The variations of S in formula (9.2.3) are given by

$$\delta S = \int_a^b \Big\langle \frac{\delta l}{\delta \xi} - \mu \,,\, \delta\xi \Big\rangle + \Big\langle \delta\mu \,,\, (g^{-1}\dot{g} - \xi) \Big\rangle + \Big\langle \mu \,,\, \delta(g^{-1}\dot{g}) \Big\rangle\, dt\,.$$

Substituting $\delta(g^{-1}\dot{g})$ from (9.1.4) into the last term produces

$$\begin{aligned} \int_a^b \Big\langle \mu \,,\, \delta(g^{-1}\dot{g}) \Big\rangle\, dt &= \int_a^b \Big\langle \mu \,,\, \dot{\eta} + \mathrm{ad}_\xi\, \eta \Big\rangle\, dt \\ &= \int_a^b \Big\langle -\dot{\mu} + \mathrm{ad}^*_\xi\, \mu \,,\, \eta \Big\rangle\, dt + \Big\langle \mu \,,\, \eta \Big\rangle \Big|_a^b\,, \end{aligned}$$

where $\eta = g^{-1}\delta g$ vanishes at the endpoints in time. Thus, stationarity $\delta S = 0$ of the Hamilton–Pontryagin variational principle yields the following set of equations:

$$\frac{\delta l}{\delta \xi} = \mu \,, \quad g^{-1}\dot{g} = \xi \,, \quad \dot{\mu} = \mathrm{ad}^*_\xi \, \mu \,. \tag{9.2.4}$$

■

Remark 9.2.1 (Interpreting variational formulas (9.2.4)) The first formula in (9.2.4) is the fibre derivative needed in the Legendre transformation $\mathfrak{g} \mapsto \mathfrak{g}^*$, for passing to the Hamiltonian formulation. The second is the reconstruction formula for obtaining the solution curve $g(t) \in G$ on the Lie group G given the solution $\xi(t) = g^{-1}\dot{g} \in \mathfrak{g}$. The third formula in (9.2.4) is the Euler–Poincaré equation on $\mathfrak{g}^*$. The interpretation of Noether's theorem in Corollary 9.1.2 transfers to the Hamilton–Pontryagin variational principle as preservation of the quantity

$$\Big(\mathrm{Ad}^*_{g^{-1}(t)}\mu(t)\Big) = \mu(0) = \text{constant},$$

under the Euler–Poincaré dynamics.

This Hamilton's principle is said to be ***implicit*** because the definitions of the quantities describing the motion emerge only *after* the variations have been taken. See [YoMa2006] for discussions of recent developments in the theory of implicit variational principles.

□

Exercise. Compute the Euler–Poincaré equation on $\mathfrak{g}^*$ when $\xi(t) = \dot{g}g^{-1} \in \mathfrak{g}$ is *right-invariant.* ★

9.3 Clebsch approach to Euler–Poincaré

The Hamilton–Pontryagin (HP) Theorem 9.2.1 elegantly delivers the three key formulas in (9.2.4) needed for deriving the Lie–Poisson

Hamiltonian formulation of the Euler–Poincaré equation. Perhaps surprisingly, the HP theorem accomplishes this without invoking any properties of how the invariance group of the Lagrangian G acts on the configuration space M.

An alternative derivation of these formulas exists that uses the Clebsch approach and does invoke the action $G \times M \to M$ of the Lie group on the configuration space, M, which is assumed to be a manifold. This alternative derivation is a bit more elaborate than the HP theorem. However, invoking the Lie group action on the configuration space provides additional valuable information. In particular, the alternative approach will yield information about the momentum map $T^*M \mapsto \mathfrak{g}^*$ which explains precisely how the canonical phase space T^*M maps to the Poisson manifold of the dual Lie algebra $\mathfrak{g}^*$.

Proposition 9.3.1 (Clebsch Euler–Poincaré principle) *The Euler–Poincaré equation*

$$\frac{d}{dt}\frac{\delta l}{\delta \xi} = \mathrm{ad}^*_\xi \frac{\delta l}{\delta \xi} \tag{9.3.1}$$

on the dual Lie algebra $\mathfrak{g}^*$ is equivalent to the following implicit variational principle,

$$\delta S(\xi, q, \dot{q}, p) = \delta \int_a^b l(\xi, q, \dot{q}, p)\, dt = 0, \tag{9.3.2}$$

for an action constrained by the reconstruction formula

$$\begin{aligned} S(\xi, q, \dot{q}, p) &= \int_a^b l(\xi, q, \dot{q}, p)\, dt \\ &= \int_a^b \Big[l(\xi) + \Big\langle\!\Big\langle\, p\,,\, \dot{q} + \pounds_\xi q \,\Big\rangle\!\Big\rangle \Big]\, dt\,, \end{aligned} \tag{9.3.3}$$

in which the pairing $\langle\!\langle \cdot\,, \cdot \rangle\!\rangle : T^*M \times TM \mapsto \mathbb{R}$ maps an element of the cotangent space (a momentum covector) and an element from the tangent space (a velocity vector) to a real number. This is the natural pairing for an action integrand and it also occurs in the Legendre transformation.

Remark 9.3.1 The Lagrange multiplier p in the second term of (9.3.3) imposes the constraint

$$\dot{q} + \pounds_\xi q = 0\,. \tag{9.3.4}$$

This is the formula for the evolution of the quantity $q(t) = g^{-1}(t)q(0)$ under the left action of the Lie algebra element $\xi \in \mathfrak{g}$ on it by the Lie derivative $\pounds_\xi$ along ξ. (For right action by g so that $q(t) = q(0)g(t)$, the formula is $\dot{q} - \pounds_\xi q = 0$.) □

9.3.1 Defining the Lie derivative

One assumes the motion follows a trajectory $q(t) \in M$ in the configuration space M given by $q(t) = g(t)q(0)$, where $g(t) \in G$ is a time-dependent curve in the Lie group G which operates on the configuration space M by a flow $\phi_t : G \times M \mapsto M$. The flow property of the map $\phi_t \circ \phi_s = \phi_{s+t}$ is guaranteed by the group composition law.

Just as for the free rotations, one defines the left-invariant and right-invariant velocity vectors. Namely, as for the body angular velocity,

$$\xi_L(t) = g^{-1}\dot{g}(t) \quad \text{is left-invariant under } g(t) \to hg(t),$$

and as for the spatial angular velocity,

$$\xi_R(t) = \dot{g}g^{-1}(t) \quad \text{is right-invariant under } g(t) \to g(t)h,$$

for any choice of matrix $h \in G$. This means neither of these velocities depends on the initial configuration.

Right-invariant velocity vector

The Lie derivative $\pounds_\xi$ appearing in the reconstruction relation $\dot{q} = -\pounds_\xi q$ in (9.3.4) is defined via the Lie group operation on the configuration space exactly as for free rotation. For example, one computes the tangent vectors to the motion induced by the group operation acting from the left as $q(t) = g(t)q(0)$ by differentiating with respect to time t,

$$\dot{q}(t) = \dot{g}(t)q(0) = \dot{g}g^{-1}(t)q(t) =: \pounds_{\xi_R} q(t)\,,$$

where $\xi_R = \dot{g}g^{-1}(t)$ is right-invariant. This is the analogue of the spatial angular velocity of a freely rotating rigid body.

Left-invariant velocity vector

Likewise, differentiating the right action $q(t) = q(0)g(t)$ of the group on the configuration manifold yields

$$\dot{q}(t) = q(t)g^{-1}\dot{g}(t) =: \pounds_{\xi_L} q(t)\,,$$

in which the quantity

$$\xi_L(t) = g^{-1}\dot{g}(t) = \mathrm{Ad}_{g^{-1}(t)}\xi_R(t)$$

is the left-invariant tangent vector.

This analogy with free rotation dynamics should be a good guide for understanding the following manipulations, at least until we have a chance to illustrate the ideas with further examples.

> **Exercise.** Compute the time derivatives and thus the forms of the right- and left-invariant velocity vectors for the group operations by the inverse $q(t) = q(0)g^{-1}(t)$ and $q(t) = g^{-1}(t)q(0)$. Observe the equivalence (up to a sign) of these velocity vectors with the vectors ξ_R and ξ_L, respectively. Note that the reconstruction formula (9.3.4) arises from the latter choice. ★

9.3.2 Clebsch Euler–Poincaré principle

Let us first define the concepts and notation that will arise in the course of the proof of Proposition 9.3.1.

Definition 9.3.1 (The diamond operation $\diamond$) *The diamond operation ($\diamond$) in Equation (9.3.8) is defined as minus the dual of the Lie derivative with respect to the pairing induced by the variational derivative in q, namely,*

$$\Big\langle p \diamond q \,,\, \xi \Big\rangle = \Big\langle\!\Big\langle p \,,\, -\pounds_\xi q \Big\rangle\!\Big\rangle\,. \tag{9.3.5}$$

Definition 9.3.2 (Transpose of the Lie derivative) *The **transpose** of the Lie derivative* $\pounds_\xi^T p$ *is defined via the pairing* $\langle\!\langle\,\cdot\,,\,\cdot\,\rangle\!\rangle$ *between* $(q,p)\in T^*M$ *and* $(q,\dot q)\in TM$ *as*

$$\Big\langle\!\!\Big\langle\, \pounds_\xi^T p\,,\, q\,\Big\rangle\!\!\Big\rangle = \Big\langle\!\!\Big\langle\, p\,,\, \pounds_\xi q\,\Big\rangle\!\!\Big\rangle\,. \tag{9.3.6}$$

Proof. The variations of the action integral

$$S(\xi,q,\dot q,p) = \int_a^b \Big[l(\xi) + \Big\langle\!\!\Big\langle\, p\,,\, \dot q + \pounds_\xi q\,\Big\rangle\!\!\Big\rangle \Big]\, dt \tag{9.3.7}$$

from formula (9.3.3) are given by

$$\begin{aligned}\delta S &= \int_a^b \Big\langle \frac{\delta l}{\delta \xi}\,,\,\delta\xi\Big\rangle + \Big\langle\!\!\Big\langle \frac{\delta l}{\delta p}\,,\,\delta p\Big\rangle\!\!\Big\rangle + \Big\langle\!\!\Big\langle \frac{\delta l}{\delta q}\,,\,\delta q\Big\rangle\!\!\Big\rangle + \Big\langle\!\!\Big\langle\, p\,,\, \pounds_{\delta\xi} q\Big\rangle\!\!\Big\rangle\, dt \\ &= \int_a^b \Big\langle \frac{\delta l}{\delta \xi} - p\diamond q\,,\,\delta\xi\Big\rangle + \Big\langle\!\!\Big\langle\, \delta p\,,\, \dot q + \pounds_\xi q\Big\rangle\!\!\Big\rangle - \Big\langle\!\!\Big\langle\, \dot p - \pounds_\xi^T p\,,\,\delta q\Big\rangle\!\!\Big\rangle\, dt\,.\end{aligned}$$

Thus, stationarity of this implicit variational principle implies the following set of equations:

$$\frac{\delta l}{\delta \xi} = p\diamond q\,,\quad \dot q = -\,\pounds_\xi q\,,\quad \dot p = \pounds_\xi^T p\,. \tag{9.3.8}$$

In these formulas, the notation distinguishes between the two types of pairings,

$$\langle\,\cdot\,,\,\cdot\,\rangle : \mathfrak{g}^*\times\mathfrak{g}\mapsto\mathbb{R} \quad\text{and}\quad \langle\!\langle\,\cdot\,,\,\cdot\,\rangle\!\rangle : T^*M\times TM\mapsto\mathbb{R}\,. \tag{9.3.9}$$

(The third pairing in the formula for δS is not distinguished because it is equivalent to the second one under integration by parts in time.)

The Euler–Poincaré equation emerges from elimination of (q,p) using these formulas and the properties of the diamond operation that arise from its definition, as follows, for any vector $\eta \in \mathfrak{g}$:

$$
\begin{aligned}
\Big\langle \frac{d}{dt}\frac{\delta l}{\delta \xi}\,,\,\eta \Big\rangle &= \frac{d}{dt}\Big\langle \frac{\delta l}{\delta \xi}\,,\,\eta \Big\rangle, \\
\text{[Definition of } \diamond\text{]} &= \frac{d}{dt}\Big\langle p\diamond q\,,\,\eta \Big\rangle = \frac{d}{dt}\Big\langle\!\Big\langle p\,,\,-\pounds_\eta q \Big\rangle\!\Big\rangle, \\
\text{[Equations (9.3.8)]} &= \Big\langle\!\Big\langle \pounds_\xi^T p\,,\,-\pounds_\eta q \Big\rangle\!\Big\rangle + \Big\langle\!\Big\langle p\,,\,\pounds_\eta \pounds_\xi q \Big\rangle\!\Big\rangle, \\
\text{[Transpose, } \diamond \text{ and ad]} &= \Big\langle\!\Big\langle p\,,\,-\pounds_{[\xi,\,\eta]} q \Big\rangle\!\Big\rangle = \Big\langle p\diamond q\,,\,\mathrm{ad}_\xi \eta \Big\rangle, \\
\text{[Definition of ad}^*\text{]} &= \Big\langle\, \mathrm{ad}^*_\xi \frac{\delta l}{\delta \xi}\,,\,\eta \Big\rangle.
\end{aligned}
$$

This is the Euler–Poincaré Equation (9.3.1). ■

Exercise. Show that the diamond operation defined in Equation (9.3.5) is antisymmetric,

$$
\Big\langle p\diamond q\,,\,\xi \Big\rangle = -\Big\langle q\diamond p\,,\,\xi \Big\rangle. \tag{9.3.10}
$$

★

Exercise. (Euler–Poincaré equation for right action) Compute the Euler–Poincaré equation for the Lie group action $G\times M \mapsto M : q(t) = q(0)g(t)$ in which the group acts from the right on a point $q(0)$ in the configuration manifold M along a time-dependent curve $g(t) \in G$. Explain why the result differs in sign from the case of left G-action on manifold M. ★

Exercise. (Clebsch approach for motion on $T^*(G\times V)$) Often the Lagrangian will contain a parameter taking values in a vector space V that represents a feature of the potential energy of the motion. We have encountered this situation already with the heavy top, in which the parameter is the vector in the body pointing from the contact point to the centre of mass. Since the potential energy will affect the motion we assume an action $G \times V \to V$ of the Lie group G on the vector space V. The Lagrangian then takes the form $L : TG \times V \to \mathbb{R}$.

Compute the variations of the action integral

$$S(\xi, q, \dot{q}, p) = \int_a^b \Big[\tilde{l}(\xi, q) + \Big\langle\!\Big\langle\, p\,,\, \dot{q} + \pounds_\xi q \,\Big\rangle\!\Big\rangle \Big]\, dt$$

and determine the effects in the Euler–Poincaré equation of having $q \in V$ appear in the Lagrangian $\tilde{l}(\xi, q)$.

Show first that stationarity of S implies the following set of equations:

$$\frac{\delta \tilde{l}}{\delta \xi} = p \diamond q\,, \quad \dot{q} = -\,\pounds_\xi q\,, \quad \dot{p} = \pounds_\xi^T p + \frac{\delta \tilde{l}}{\delta q}\,.$$

Then transform to the variable $\delta l/\delta\xi$ to find the associated Euler–Poincaré equations on the space $\mathfrak{g}^* \times V$,

$$\begin{aligned} \frac{d}{dt}\frac{\delta \tilde{l}}{\delta \xi} &= \operatorname{ad}^*_\xi \frac{\delta \tilde{l}}{\delta \xi} + \frac{\delta \tilde{l}}{\delta q} \diamond q\,, \\ \frac{dq}{dt} &= -\,\pounds_\xi q\,. \end{aligned}$$

Perform the Legendre transformation to derive the Lie–Poisson Hamiltonian formulation corresponding to $\tilde{l}(\xi, q)$.

★

9.4 Lie–Poisson Hamiltonian formulation

The Clebsch variational principle for the Euler–Poincaré equation provides a natural path to its canonical and Lie–Poisson Hamiltonian formulations. The Legendre transform takes the Lagrangian

$$l(p,q,\dot q,\xi) = l(\xi) + \left\langle\!\left\langle\, p\,,\, \dot q + \pounds_\xi q \,\right\rangle\!\right\rangle$$

in the action (9.3.7) to the Hamiltonian,

$$H(p,q) = \left\langle\!\left\langle\, p\,,\, \dot q \,\right\rangle\!\right\rangle - l(p,q,\dot q,\xi) = \left\langle\!\left\langle\, p\,,\, -\pounds_\xi q \,\right\rangle\!\right\rangle - l(\xi)\,,$$

whose variations are given by

$$\begin{aligned}\delta H(p,q) &= \left\langle\!\left\langle\, \delta p\,,\, -\pounds_\xi q \,\right\rangle\!\right\rangle + \left\langle\!\left\langle\, p\,,\, -\pounds_\xi \delta q \,\right\rangle\!\right\rangle \\ &\quad + \left\langle\!\left\langle\, p\,,\, -\pounds_{\delta\xi} q \,\right\rangle\!\right\rangle - \left\langle\, \frac{\delta l}{\delta \xi}\,,\, \delta\xi \,\right\rangle \\ &= \left\langle\!\left\langle\, \delta p\,,\, -\pounds_\xi q \,\right\rangle\!\right\rangle + \left\langle\!\left\langle\, -\pounds_\xi^T p\,,\, \delta q \,\right\rangle\!\right\rangle + \left\langle\, p\diamond q - \frac{\delta l}{\delta \xi}\,,\, \delta\xi \,\right\rangle .\end{aligned}$$

These variational derivatives recover Equations (9.3.8) in canonical Hamiltonian form,

$$\dot q = \delta H/\delta p = -\pounds_\xi q \quad \text{and} \quad \dot p = -\delta H/\delta q = \pounds_\xi^T p\,.$$

Moreover, independence of H from ξ yields the momentum relation,

$$\frac{\delta l}{\delta \xi} = p\diamond q\,. \tag{9.4.1}$$

The Legendre transformation of the Euler–Poincaré equations using the Clebsch canonical variables leads to the ***Lie–Poisson Hamiltonian form*** of these equations,

$$\frac{d\mu}{dt} = \{\mu, h\} = \mathrm{ad}^*_{\delta h/\delta\mu}\,\mu\,, \tag{9.4.2}$$

with

$$\mu = p \diamond q = \frac{\delta l}{\delta \xi}, \quad h(\mu) = \langle \mu, \xi \rangle - l(\xi), \quad \xi = \frac{\delta h}{\delta \mu}. \qquad (9.4.3)$$

By Equation (9.4.3), the evolution of a smooth real function $f : \mathfrak{g}^* \to \mathbb{R}$ is governed by

$$\begin{aligned} \frac{df}{dt} &= \left\langle \frac{\delta f}{\delta \mu}, \frac{d\mu}{dt} \right\rangle \\ &= \left\langle \frac{\delta f}{\delta \mu}, \mathrm{ad}^*_{\delta h/\delta \mu}\, \mu \right\rangle \\ &= \left\langle \mathrm{ad}_{\delta h/\delta \mu} \frac{\delta f}{\delta \mu}, \mu \right\rangle \\ &= -\left\langle \mu, \left[\frac{\delta f}{\delta \mu}, \frac{\delta h}{\delta \mu} \right] \right\rangle \\ &=: \{f, h\}. \end{aligned} \qquad (9.4.4)$$

The last equality defines the ***Lie–Poisson bracket*** $\{f, h\}$ for smooth real functions f and h on the dual Lie algebra $\mathfrak{g}^*$. One may check directly that this bracket operation is a bilinear, skew-symmetric derivation that satisfies the Jacobi identity. Thus, it defines a proper Poisson bracket on $\mathfrak{g}^*$.

9.4.1 Cotangent-lift momentum maps

Although it is more elaborate than the Hamilton–Pontryagin principle and it requires input about the action of a Lie algebra on the configuration space, the Clebsch variational principle for the Euler–Poincaré equation reveals useful information.

As we shall see, the Clebsch approach provides a direct means of computing the ***momentum map*** for the specified Lie algebra action on a given configuration manifold M. In fact, the first equation in (9.4.3) is the standard example of the momentum map obtained by the ***cotangent lift*** of a Lie algebra action on a configuration manifold.

Momentum maps will be discussed later, in Chapter 11. For now, the reader may wish to notice that the formulas (9.4.3) and (11.2.1) involving the diamond operation have remarkable similarities. In particular, the term $q \diamond p$ in these formulas has the same meaning. Consequently, we may state the following proposition.

Proposition 9.4.1 (Momentum maps) *The Lie–Poisson form (9.4.2) of the Euler–Poincaré Equation (9.3.1) governs the evolution of the momentum map derived from the cotangent lift of the Lie algebra action on the configuration manifold.*

The remainder of the present text should provide the means to fully understand this statement.

10 LIE–POISSON HAMILTONIAN FORM OF A CONTINUUM SPIN CHAIN

Contents

10.1 Formulating continuum spin chain equations **210**

10.2 Euler–Poincaré equations **212**

10.3 Hamiltonian formulation **213**

In this chapter we will begin thinking in terms of Hamiltonian partial differential equations in the specific example of *G-strands,* which are evolutionary maps into a Lie group $g(t,x) : \mathbb{R}\times\mathbb{R} \to G$ that follow from Hamilton's principle for a certain class of G-invariant Lagrangians. The case when $G = SO(3)$ may be regarded physically as a smooth distribution of $so(3)$-valued spins attached to a one-dimensional straight strand lying along the x-axis. We will investigate its three-dimensional orientation dynamics at each point along the strand. For no additional cost, we may begin with the Euler–Poincaré theorem for a left-invariant Lagrangian defined on the tangent space of an *arbitrary* Lie group G and later specialise to the case where G is the rotation group $SO(3)$.

The Lie–Poisson Hamiltonian formulation of the Euler–Poincaré Equation (9.3.1) for this problem will be derived via the Legendre Transformation by following calculations similar to those done previously for the rigid body in Section 2.5. To emphasise the systematic nature of the Legendre transformation from the Euler–Poincaré picture to the Lie–Poisson picture, we will lay out the procedure in well-defined steps.

10.1 Formulating continuum spin chain equations

We shall consider Hamilton's principle $\delta S = 0$ for a left-invariant Lagrangian,

$$S = \int_a^b \int_{-\infty}^{\infty} \ell(\Omega, \Xi)\, dx\, dt\,, \qquad (10.1.1)$$

with the following definitions of the tangent vectors Ω and Ξ,

$$\Omega(t,x) = g^{-1}\partial_t g(t,x) \quad \text{and} \quad \Xi(t,x) = g^{-1}\partial_x g(t,x)\,, \qquad (10.1.2)$$

where $g(t,x) \in G$ is a real-valued map $g : \mathbb{R}\times\mathbb{R} \to G$ for a Lie group G. Later, we shall specialise to the case where G is the rotation group $SO(3)$. We shall apply the by now standard Euler–Poincaré procedure, modulo the partial spatial derivative in the definition of $\Xi(t,x) = g^{-1}\partial_x g(t,x) \in \mathfrak{g}$. This procedure takes the following steps:

(i) Write the auxiliary equation for the evolution of $\Xi : \mathbb{R} \times \mathbb{R} \to \mathfrak{g}$, obtained by differentiating its definition with respect to time and invoking equality of cross derivatives.

(ii) Use the Euler–Poincaré theorem for left-invariant Lagrangians to obtain the equation of motion for the momentum variable $\partial\ell/\partial\Omega : \mathbb{R} \times \mathbb{R} \to \mathfrak{g}^*$, where $\mathfrak{g}^*$ is the dual Lie algebra. Use the L^2 pairing defined by the spatial integration.

(These will be partial differential equations. Assume homogeneous boundary conditions on $\Omega(t,x)$, $\Xi(t,x)$ and vanishing endpoint conditions on the variation $\eta = g^{-1}\delta g(t,x) \in \mathfrak{g}$ when integrating by parts.)

(iii) Legendre-transform this Lagrangian to obtain the corresponding Hamiltonian. Differentiate the Hamiltonian and determine its partial derivatives. Write the Euler–Poincaré equation in terms of the new momentum variable $\Pi = \delta\ell/\delta\Omega \in \mathfrak{g}^*$.

(iv) Determine the Lie–Poisson bracket implied by the Euler–Poincaré equation in terms of the Legendre-transformed quantities $\Pi = \delta\ell/\delta\Omega$, by rearranging the time derivative of a smooth function $f(\Pi, \Xi) : \mathfrak{g}^* \times \mathfrak{g} \to \mathbb{R}$.

(v) Specialise to $G = SO(3)$ and write the Lie–Poisson Hamiltonian form in terms of vector operations in $\mathbb{R}^3$.

(vi) For $G = SO(3)$ choose the Lagrangian

$$\begin{aligned} \ell &= \frac{1}{2}\int_{-\infty}^{\infty} \mathrm{Tr}\Big([g^{-1}\partial_t g, g^{-1}\partial_x g]^2 \Big)\, dx \\ &= \frac{1}{2}\int_{-\infty}^{\infty} \mathrm{Tr}\Big([\Omega, \Xi]^2 \Big)\, dx\,, \end{aligned} \qquad (10.1.3)$$

where $[\Omega, \Xi] = \Omega\Xi - \Xi\Omega$ is the commutator in the Lie algebra $\mathfrak{g}$. Use the hat map to write the Euler–Poincaré equation and its Lie–Poisson Hamiltonian form in terms of vector operations in $\mathbb{R}^3$.

10.2 Euler–Poincaré equations

The Euler–Poincaré procedure systematically produces the following results.

Auxiliary equations By definition, $\Omega(t,x) = g^{-1}\partial_t g(t,x)$ and $\Xi(t,x) = g^{-1}\partial_x g(t,x)$ are Lie-algebra-valued functions over $\mathbb{R} \times \mathbb{R}$. The evolution of Ξ is obtained from these definitions by taking the difference of the two equations for the partial derivatives

$$\begin{aligned} \partial_t\Xi(t,x) &= -(g^{-1}\partial_t g)(g^{-1}\partial_x g) + g^{-1}\partial_t\partial_x g(t,x)\,, \\ \partial_x\Omega(t,x) &= -(g^{-1}\partial_x g)(g^{-1}\partial_t g) + g^{-1}\partial_x\partial_t g(t,x)\,, \end{aligned}$$

and invoking equality of cross derivatives. Hence, Ξ evolves by the adjoint operation, much like in the derivation of the variational derivative of Ω,

$$\partial_t\Xi(t,x) - \partial_x\Omega(t,x) = \Xi\,\Omega - \Omega\,\Xi = [\Xi, \Omega] =: -\,\mathrm{ad}_\Omega \Xi\,. \qquad (10.2.1)$$

This is the auxiliary equation for $\Xi(t,x)$. In differential geometry, this relation is called a ***zero curvature relation***, because it implies that the curvature vanishes for the Lie-algebra-valued connection one-form $A = \Omega dt + \Xi dx$ [doCa1976].

Hamilton's principle For $\eta = g^{-1}\delta g(t,x) \in \mathfrak{g}$, Hamilton's principle $\delta S = 0$ for $S = \int_a^b \ell(\Omega, \Xi)\, dt$ leads to

$$\delta S = \int_a^b \Big\langle \frac{\delta \ell}{\delta \Omega}\,,\, \delta\Omega \Big\rangle + \Big\langle \frac{\delta \ell}{\delta \Xi}\,,\, \delta\Xi \Big\rangle\, dt$$

$$= \int_a^b \Big\langle \frac{\delta \ell}{\delta \Omega}\,,\, \partial_t \eta + \mathrm{ad}_\Omega \eta \Big\rangle + \Big\langle \frac{\delta \ell}{\delta \Xi}\,,\, \partial_x \eta + \mathrm{ad}_\Xi \eta \Big\rangle \, dt$$

$$= \int_a^b \Big\langle - \partial_t \frac{\delta \ell}{\delta \Omega} + \mathrm{ad}^*_\Omega \frac{\delta \ell}{\delta \Omega}\,,\, \eta \Big\rangle + \Big\langle - \partial_x \frac{\delta \ell}{\delta \Xi} + \mathrm{ad}^*_\Xi \frac{\delta \ell}{\delta \Xi}\,,\, \eta \Big\rangle \, dt$$

$$= \int_a^b \Big\langle - \frac{\partial}{\partial t} \frac{\delta \ell}{\delta \Omega} + \mathrm{ad}^*_\Omega \frac{\delta \ell}{\delta \Omega} - \frac{\partial}{\partial x} \frac{\delta \ell}{\delta \Xi} + \mathrm{ad}^*_\Xi \frac{\delta \ell}{\delta \Xi}\,,\, \eta \Big\rangle \, dt\,,$$

where the formulas for the variations $\delta\Omega$ and $\delta\Xi$ are obtained by essentially the same calculation as in part (i). Hence, $\delta S = 0$ yields

$$\frac{\partial}{\partial t} \frac{\delta \ell}{\delta \Omega} = \mathrm{ad}^*_\Omega \frac{\delta \ell}{\delta \Omega} - \frac{\partial}{\partial x} \frac{\delta \ell}{\delta \Xi} + \mathrm{ad}^*_\Xi \frac{\delta \ell}{\delta \Xi}\,. \tag{10.2.2}$$

This is the Euler–Poincaré equation for $\delta\ell/\delta\Omega \in \mathfrak{g}^*$.

Exercise. Use Equation (4.2.9) in Proposition 4.2.2 to show that the Euler–Poincaré Equation (10.2.2) is a ***conservation law*** for spin angular momentum $\Pi = \delta\ell/\delta\Omega$,

$$\frac{\partial}{\partial t}\left(\mathrm{Ad}^*_{g(t,x)^{-1}} \frac{\delta l}{\delta \Omega} \right) = - \frac{\partial}{\partial x}\left(\mathrm{Ad}^*_{g(t,x)^{-1}} \frac{\delta l}{\delta \Xi} \right). \tag{10.2.3}$$

★

10.3 Hamiltonian formulation

Legendre transform Legendre-transforming the Lagrangian $\ell(\Omega, \Xi)$: $\mathfrak{g} \times V \to \mathbb{R}$ yields the Hamiltonian $h(\Pi, \Xi) : \mathfrak{g}^* \times V \to \mathbb{R}$,

$$h(\Pi, \Xi) = \big\langle \Pi\,,\, \Omega \big\rangle - \ell(\Omega, \Xi)\,. \tag{10.3.1}$$

Differentiating the Hamiltonian determines its partial derivatives:

$$\begin{aligned}
\delta h &= \Big\langle \delta\Pi\,,\, \frac{\delta h}{\delta \Pi} \Big\rangle + \Big\langle \frac{\delta h}{\delta \Xi}\,,\, \delta\Xi \Big\rangle \\
&= \Big\langle \delta\Pi\,,\, \Omega \Big\rangle + \Big\langle \Pi - \frac{\delta l}{\delta \Omega}\,,\, \delta\Omega \Big\rangle - \Big\langle \frac{\delta \ell}{\delta \Xi}\,,\, \delta\Xi \Big\rangle \\
&\Rightarrow\ \frac{\delta l}{\delta \Omega} = \Pi\,, \quad \frac{\delta h}{\delta \Pi} = \Omega \quad \text{and} \quad \frac{\delta h}{\delta \Xi} = -\,\frac{\delta \ell}{\delta \Xi}\,.
\end{aligned}$$

The middle term vanishes because $\Pi - \delta l/\delta\Omega = 0$ defines Π. These derivatives allow one to rewrite the Euler–Poincaré equation solely in terms of momentum Π as

$$\begin{aligned}
\partial_t \Pi &= \mathrm{ad}^*_{\delta h/\delta \Pi}\,\Pi + \partial_x \frac{\delta h}{\delta \Xi} - \mathrm{ad}^*_{\Xi}\,\frac{\delta h}{\delta \Xi}\,, \\
\partial_t \Xi &= \partial_x \frac{\delta h}{\delta \Pi} - \mathrm{ad}_{\delta h/\delta \Pi}\,\Xi\,.
\end{aligned} \tag{10.3.2}$$

Hamiltonian equations The corresponding Hamiltonian equation for any functional of $f(\Pi, \Xi)$ is then

$$\begin{aligned}
\frac{\partial}{\partial t} f(\Pi, \Xi) &= \Big\langle \partial_t \Pi\,,\, \frac{\delta f}{\delta \Pi} \Big\rangle + \Big\langle \partial_t \Xi\,,\, \frac{\delta f}{\delta \Xi} \Big\rangle \\
&= \Big\langle \mathrm{ad}^*_{\delta h/\delta \Pi}\Pi + \partial_x \frac{\delta h}{\delta \Xi} - \mathrm{ad}^*_{\Xi}\frac{\delta h}{\delta \Xi}\,,\, \frac{\delta f}{\delta \Pi} \Big\rangle \\
&\quad + \Big\langle \partial_x \frac{\delta h}{\delta \Pi} - \mathrm{ad}_{\delta h/\delta \Pi}\Xi\,,\, \frac{\delta f}{\delta \Xi} \Big\rangle \\
&= -\Big\langle \Pi\,,\, \Big[\frac{\delta f}{\delta \Pi}\,,\, \frac{\delta h}{\delta \Pi}\Big] \Big\rangle \\
&\quad + \Big\langle \partial_x \frac{\delta h}{\delta \Xi}\,,\, \frac{\delta f}{\delta \Pi} \Big\rangle - \Big\langle \partial_x \frac{\delta f}{\delta \Xi}\,,\, \frac{\delta h}{\delta \Pi} \Big\rangle \\
&\quad + \Big\langle \Xi\,,\, \mathrm{ad}^*_{\delta f/\delta \Pi}\,\frac{\delta h}{\delta \Xi} - \mathrm{ad}^*_{\delta h/\delta \Pi}\,\frac{\delta f}{\delta \Xi} \Big\rangle \\
&=: \{f\,,\,h\}(\Pi, \Xi)\,.
\end{aligned}$$

Assembling these equations into Hamiltonian form gives, symbolically,

$$\frac{\partial}{\partial t}\begin{bmatrix} \Pi \\ \Xi \end{bmatrix} = \begin{bmatrix} \mathrm{ad}^*_{\square}\Pi & (\mathrm{div} - \mathrm{ad}^*_{\Xi})\square \\ (\mathrm{grad} - \mathrm{ad}_{\square})\Xi & 0 \end{bmatrix} \begin{bmatrix} \delta h/\delta \Pi \\ \delta h/\delta \Xi \end{bmatrix} \tag{10.3.3}$$

The boxes $\square$ in Equation (10.3.3) indicate how the ad and ad* operations are applied in the matrix multiplication. For example,

$$\mathrm{ad}^*_\square \Pi(\delta h/\delta \Pi) = \mathrm{ad}^*_{\delta h/\delta \Pi}\Pi\,,$$

so each matrix entry acts on its corresponding vector component.[1]

Higher dimensions Although it is beyond the scope of the present text, we shall make a few short comments about the meaning of the terms appearing in the Hamiltonian matrix (10.3.3). First, the notation indicates that the natural jump to higher dimensions has been made. This is done by using the spatial gradient to define the left-invariant auxiliary variable $\Xi \equiv g^{-1}\nabla g$ in higher dimensions. The lower left entry of the matrix (10.3.3) defines the ***covariant spatial gradient,*** and its upper right entry defines the adjoint operator, the ***covariant spatial divergence.*** More explicitly, in terms of indices and partial differential operators, this Hamiltonian matrix becomes,

$$\frac{\partial}{\partial t}\begin{bmatrix}\Pi_\alpha \\ \Xi^\alpha_i\end{bmatrix} = B_{\alpha\beta}\begin{bmatrix}\delta h/\delta \Pi_\beta \\ \delta h/\delta \Xi^\beta_j\end{bmatrix}, \tag{10.3.4}$$

where the Hamiltonian structure matrix $B_{\alpha\beta}$ is given explicitly as

$$B_{\alpha\beta} = \begin{bmatrix} -\Pi_\kappa\, t^\kappa_{\alpha\beta} & \delta^\beta_\alpha \partial_j + t^\beta_{\alpha\kappa}\Xi^\kappa_j \\ \delta^\alpha_\beta \partial_i - t^\alpha_{\beta\kappa}\Xi^\kappa_i & 0 \end{bmatrix}. \tag{10.3.5}$$

Here, the summation convention is enforced on repeated indices. Superscript Greek indices refer to the Lie algebraic basis set, subscript Greek indices refer to the dual basis and Latin indices refer to the spatial reference frame. The partial derivative $\partial_j = \partial/\partial x_j$, say, acts to the right on all terms in a product by the chain rule.

[1]This is the lower right corner of the Hamiltonian matrix for a perfect complex fluid [Ho2002, GBRa2008]. It also appears in the Lie–Poisson brackets for Yang–Mills fluids [GiHoKu1982] and for spin glasses [HoKu1988].

Lie–Poisson bracket For the case that $t^{\alpha}_{\beta\kappa}$ are structure constants for the Lie algebra $so(3)$, then $t^{\alpha}_{\beta\kappa} = \epsilon_{\alpha\beta\kappa}$ with $\epsilon_{123} = +1$. By using the hat map (2.1.11), the Lie–Poisson Hamiltonian matrix in (10.3.5) may be rewritten for the $so(3)$ case in $\mathbb{R}^3$ vector form as

$$\frac{\partial}{\partial t}\begin{bmatrix} \mathbf{\Pi} \\ \mathbf{\Xi}_i \end{bmatrix} = \begin{bmatrix} \mathbf{\Pi}\times & \partial_j + \mathbf{\Xi}_j\times \\ \partial_i + \mathbf{\Xi}_i\times & 0 \end{bmatrix} \begin{bmatrix} \delta h/\delta\mathbf{\Pi} \\ \delta h/\delta\mathbf{\Xi}_j \end{bmatrix}. \tag{10.3.6}$$

Returning to one dimension, stationary solutions $\partial_t \to 0$ and spatially independent solutions $\partial_x \to 0$ both satisfy equations of the same $se(3)$ form as the heavy top. For example, the time-independent solutions satisfy, with $\mathbf{\Omega} = \delta h/\delta\mathbf{\Pi}$ and $\mathbf{\Lambda} = \delta h/\delta\mathbf{\Xi}$,

$$\frac{d}{dx}\mathbf{\Lambda} = -\,\mathbf{\Xi}\times\mathbf{\Lambda} - \mathbf{\Pi}\times\mathbf{\Omega} \quad \text{and} \quad \frac{d}{dx}\mathbf{\Omega} = -\,\mathbf{\Xi}\times\mathbf{\Omega}\,.$$

That the equations have the same form is to be expected because of the exchange symmetry under $t \leftrightarrow x$ and $\mathbf{\Omega} \leftrightarrow \mathbf{\Xi}$. Perhaps less expected is that the heavy-top form reappears.

For $G = SO(3)$ and the Lagrangian $\mathbb{R}^3 \times \mathbb{R}^3 \to \mathbb{R}$ in one spatial dimension $\ell(\mathbf{\Omega},\,\mathbf{\Xi})$ the Euler–Poincaré equation and its Hamiltonian form are given in terms of vector operations in $\mathbb{R}^3$, as follows. First, the Euler–Poincaré Equation (10.2.2) becomes

$$\frac{\partial}{\partial t}\frac{\delta\ell}{\delta\mathbf{\Omega}} = -\,\mathbf{\Omega}\times\frac{\delta\ell}{\delta\mathbf{\Omega}} - \frac{\partial}{\partial x}\frac{\delta\ell}{\delta\mathbf{\Xi}} - \mathbf{\Xi}\times\frac{\delta\ell}{\delta\mathbf{\Xi}}. \tag{10.3.7}$$

Choices for the Lagrangian

- Interesting choices for the Lagrangian include those symmetric under exchange of $\mathbf{\Omega}$ and $\mathbf{\Xi}$, such as

$$\ell_{\perp} = |\mathbf{\Omega}\times\mathbf{\Xi}|^2/2 \quad \text{and} \quad \ell_{\parallel} = (\mathbf{\Omega}\cdot\mathbf{\Xi})^2/2\,,$$

  for which the variational derivatives are, respectively,

$$\frac{\delta\ell_{\perp}}{\delta\mathbf{\Omega}} = \mathbf{\Xi}\times(\mathbf{\Omega}\times\mathbf{\Xi}) =: |\mathbf{\Xi}|^2\mathbf{\Omega}_{\perp}\,,$$
$$\frac{\delta\ell_{\perp}}{\delta\mathbf{\Xi}} = \mathbf{\Omega}\times(\mathbf{\Xi}\times\mathbf{\Omega}) =: |\mathbf{\Omega}|^2\mathbf{\Xi}_{\perp}\,,$$

for $\ell_\perp$ and the complementary quantities,

$$\frac{\delta \ell_\parallel}{\delta \boldsymbol{\Omega}} = (\boldsymbol{\Omega} \cdot \boldsymbol{\Xi})\boldsymbol{\Xi} =: |\boldsymbol{\Xi}|^2 \boldsymbol{\Omega}_\parallel ,$$
$$\frac{\delta \ell_\parallel}{\delta \boldsymbol{\Xi}} = (\boldsymbol{\Omega} \cdot \boldsymbol{\Xi})\boldsymbol{\Omega} =: |\boldsymbol{\Omega}|^2 \boldsymbol{\Xi}_\parallel ,$$

for $\ell_\parallel$. With either of these choices, $\ell_\perp$ or $\ell_\parallel$, Equation (10.3.7) becomes a local conservation law for spin angular momentum

$$\frac{\partial}{\partial t}\frac{\delta \ell}{\delta \boldsymbol{\Omega}} = -\frac{\partial}{\partial x}\frac{\delta \ell}{\delta \boldsymbol{\Xi}} .$$

The case $\ell_\perp$ is reminiscent of the ***Skyrme model*** [Sk1961], a nonlinear topological model of pions in nuclear physics.

- Another interesting choice for $G = SO(3)$ and the Lagrangian $\mathbb{R}^3 \times \mathbb{R}^3 \to \mathbb{R}$ in one spatial dimension is

$$\ell(\boldsymbol{\Omega}, \boldsymbol{\Xi}) = \frac{1}{2}\int_{-\infty}^{\infty} \boldsymbol{\Omega} \cdot \mathbb{A}\boldsymbol{\Omega} + \boldsymbol{\Xi} \cdot \mathbb{B}\boldsymbol{\Xi}\, dx ,$$

for symmetric matrices $\mathbb{A}$ and $\mathbb{B}$, which may also be L^2-symmetric differential operators. In this case the variational derivatives are given by

$$\delta \ell(\boldsymbol{\Omega}, \boldsymbol{\Xi}) = \int_{-\infty}^{\infty} \delta\boldsymbol{\Omega} \cdot \mathbb{A}\boldsymbol{\Omega} + \delta\boldsymbol{\Xi} \cdot \mathbb{B}\boldsymbol{\Xi}\, dx ,$$

and the Euler–Poincaré Equation (10.2.2) becomes

$$\frac{\partial}{\partial t}\mathbb{A}\boldsymbol{\Omega} + \boldsymbol{\Omega} \times \mathbb{A}\boldsymbol{\Omega} + \frac{\partial}{\partial x}\mathbb{B}\boldsymbol{\Xi} + \boldsymbol{\Xi} \times \mathbb{B}\boldsymbol{\Xi} = 0 . \tag{10.3.8}$$

This is the sum of two coupled rotors, one in space and one in time, again suggesting the one-dimensional spin glass, or spin chain. When $\mathbb{A}$ and $\mathbb{B}$ are taken to be the identity, Equation (10.3.8) recovers the ***chiral model***, or ***sigma model***, which is completely integrable, cf. [Wi1984, ZaMi1980].

Hamiltonian structures The Hamiltonian structures of these equations on $so(3)^*$ are obtained from the Legendre-transform relations

$$\frac{\delta \ell}{\delta \boldsymbol{\Omega}} = \boldsymbol{\Pi}\,, \quad \frac{\delta h}{\delta \boldsymbol{\Pi}} = \boldsymbol{\Omega} \quad \text{and} \quad \frac{\delta h}{\delta \boldsymbol{\Xi}} = -\frac{\delta \ell}{\delta \boldsymbol{\Xi}}\,.$$

Hence, the Euler–Poincaré Equation (10.2.2) becomes

$$\frac{\partial}{\partial t}\boldsymbol{\Pi} = \boldsymbol{\Pi} \times \frac{\delta h}{\delta \boldsymbol{\Pi}} + \frac{\partial}{\partial x}\frac{\delta h}{\delta \boldsymbol{\Xi}} + \boldsymbol{\Xi} \times \frac{\delta h}{\delta \boldsymbol{\Xi}}\,, \qquad (10.3.9)$$

and the auxiliary Equation (10.3.10) becomes

$$\frac{\partial}{\partial t}\boldsymbol{\Xi} = \frac{\partial}{\partial x}\frac{\delta h}{\delta \boldsymbol{\Pi}} + \boldsymbol{\Xi} \times \frac{\delta h}{\delta \boldsymbol{\Pi}}\,, \qquad (10.3.10)$$

which recovers the Lie–Poisson structure in Equation (10.3.6).

Finally, the reconstruction equations may be expressed using the hat map as

$$\partial_t O(t,x) = O(t,x)\widehat{\Omega}(t,x) \quad \text{and}$$
$$\partial_x O(t,x) = O(t,x)\widehat{\Xi}(t,x)\,. \qquad (10.3.11)$$

Remark 10.3.1 The Euler–Poincaré equations for the continuum spin chain discussed here and their Lie–Poisson Hamiltonian formulation provide a framework for systematically investigating three-dimensional orientation dynamics along a one-dimensional strand. These partial differential equations are interesting in their own right and they have many possible applications. For an idea of where the applications of these equations could lead, consult [SiMaKr1988,EGHPR2010].

□

Exercise. Write the Euler–Poincaré equations of the continuum spin chain for $SE(3)$, in which each point is both rotating and translating. Recall that

$$\left(\frac{d}{dt}\frac{\delta l}{\delta \xi}\,, \frac{d}{dt}\frac{\delta l}{\delta \alpha}\right) = \mathrm{ad}^*_{(\xi\,,\,\alpha)}\left(\frac{\delta l}{\delta \xi}\,, \frac{\delta l}{\delta \alpha}\right), \qquad (10.3.12)$$

where the ad^* operation of the Lie algebra $se(3)$ on its dual $se(3)^*$ is given in Equations (6.3.2) and (6.3.3).

Apply formula (10.3.12) to express the space-time Euler–Poincaré Equation (10.2.2) for $SE(3)$ in vector form.

Complete the computation of the Lie–Poisson Hamiltonian form for the continuum spin chain on $SE(3)$. ★

Exercise. Let the set of 2×2 matrices M_i with $i = 1, 2, 3$ satisfy the defining relation for the symplectic Lie group $Sp(2)$,

$$M_i J M_i^T = J \quad \text{with} \quad J = \begin{pmatrix} 0 & -1 \\ 1 & 0 \end{pmatrix}. \tag{10.3.13}$$

The corresponding elements of its Lie algebra $m_i = \dot{M}_i M_i^{-1} \in sp(2)$ satisfy $(Jm_i)^T = Jm_i$ for each $i = 1, 2, 3$. Thus, $\mathsf{X}_i = Jm_i$ satisfying $\mathsf{X}_i^T = \mathsf{X}_i$ is a set of three symmetric 2×2 matrices. Define $\mathsf{X} = J\dot{M}M^{-1}$ with time derivative $\dot{M} = \partial M(t,x)/\partial t$ and $\mathsf{Y} = JM'M^{-1}$ with space derivative $M' = \partial M(t,x)/\partial x$. Then show that

$$\mathsf{X}' = \dot{\mathsf{Y}} + [\mathsf{X}, \mathsf{Y}]_J\,, \tag{10.3.14}$$

for the J-bracket defined by

$$[\mathsf{X}, \mathsf{Y}]_J := \mathsf{X}J\mathsf{Y} - \mathsf{Y}J\mathsf{X} =: 2\text{sym}(\mathsf{X}J\mathsf{Y}) =: \text{ad}^J_{\mathsf{X}}\mathsf{Y}\,.$$

In terms of the J-bracket, compute the continuum Euler–Poincaré equations for a Lagrangian $\ell(\mathsf{X}, \mathsf{Y})$ defined on the symplectic Lie algebra $\mathfrak{sp}(2)$.

Compute the Lie–Poisson Hamiltonian form of the system comprising the continuum Euler–Poincaré equations on $\mathfrak{sp}(2)^*$ and the compatibility equation (10.3.14) on $\mathfrak{sp}(2)$. ★

11 MOMENTUM MAPS

Contents

11.1 The momentum map 222

11.2 Cotangent lift 224

11.3 Examples of momentum maps 226

11.3.1 The Poincaré sphere $S^2 \in S^3$ 237

11.3.2 Overview 242

11.1 The momentum map

A momentum map $J : M \mapsto \mathfrak{g}^*$ arises when the smooth Lie group action of G on a manifold M preserves either the symplectic structure, or the Poisson structure on M. Here $\mathfrak{g}$ is the Lie algebra of G and $\mathfrak{g}^*$ is its dual. We concentrate on the situation in which $M = T^*Q$ is the cotangent bundle of a configuration manifold Q on which the Lie group G acts smoothly.

Example 11.1.1 *An example of a momentum map $J : M \mapsto \mathfrak{g}^*$ is the quantity $J(p, q)$ defined by*

$$\begin{aligned} J^\xi(p,q) &:= \Big\langle J(p,q),\, \xi \Big\rangle_{\mathfrak{g}^*\times\mathfrak{g}} \\ &= \Big\langle\!\Big\langle (p,q),\, \xi_Q(q) \Big\rangle\!\Big\rangle_{T^*Q\times TQ} \\ &= \Big\langle\!\Big\langle p,\, \pounds_\xi q \Big\rangle\!\Big\rangle_{T^*Q\times TQ} . \end{aligned} \tag{11.1.1}$$

In this formula, the infinitesimal action of a Lie group G by an element $\xi \in \mathfrak{g}$ of its Lie algebra is expressed as a Lie derivative $\pounds_\xi q = \xi_Q(q)$ on the configuration space Q. The momentum map lifts this expression into phase space $M = T^*Q$ by expressing it as a Hamiltonian vector field $X_{J^\xi(p,q)} = \{\,\cdot\,,\, J^\xi(p,q)\}$ on T^*Q.

Denote the action of the Lie group G on the configuration manifold Q as $q(s) = g(s)q(0)$ for $g(s) \in G$, $s \in \mathbb{R}$ and $q \in Q$. As usual, a vector field $\xi_Q(q) \in TQ$ at a point $q \in Q$ is obtained by differentiating $q(s) = g(s)q(0)$ with respect to s in the direction ξ at the identity $s = 0$, where $g(0) = e$. That is,

$$\xi_Q(q) = q'(s)\big|_{s=0} = g'(s)q(0)\big|_{s=0} = (g'g^{-1})\big|_{s=0}q(0) =: \xi q\,,$$

for $q = q(0)$. In other notation, the vector field $\xi_Q(q) \in TQ$ may be expressed as a Lie derivative,

$$\xi_Q(q) = \frac{d}{ds}\big[\exp(s\xi)q\big]\Big|_{s=0} =: \pounds_\xi q = \xi q \in TQ\,. \tag{11.1.2}$$

Remark 11.1.1 The formula for $\xi_Q(q) \in TQ$ is the ***tangent lift*** of the action of G on Q at $q \in Q$. The tangent lift action of G on TQ induces an action of G on T^*Q by the ***cotangent lift*** (the inverse transpose of the tangent lift). The cotangent lift action of G on T^*Q is always symplectic; so it may be written using canonical coordinates $(p,q) \in T^*Q$ as a Hamiltonian vector field $X_{J^\xi} = \{\,\cdot\,,\, J^\xi(p,q)\,\}$. □

For the case when the symplectic manifold M is the cotangent bundle T^*Q of a configuration manifold Q, the quantity

$$J^\xi(p,q) := \Big\langle J(p,q),\, \xi \Big\rangle_{\mathfrak{g}^*\times\mathfrak{g}}$$

is the Hamiltonian on T^*Q. The canonical Poisson bracket

$$X_{J^\xi} = \{\,\cdot\,,\, J^\xi(p,q)\,\}$$

is the Hamiltonian vector field X_{J^ξ} for the infinitesimal action of the Lie group G on Q (configuration space), lifted to the cotangent bundle T^*Q (phase space) with symplectic form $\omega = dq \wedge dp$. Equivalently,

$$dJ^\xi(p,q) = X_{J^\xi} \rfloor \omega = \omega(X_{J^\xi}, \cdot\,)\,.$$

This property defines the standard momentum map.

Definition 11.1.1 (Standard momentum map) *The standard momentum map $J : T^*Q \mapsto \mathfrak{g}^*$ is defined by requiring*

$$X_{J^\xi(p,q)} = \xi_{(p,q)} \quad \textit{for each} \quad \xi \in \mathfrak{g}\,. \tag{11.1.3}$$

That is, the infinitesimal generator $\xi_{(p,q)}$ of the cotangent lift action of G for each element $\xi \in \mathfrak{g}$ is equal to the Hamiltonian vector field

$$X_{J^\xi(p,q)} = \{\,\cdot\,,\, J^\xi(p,q)\,\}\,, \tag{11.1.4}$$

of the function $J^\xi : T^*Q \mapsto \mathbb{R}$. The momentum map $J : T^*Q \mapsto \mathfrak{g}^*$ is defined as $J(p,q)$, satisfying

- *$J^\xi(p,q) := \langle J(p,q),\, \xi\rangle$ for each $\xi \in \mathfrak{g}$ and any $(p,q) \in T^*Q$;*
- *$\langle\,\cdot\,,\,\cdot\,\rangle$ is the natural pairing $\mathfrak{g}^* \times \mathfrak{g} \mapsto \mathbb{R}$; and*
- *$\{\,\cdot\,,\,\cdot\,\} : \mathcal{F}(p,q) \times \mathcal{F}(p,q) \mapsto \mathcal{F}(p,q)$ is the canonical Poisson bracket on T^*Q.*

11.2 Cotangent lift

Theorem 11.2.1 (Cotangent-lift momentum map) *The cotangent-lift momentum map $J : T^*Q \mapsto \mathfrak{g}^*$ satisfies*

$$
\begin{aligned}
J^\xi(p,q) &:= \Big\langle J(p,q),\, \xi \Big\rangle \\
&= \Big\langle\!\Big\langle (p,q),\, \xi_Q(q) \Big\rangle\!\Big\rangle \\
&= \Big\langle\!\Big\langle p,\, \pounds_\xi q \Big\rangle\!\Big\rangle \\
&=: \Big\langle q \diamond p,\, \xi \Big\rangle . \qquad (11.2.1)
\end{aligned}
$$

Proof. The first equality repeats the definition of $J^\xi(p,q)$. The second equality inserts the definition of the infinitesimal action $\xi_Q(q) \in TQ$ of the Lie group G on Q at the point $q \in Q$. The pairing $\langle\!\langle \cdot\,, \cdot \rangle\!\rangle : T^*Q \times TQ \mapsto \mathbb{R}$ in this equality is between the tangent and cotangent spaces of the configuration Q. The third equality inserts the definition of the infinitesimal action $\xi_Q(q) = \pounds_\xi q$ in terms of the Lie derivative. The last equality provides the required Hamiltonian $J^\xi(p,q) = \langle q \diamond p,\, \xi\rangle$ for the Hamiltonian vector field $X_{J^\xi} = \{\,\cdot\,,\, J^\xi(p,q)\,\}$ in the cotangent lift action of G on T^*Q by defining the diamond operation ($\diamond$) in terms of the two pairings and the Lie derivative. ■

Remark 11.2.1 The diamond operation was introduced in (6.2.9) and in the Clebsch Equation (9.3.8), where it was defined using the dual of the Lie derivative with respect to the $T^*Q \times TQ$ pairing induced by the variational derivative in q, namely,

$$
\Big\langle q \diamond p\,,\, \xi \Big\rangle_{\mathfrak{g}^* \times \mathfrak{g}} = \Big\langle\!\Big\langle p\,,\, \pounds_\xi q \Big\rangle\!\Big\rangle_{T^*Q \times TQ} . \qquad (11.2.2)
$$

Thus, as discussed in Section 9.3.2, the variational relation in the Clebsch procedure associated with the dynamical Clebsch constraint, or reconstruction Equation (9.3.4), defines the corresponding cotangent-lift momentum map in Equation (9.4.3). (The sign is correct because $(\diamond)$ is antisymmetric.) □

Theorem 11.2.2 (Hamiltonian Noether's theorem) *If the Hamiltonian $H(p,q)$ on T^*Q is invariant under the action of the Lie group G, then $J^\xi(p,q)$ is conserved on trajectories of the corresponding Hamiltonian vector field,*

$$X_H = \{\,\cdot\,,\, H(p,q)\}\,.$$

Proof. Differentiating the invariance condition $H(gp, gq) = H(p,q)$ with respect to g for fixed $(p,q) \in T^*Q$ yields

$$\begin{aligned} \pounds_\xi H(p,q) &= dH(p,q)\cdot \xi_{(p,q)} = 0 = X_{J^\xi(p,q)} H(p,q) \\ &= -\{J^\xi\,,\,H\}(p,q) = -\,X_{H(p,q)} J^\xi(p,q)\,. \end{aligned}$$

Consequently, the momentum map $J^\xi(p,q)$ is conserved on trajectories of the Hamiltonian vector field $X_H = \{\,\cdot\,,\, H(p,q)\}$ for a G-invariant Hamiltonian. ■

Proposition 11.2.1 (Equivariant group actions)

- *A group action $\Phi_g : G \times T^*Q \mapsto T^*Q$ is said to be* ***equivariant*** *if it satisfies*

$$J \circ \Phi_g = \mathrm{Ad}^*_{g^{-1}} \circ J\,.$$

This means the following diagram commutes:

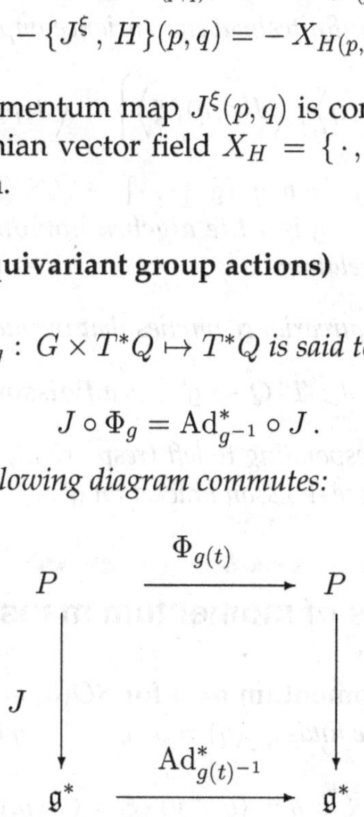

- *Equivariance implies* ***infinitesimal equivariance****. Namely,*

$$\frac{d}{dt}\bigg|_{t=0} J(\Phi_{g(t)}(z)) = \frac{d}{dt}\bigg|_{t=0} \mathrm{Ad}^*_{g^{-1}} \circ J(z)$$

implies

$$dJ(z) \cdot \xi_P(z) = -\mathrm{ad}^*_\xi J(z)\,,$$

with $z = (p, q)$. Setting $dJ(z) \cdot \xi_P(z) = X_{J^\xi} J$ and pairing with a fixed Lie algebra element η yields the η-component:

$$\begin{aligned}\Big\langle dJ(z) \cdot \xi_P(z)\,,\, \eta \Big\rangle &= \Big\langle -\mathrm{ad}^*_\xi J(z)\,,\, \eta \Big\rangle\,, \\ X_{J^\xi} J^\eta &= \Big\langle J(z)\,,\, -\mathrm{ad}_\xi \eta \Big\rangle\,, \\ \Big\{ J^\eta(z)\,,\, J^\xi(z) \Big\} &= \Big\langle J(z)\,,\, [\eta\,,\, \xi] \Big\rangle\,.\end{aligned} \tag{11.2.3}$$

- *Consequently, infinitesimal equivariance implies*

$$\Big\{ \Big\langle J(p,q),\, \eta \Big\rangle,\, \Big\langle J(p,q),\, \xi \Big\rangle \Big\} = \Big\langle J(p,q),\, [\eta\,,\, \xi] \Big\rangle. \tag{11.2.4}$$

This means that the map $(\mathfrak{g}, [\,\cdot\,,\,\cdot\,] \to (C^\infty(T^*Q), \{\,\cdot\,,\,\cdot\,\})$ defined by $\xi \mapsto J^\xi$, $\xi \in \mathfrak{g}$ is a ***Lie algebra homomorphism*** *(i.e., it preserves bracket relations).*

- *Infinitesimal equivariance implies that the momentum map*

$$J : T^*Q \mapsto \mathfrak{g}^* \quad \text{is a } \textbf{Poisson map}.$$

That is, J corresponding to left (resp., right) group action produces a $+$ (resp., $-$) Lie–Poisson bracket on $\mathfrak{g}^*$.

11.3 Examples of momentum maps

Example 11.3.1 (Momentum map for $SO(3)$ acting on $\mathbb{R}^3$) *For $Q = \mathbb{R}^3$ and $\mathfrak{g} = so(3)$ one finds $\xi_Q(q) = \pounds_\xi q = \xi \times q$ by the hat map and*

$$\Big\langle\!\Big\langle p,\, \pounds_\xi q \Big\rangle\!\Big\rangle = p \cdot \xi \times q = (q \times p) \cdot \xi = \Big\langle J(p,q),\, \xi \Big\rangle = J^\xi(p,q)\,,$$

which is the Hamiltonian for an infinitesimal rotation around ξ in $\mathbb{R}^3$. In the case that $\mathfrak{g} = so(3)$, the pairings $\langle\,\cdot\,,\,\cdot\,\rangle$ and $\langle\!\langle\,\cdot\,,\,\cdot\,\rangle\!\rangle$ may both be taken as dot products of vectors in $\mathbb{R}^3$, the momentum map $J(p,q) = q \diamond p = q \times p \in \mathbb{R}^3$ is the phase-space expression for angular momentum and the $\diamond$ operation is $\times$, the cross product of vectors in $\mathbb{R}^3$. This is an example of a ***cotangent-lift*** *momentum map.*

Example 11.3.2 (Momentum map for $SU(2)$ acting on $\mathbb{C}^2$) *The Lie group $SU(2)$ of complex 2×2 unitary matrices $U(s)$ with unit determinant acts on $\mathbf{a}\in\mathbb{C}^2$ by matrix multiplication as*

$$\mathbf{a}(s) = U(s)\mathbf{a}(0) = \exp(is\xi)\mathbf{a}(0)\,,$$

in which $i\xi = U'U^{-1}|_{s=0}$ is a 2×2 traceless skew-Hermitian matrix, as seen from the following:

$$UU^\dagger = Id \quad implies \quad U'U^\dagger + UU'^{\,\dagger} = 0 = U'U^\dagger + (U'U^\dagger)^\dagger\,.$$

Likewise, ξ alone (that is, not multiplied by i) is a 2×2 traceless Hermitian matrix.

The infinitesimal generator $\xi(\mathbf{a}) \in \mathbb{C}^2$ may be expressed as a linear transformation,

$$\xi(\mathbf{a}) = \frac{d}{ds}\Big[\exp(is\xi)\mathbf{a}\Big]\Big|_{s=0} = i\xi\mathbf{a}\,,$$

in which the product $(\xi\mathbf{a})$ of the Hermitian matrix (ξ) and the two-component complex vector $(\mathbf{a})$ has components $\xi_{kl}a_l$, with $k,l = 1,2$.

To be a momentum map, $J : \mathbb{C}^2 \mapsto su(2)^*$ must satisfy the defining relation (11.2.1),

$$\begin{aligned} J^\xi(\mathbf{a}) &:= \Big\langle J(\mathbf{a}),\,\xi\Big\rangle_{su(2)^*\times su(2)} = \Big\langle\!\Big\langle \mathbf{a},\,\xi(\mathbf{a})\Big\rangle\!\Big\rangle_{\mathbb{C}^2} = \Big\langle\!\Big\langle \mathbf{a},\,i\xi\mathbf{a}\Big\rangle\!\Big\rangle_{\mathbb{C}^2} \\ &= \mathrm{Im}(a_k^*(i\xi)_{kl}a_l) = a_k^*\xi_{kl}a_l = \mathrm{tr}\big((\mathbf{a}\otimes\mathbf{a}^*)\xi\big) = \mathrm{tr}\big(Q^\dagger\xi\big)\,. \end{aligned}$$

Being traceless, ξ has zero pairing with any multiple of the identity; so one may subtract the trace of $Q = \mathbf{a}\otimes\mathbf{a}^*$. Thus, the traceless Hermitian quantity

$$J(\mathbf{a}) = Q - \frac{1}{2}\,(\mathrm{tr}\;Q)\,\mathrm{Id} = \mathbf{a}\otimes\mathbf{a}^* - \frac{1}{2}\,\mathrm{Id}\;|\mathbf{a}|^2 \in su(2)^* \qquad (11.3.1)$$

defines a momentum map $J : \mathbb{C}^2 \mapsto su(2)^*$. *That is,* J *maps* $\mathbf{a} \in \mathbb{C}^2$ *to the traceless Hermitian matrix* $J(\mathbf{a})$, *which is an element of* $su(2)^*$, *the dual space to* $su(2)$ *under the pairing* $\langle \cdot\,, \cdot \rangle : su(2)^* \times su(2) \mapsto \mathbb{R}$ *given by the trace of the matrix product,*

$$\Big\langle J,\, \xi \Big\rangle_{su(2)^* \times su(2)} = \mathrm{tr}\big(J(\mathbf{a})^\dagger \xi\big)\,, \tag{11.3.2}$$

$$\textit{for}\quad J(\mathbf{a}) = J(\mathbf{a})^\dagger \in su(2)^* \quad \textit{and} \quad i\xi \in su(2)\,. \tag{11.3.3}$$

Proposition 11.3.1 (Momentum map equivariance) *Let* $U \in SU(2)$ *and* $\mathbf{a} \in \mathbb{C}^2$. *The momentum map for* $SU(2)$ *acting on* $\mathbb{C}^2$ *defined by*

$$\Big\langle J(\mathbf{a}),\, \xi \Big\rangle_{su(2)^* \times su(2)} = \Big\langle\!\Big\langle \mathbf{a},\, i\xi\mathbf{a} \Big\rangle\!\Big\rangle_{\mathbb{C}^2} \tag{11.3.4}$$

is equivariant. That is,

$$J(U\mathbf{a}) = \mathrm{Ad}^*_{U^{-1}} J(\mathbf{a}).$$

Proof. Substitute $\mathrm{Ad}_{U^{-1}}\xi$ into the momentum map definition,

$$\begin{aligned}
\Big\langle \mathrm{Ad}^*_{U^{-1}} J(\mathbf{a})\,,\, \xi \Big\rangle_{su(2)^* \times su(2)} &= \Big\langle J(\mathbf{a})\,,\, \mathrm{Ad}_{U^{-1}}\xi \Big\rangle_{su(2)^* \times su(2)} \\
&= \Big\langle\!\Big\langle \mathbf{a},\, U^\dagger i\xi U\mathbf{a} \Big\rangle\!\Big\rangle_{\mathbb{C}^2} \\
&= \Big\langle\!\Big\langle (U\mathbf{a})\,,\, i\xi(U\mathbf{a}) \Big\rangle\!\Big\rangle_{\mathbb{C}^2} \\
&= \Big\langle J(U\mathbf{a})\,,\, \xi \Big\rangle_{su(2)^* \times su(2)}.
\end{aligned}$$

Therefore, $J(U\mathbf{a}) = \mathrm{Ad}^*_{U^{-1}} J(\mathbf{a})$, as claimed. ∎

Remark 11.3.1 (Poincaré sphere momentum map) Looking at Equation (3.2.16) reveals that the momentum map $\mathbb{C}^2 \mapsto su(2)^*$ for the action of $SU(2)$ acting on $\mathbb{C}^2$ in Equation (11.3.1) is a component of the map $\mathbb{C}^2 \mapsto S^2$ to the Poincaré sphere, which defines the Hopf fibration $S^3 \simeq S^2 \times S^1$. To see this, one simply replaces $\xi \in su(2)$

with the vector of Pauli matrices σ in Equation (3.2.18) to find

$$J(\mathrm{a}) = \frac{1}{2}\mathrm{n}\cdot\boldsymbol{\sigma} \quad \text{in which} \quad \mathrm{n} = \operatorname{tr} Q\,\boldsymbol{\sigma} = a_k^*\boldsymbol{\sigma}_{kl}a_l\,. \qquad (11.3.5)$$

Thus, the quadratic S^1-invariant quantities comprising the components of the unit vector $\mathrm{n} = (n_1, n_2, n_3)$ given by

$$n_1 + i\,n_2 = 2a_1a_2^*,\ n_3 = |a_1|^2 - |a_2|^2 \qquad (11.3.6)$$

on the Poincaré sphere $|\mathrm{n}|^2 = n_0^2$ in (3.2.20) are precisely the components of the momentum map in Equation (11.3.1). □

Exercise. Compute the Lie–Poisson brackets among the components of the unit vector $\mathrm{n} = (n_1, n_2, n_3)$ by using their definitions in terms of $\mathrm{a} \in \mathbb{C}^2$ and applying the canonical Poisson brackets $\{a_k^*, a_l\} = 2i\delta_{kl}$. ★

Answer.

$$\{n_i, n_j\} = 4\epsilon_{ijk}n_k\,.$$

▲

Remark 11.3.2 (Cotangent-lift momentum maps) The formula determining the momentum map for the cotangent lift action of a Lie group G on a smooth manifold Q may be expressed in terms of the pairings

$$\langle\,\cdot\,,\cdot\,\rangle : \mathfrak{g}^*\times\mathfrak{g} \mapsto \mathbb{R} \quad \text{and} \quad \langle\!\langle\,\cdot\,,\cdot\,\rangle\!\rangle : T^*M \times TM \mapsto \mathbb{R}\,,$$

as

$$\Big\langle J(p,q)\,,\,\xi\Big\rangle_{\mathfrak{g}^*\times\mathfrak{g}} = \Big\langle q\diamond p,\,\xi\Big\rangle_{\mathfrak{g}^*\times\mathfrak{g}} = \Big\langle\!\!\Big\langle\, p\,,\,\pounds_\xi q\,\Big\rangle\!\!\Big\rangle_{T^*M\times TM}, \qquad (11.3.7)$$

where $(q,p) \in T_q^*M$ and $\pounds_\xi q \in T_qM$ is the infinitesimal generator of the action of the Lie algebra element ξ on the coordinate q. □

Proposition 11.3.2 (Equivariance of cotangent lifts) *Cotangent-lift momentum maps (11.3.7) are equivariant. That is,*

$$J(g \cdot p, g \cdot q) = \mathrm{Ad}^*_{g^{-1}} J(p, q) \,, \qquad (11.3.8)$$

where $(g \cdot p, g \cdot q)$ denotes the cotangent lift to T^*M of the action of G on manifold M.

Proof. The proof follows from Remark 4.2.2, that $\mathrm{Ad}^*_{g^{-1}}$ is a representation of the coAdjoint action Φ^*_g of the group G on its dual Lie algebra $\mathfrak{g}^*$. This means that $\mathrm{Ad}^*_{g^{-1}}(q \diamond p) = (g \cdot q \diamond g \cdot p)$, and we have

$$\begin{aligned}
\Big\langle \mathrm{Ad}^*_{g^{-1}} J(p, q) \,,\, \xi \Big\rangle_{\mathfrak{g}^* \times \mathfrak{g}} &= \Big\langle \mathrm{Ad}^*_{g^{-1}}(q \diamond p) \,,\, \xi \Big\rangle_{\mathfrak{g}^* \times \mathfrak{g}} \\
&= \Big\langle g \cdot q \diamond g \cdot p \,,\, \xi \Big\rangle_{\mathfrak{g}^* \times \mathfrak{g}} \\
&= \Big\langle J(g \cdot p, g \cdot q) \,,\, \xi \Big\rangle_{\mathfrak{g}^* \times \mathfrak{g}} \,. \qquad (11.3.9)
\end{aligned}$$

Thus, Equation (11.3.8) holds and cotangent-lift momentum maps are equivariant. ■

Importance of equivariance

Equivariance of a momentum map is important, because Poisson brackets among the components of an equivariant momentum map close among themselves and satisfy the Jacobi identity. That is, the following theorem holds.

Theorem 11.3.1 *Equivariant momentum maps are Poisson.*

Proof. As we know, a momentum map $J : P \to \mathfrak{g}^*$ is equivariant, if

$$J \circ \Phi_{g(t)} = \mathrm{Ad}^*_{g(t)^{-1}} \circ J \,,$$

for any curve $g(t) \in G$. As discussed earlier, the time derivative of the equivariance relation leads to the infinitesimal equivariance relation,

$$\{\langle J\,,\,\xi\rangle\,,\,\langle J\,,\,\eta\rangle\} = \big\langle J\,,\,[\xi,\,\eta]\big\rangle\,, \tag{11.3.10}$$

where $\xi,\,\eta \in \mathfrak{g}$ and $\{\,\cdot\,,\,\cdot\,\}$ denotes the Poisson bracket on the manifold P. This in turn implies that the momentum map preserves Poisson brackets in the sense that

$$\{F_1 \circ J\,,\,F_2 \circ J\} = \{F_1\,,\,F_2\}_{LP} \circ J\,, \tag{11.3.11}$$

for all $F_1, F_2 \in \mathcal{F}(\mathfrak{g}^*)$, where $\{F_1\,,\,F_2\}_{LP}$ denotes the Lie–Poisson bracket for the appropriate left or right action of $\mathfrak{g}$ on P. That is, equivariance implies infinitesimal equivariance, which is sufficient for the momentum map to be Poisson. ■

Exercise. (Compute N-dimensional momentum maps) Define appropriate pairings and determine the momentum maps explicitly for the following actions:

(i) $\pounds_\xi q = \xi \times q$ for $\mathbb{R}^3 \times \mathbb{R}^3 \mapsto \mathbb{R}^3$.

(ii) $\pounds_\xi q = \mathrm{ad}_\xi q$ for adjoint action $\mathrm{ad} : \mathfrak{g} \times \mathfrak{g} \mapsto \mathfrak{g}$ in a Lie algebra $\mathfrak{g}$.

(iii) AqA^{-1} for $A \in GL(3,R)$ acting on $q \in GL(3,R)$ by matrix conjugation.

(iv) Aq for left action of $A \in SO(3)$ on $q \in SO(3)$.

(v) AqA^T for $A \in GL(3,R)$ acting on $q \in Sym(3)$, that is $q = q^T$.

(vi) Adjoint action of the Lie algebra of the semidirect-product group $SL(2,\mathbb{R})Ⓢ\mathbb{R}^2$ on itself. See Section 6.5 for notation and coadjoint actions.

★

Answer.

(i) For the pairing by scalar product of vectors, one writes

$$\left\langle\!\left\langle p\,,\, \pounds_\xi q \right\rangle\!\right\rangle_{T^*M\times TM} = p\cdot\xi\times q = q\times p\cdot\xi\,,$$

so that the momentum map for the spatial rotation (2.1.13) is

$$J = q\times p\,. \qquad (11.3.12)$$

(ii) Similarly, for the pairing $\langle\cdot\,,\,\cdot\rangle : \mathfrak{g}^*\times\mathfrak{g}\mapsto\mathbb{R}$,

$$\langle p,\, \mathrm{ad}_\xi q\rangle = -\langle\, \mathrm{ad}^*_q p\,,\,\xi\rangle \quad\Rightarrow\quad J = -\,\mathrm{ad}^*_q p\,.$$

(iii) Compute the ad action for $GL(3,R)$ conjugation as

$$T_e(AqA^{-1}) = \xi q - q\xi = [\xi, q]\,,$$

for $\xi = A'(0)\in gl(3,R)$ acting on $q\in GL(3,R)$ by matrix Lie bracket $[\cdot\,,\,\cdot]$. For the matrix pairing

$$\langle\, A\,,\,B\,\rangle = \mathrm{tr}(A^TB)\,,$$

one finds the momentum map,

$$\mathrm{tr}(p^T[\xi,q]) = \mathrm{tr}\Big((pq^T - q^Tp)^T\xi\Big) \;\Rightarrow\; J = pq^T - q^Tp\,.$$

(iv) Compute $T_e(Aq) = \xi q$ for $\xi = A'(0)\in so(3)$ acting on $q\in SO(3)$ by left matrix multiplication. For the matrix pairing $\langle\, A\,,\,B\,\rangle = \mathrm{trace}(A^TB)$, one finds the following expression for the momentum map,

$$\mathrm{trace}(p^T\xi q) = \mathrm{trace}((pq^T)^T\xi) \;\Rightarrow\; J = \frac{1}{2}(pq^T - qp^T)\,,$$

upon using antisymmetry of the matrix $\xi\in so(3)$.

(v) Compute

$$T_e(AqA^T) = \xi q + q\xi^T$$

for $\xi = A'(0) \in gl(3, R)$ acting on $q \in Sym(3)$. For the matrix pairing

$$\begin{aligned} \langle\, A\,,\, B\,\rangle &= \mathrm{tr}(A^T B), \text{ one finds} \\ \mathrm{tr}(p^T(\xi q + q\xi^T)) &= \mathrm{tr}(q(p^T + p)\xi) \\ &= \mathrm{tr}(2qp)^T\xi) \Rightarrow J = 2qp \end{aligned}$$

upon using symmetry of the matrix $\xi q + q\xi^T$ to choose $p = p^T$. (The momentum canonical to the symmetric matrix $q = q^T$ should be symmetric, in order to have the correct number of components.)

(vi) For the pairing $\langle\,\cdot\,,\,\cdot\,\rangle : \mathfrak{g}^* \times \mathfrak{g} \mapsto \mathbb{R}$,

$$\langle\, p\,,\, \mathrm{ad}_\xi q\,\rangle = -\,\langle\, \mathrm{ad}^*_q\, p\,,\, \xi\,\rangle \quad \Rightarrow \quad J = -\,\mathrm{ad}^*_q\, p\,.$$

From Equation (6.5.22), this is

$$J = -\,\mathrm{ad}^*_{(A,h)}\begin{pmatrix} D \\ k \end{pmatrix} = -\begin{pmatrix} [D, A] - hk + \frac{1}{2}(\mathbf{h}\cdot\mathbf{k})\mathbb{1} \\ kA \end{pmatrix}$$

for $(A, h) \in \mathfrak{g}$ and $(D, k) \in \mathfrak{g}^*$. ▲

Exercise. (Unitary transformations of Hermitian matrices) Consider the manifold $\mathcal{Q}$ of $n \times n$ Hermitian matrices, so that $Q^\dagger = Q$ for $Q \in \mathcal{Q}$. The Poisson (symplectic) manifold is $T^*\mathcal{Q}$, whose elements are pairs (Q, P) of Hermitian matrices. The corresponding Poisson bracket is

$$\{F, H\} = \mathrm{tr}\left(\frac{\partial F}{\partial Q}\frac{\partial H}{\partial P} - \frac{\partial H}{\partial Q}\frac{\partial F}{\partial P}\right).$$

Let G be the group $U(n)$ of $n \times n$ unitary matrices. The group G acts on $T^*\mathcal{Q}$ through

$$(Q, P) \mapsto (UQU^\dagger, UPU^\dagger)\,, \quad UU^\dagger = Id\,.$$

(i) What is the linearisation of this group action?

(ii) What is its momentum map?

(iii) Is this momentum map equivariant?

(iv) Is this momentum map conserved by the Hamiltonian $H = \frac{1}{2}\mathrm{tr}\, P^2$? Prove it.

(v) What changes occur in the solution for *orthogonal* transformations of the manifold of $n \times n$ *symmetric* matrices, instead?

★

Answer. (Unitary transformations of Hermitian matrices)

(i) The linearisation of this group action with $U = \exp(t\xi)$, with skew-Hermitian $\xi^\dagger = -\xi$, yields the vector field with (Q, P) components,

$$X_\xi = \Big([\xi, Q], [\xi, P] \Big) .$$

(ii) This is the Hamiltonian vector field for

$$H_\xi = \mathrm{tr}\big([Q, P]\xi \big) ,$$

thus yielding the momentum map $J(Q, P) = [Q, P]$.

That is, the momentum map for the lifted action of the unitary transformations on the phase space $(Q, P) \in T^*\mathcal{Q}$ of the Hermitian matrices is the matrix commutator $[Q, P]$.

This is entirely natural from the viewpoint of quantum mechanics, in which the commutator $[Q, P]$ is responsible for the ***uncertainty principle***.

(iii) Being defined by a cotangent lift, this momentum map is equivariant.

(iv) For $H = \frac{1}{2}\mathrm{tr}\, P^2$,

$$\Big\{[Q,P],H\Big\} = \mathrm{tr}\left(\frac{\partial[Q,P]}{\partial Q}\frac{\partial H}{\partial P}\right) = \mathrm{tr}\Big(P^2 - P^2\Big) = 0\,,$$

so the momentum map $J(Q,P) = [Q,P]$ is conserved by this Hamiltonian.

Alternatively, one may simply observe that the map

$$(Q,P) \mapsto (UQU^\dagger, UPU^\dagger)\,, \quad UU^\dagger = Id\,,$$

preserves $\mathrm{tr}(P^2)$, since it takes

$$\mathrm{tr}(P^2) \mapsto \mathrm{tr}(UPU^\dagger UPU^\dagger) = \mathrm{tr}(P^2)\,.$$

(v) The computation for orthogonal transformations of the manifold of $n \times n$ symmetric matrices is entirely analogous, except for minor changes in interpretation.

▲

Example 11.3.3 (The $1\!:\!1$ **resonance [Ku1978])** *This example extends the* $\mathbb{C}^2 \mapsto su(2)^*$ *momentum map (11.3.1) to* $\mathbb{C}^2 \mapsto u(2)^*$ *and thereby completes the relation of the momentum map to the Poincaré sphere and Hopf fibration.*

A unitary 2×2 *matrix* $U(s)$ *acts on a complex two-vector* $\mathbf{a} \in \mathbb{C}^2$ *by matrix multiplication as*

$$\mathbf{a}(s) = U(s)\mathbf{a}(0) = \exp(is\xi)\mathbf{a}(0)\,,$$

in which $i\xi = U'U^{-1}|_{s=0}$ *is a* 2×2 *skew-Hermitian matrix. Therefore, the infinitesimal generator* $\xi(\mathbf{a}) \in \mathbb{C}^2$ *may be expressed as a linear transformation,*

$$\xi(\mathbf{a}) = \frac{d}{ds}\big[\exp(is\xi)\mathbf{a}\big]\Big|_{s=0} = i\xi\mathbf{a}\,,$$

in which the matrix $\xi^\dagger = \xi$ *is Hermitian.*

Definition 11.3.1 (Momentum map $J: \mathbb{C}^2 \mapsto u(2)^*$) *The momentum map $J(\mathbf{a}) : \mathbb{C}^2 \mapsto u(2)^*$ for the matrix action of $U(2)$ on $\mathbb{C}^2$ is defined by*

$$\begin{aligned} J^\xi(\mathbf{a}) &:= \big\langle J(\mathbf{a}),\, \xi \big\rangle_{u(2)} = \frac{i}{2}\langle\!\langle \mathbf{a},\, \xi(\mathbf{a}) \rangle\!\rangle_{\mathbb{C}^2} \\ &= \frac{1}{2}\omega(\mathbf{a}, \xi(\mathbf{a})) \quad \textit{with} \quad \xi(\mathbf{a}) = i\xi\mathbf{a}, \end{aligned} \tag{11.3.13}$$

and $\xi^\dagger = \xi$. The $\mathbb{C}^2$ pairing $\langle\!\langle \cdot\,,\, \cdot \rangle\!\rangle_{\mathbb{C}^2}$ in this map is the Hermitian pairing, which for skew-Hermitian $\xi(\mathbf{a})^\dagger = -\,\xi(\mathbf{a})$ is also the canonical symplectic form, $\omega(\mathbf{a}, \mathbf{b}) = \mathrm{Im}(\mathbf{a}^* \cdot \mathbf{b})$ on $\mathbb{C}^2$, as discussed in [MaRa1994]. Thus,

$$\begin{aligned} 2J^\xi(\mathbf{a}) &:= \omega(\mathbf{a}, \xi(\mathbf{a})) = \omega(\mathbf{a}, i\xi\mathbf{a}) \\ &= \mathrm{Im}(a_k^*(i\xi)_{kl}a_l) \\ &= a_k^*\xi_{kl}a_l \\ &= \mathrm{tr}\big((\mathbf{a}\otimes\mathbf{a}^*)\,\xi\big) \\ &= \mathrm{tr}\big(J^\dagger(\mathbf{a}^*, \mathbf{a})\,\xi\big)\,. \end{aligned} \tag{11.3.14}$$

Consequently, the momentum map $J : \mathbb{C}^2 \mapsto u(2)^*$ is given by the Hermitian expression

$$J(\mathbf{a}) = \frac{1}{2}\mathbf{a}\otimes\mathbf{a}^*. \tag{11.3.15}$$

This conclusion may be checked by computing the differential of the Hamiltonian $dJ^\xi(\mathbf{a})$ for the momentum map, which should be canonically related to its Hamiltonian vector field $X_{J^\xi(\mathbf{a})} = \{\,\cdot\,, J^\xi(\mathbf{a})\}$. As the infinitesimal generator $\xi(\mathbf{a}) = i\xi\mathbf{a}$ is linear, we have

$$\begin{aligned} dJ^\xi(\mathbf{a}) = d\big\langle J(\mathbf{a}), \xi \big\rangle_{u(2)} &= \frac{i}{2}\langle\!\langle \mathbf{a}, \xi(d\mathbf{a}) \rangle\!\rangle_{\mathbb{C}^2} + \frac{i}{2}\langle\!\langle\, d\mathbf{a}, \xi(\mathbf{a}) \rangle\!\rangle_{\mathbb{C}^2} \\ &= \Im\langle\!\langle \xi(\mathbf{a}),\, d\mathbf{a}\,\rangle\!\rangle_{\mathbb{C}^2} = \omega(\xi(\mathbf{a}),\,\cdot\,) = X_{J^\xi(\mathbf{a})} \,\lrcorner\, \omega, \end{aligned}$$

which is the desired canonical relation.

11.3.1 The Poincaré sphere $S^2 \in S^3$

We expand the Hermitian matrix $J = \frac{1}{2}\mathrm{a} \otimes \mathrm{a}^*$ in (11.3.15) in a basis of four 2×2 unit Hermitian matrices $(\sigma_0\,, \boldsymbol{\sigma})$, with $\boldsymbol{\sigma} = (\sigma_1, \sigma_2, \sigma_3)$ given by

$$\begin{aligned} \sigma_0 &= \begin{bmatrix} 1 & 0 \\ 0 & 1 \end{bmatrix}, \quad \sigma_1 = \begin{bmatrix} 0 & 1 \\ 1 & 0 \end{bmatrix}, \\ \sigma_2 &= \begin{bmatrix} 0 & -i \\ i & 0 \end{bmatrix}, \quad \sigma_3 = \begin{bmatrix} 1 & 0 \\ 0 & -1 \end{bmatrix}. \end{aligned} \tag{11.3.16}$$

The result is the decomposition

$$J = \frac{1}{4}\big(R\sigma_0 + \mathbf{Y} \cdot \boldsymbol{\sigma}\big)\,. \tag{11.3.17}$$

Here we denote $R := \mathrm{tr}(J\,\sigma_0) = |a_1|^2 + |a_2|^2$ and

$$\mathbf{Y} = \mathrm{tr}(J\,\boldsymbol{\sigma}) = a_k^*\boldsymbol{\sigma}_{kl}a_l\,, \tag{11.3.18}$$

with vector notation $\boldsymbol{\sigma} = (\sigma_1, \sigma_2, \sigma_3)$. In components, one finds

$$J = \frac{1}{2}\begin{bmatrix} a_1^*a_1 & a_1^*a_2 \\ a_2^*a_1 & a_2^*a_2 \end{bmatrix} = \frac{1}{4}\begin{bmatrix} R + Y_3 & Y_1 - iY_2 \\ Y_1 + iY_2 & R - Y_3 \end{bmatrix}, \tag{11.3.19}$$

with trace $\mathrm{tr}\,J = R$. Thus, the decomposition (11.3.17) splits the momentum map into its trace part $R \in \mathbb{R}$ and its traceless part $\mathbf{Y} \in \mathbb{R}^3$, given by

$$\mathbf{Y} = J - \frac{1}{2}(\mathrm{tr}\,J)\,\mathrm{Id} \in su(2)^* \cong \mathbb{R}^3\,. \tag{11.3.20}$$

This formula recovers the $SU(2)$ momentum map in Equation (11.3.5) found earlier for the Poincaré sphere.

Definition 11.3.2 (Poincaré sphere) *The coefficients $R \in \mathbb{R}$ and $\mathbf{Y} \in \mathbb{R}^3$ in the expansion of the matrix J in (11.3.17) comprise the four real quadratic quantities,*

$$\begin{aligned} R &= \frac{1}{2}\big(|a_1|^2 + |a_2|^2\big)\,, \\ Y_3 &= \frac{1}{2}\big(|a_1|^2 - |a_2|^2\big) \quad \textit{and} \\ Y_1 - i\,Y_2 &= a_1^*a_2\,. \end{aligned} \tag{11.3.21}$$

These quantities are all invariant under the action $\mathbf{a} \to e^{i\phi}\mathbf{a}$ *of* $\phi \in S^1$ *on* $\mathbf{a} \in \mathbb{C}^2$. *The* S^1*-invariant coefficients in the expansion of the momentum map* $J = \mathbf{a} \otimes \mathbf{a}^*$ *(11.3.15) in the basis of sigma matrices (11.3.16) satisfy the relation*

$$4 \det J = R^2 - |\mathbf{Y}|^2 = 0 , \quad \textit{with} \quad |\mathbf{Y}|^2 \equiv Y_1^2 + Y_2^2 + Y_3^2 . \quad (11.3.22)$$

This relation defines the ***Poincaré sphere*** $S^2 \in S^3$ *of radius* R *which, in turn, is related to the Hopf fibration* $\mathbb{C}^2/S^1 \simeq S^3$. *For more information about the Poincaré sphere and the Hopf fibration, consult, e.g., [Ho2008] and references therein.*

The $U(2)$ *Lie group structure*

The Lie group $U(2) = S^1 \times SU(2)$ is the direct product of its centre,

$$Z(U(2)) = \{zI \text{ with } |z| = 1\} \equiv S^1 ,$$

and the special unitary group in two dimensions,

$$SU(2) = \left\{ \begin{bmatrix} \alpha & \beta \\ -\beta^* & \alpha^* \end{bmatrix} \text{ with } |\alpha|^2 + |\beta|^2 = 1 \right\} .$$

As a consequence, the momentum map $J(\mathbf{a}) = \frac{1}{2}\mathbf{a} \otimes \mathbf{a}^*$ in (11.3.15) for the action $U(2) \times \mathbb{C}^2 \to \mathbb{C}^2$ decomposes into two momentum maps obtained by separating J into its trace part $J_{S^1} = R \in \mathbb{R}$ and its traceless part $\mathbf{J}_{SU(2)} = \mathbf{Y} \in \mathbb{R}^3$. This decomposition may be sketched, as follows.

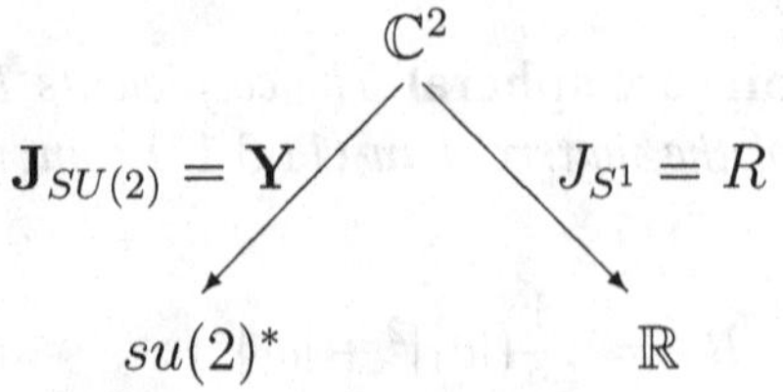

The target spaces $su(2)^*$ and $\mathbb{R}$ of the left and right legs of this pair of momentum maps are each Poisson manifolds, with coordinates $\mathbf{Y} \in su(2)^*$ and $R \in \mathbb{R}$, respectively. The corresponding Poisson brackets are given in tabular form as

$$
\begin{array}{|c|cccc|}
\hline
\{\cdot,\cdot\} & Y_1 & Y_2 & Y_3 & R \\
\hline
Y_1 & 0 & Y_3 & -Y_2 & 0 \\
Y_2 & -Y_3 & 0 & Y_1 & 0 \\
Y_3 & Y_2 & -Y_1 & 0 & 0 \\
R & 0 & 0 & 0 & 0 \\
\hline
\end{array}\;. \tag{11.3.23}
$$

In index notation, these Poisson brackets are given as

$$\{Y_k\,,\,Y_l\} = \epsilon_{klm}Y_m \quad \text{and} \quad \{Y_k\,,\,R\} = 0\,. \tag{11.3.24}$$

The last Poisson bracket relation means that the spaces with coordinates $\mathbf{Y} \in su(2)^*$ and $R \in \mathbb{R}$ are *symplectically orthogonal* in $u(2)^* = su(2)^* \times \mathbb{R}$.

Equations (11.3.24) prove the following.

Theorem 11.3.2 (Momentum map (11.3.15) is Poisson) *The direct-product structure of $U(2) = S^1 \times SU(2)$ decomposes the momentum map J in Equation (11.3.15) into two other momentum maps, $J_{SU(2)} : \mathbb{C}^2 \mapsto su(2)^*$ and $J_{S^1} : \mathbb{C}^2 \mapsto \mathbb{R}$. These other momentum maps are also Poisson maps. That is, they each satisfy the Poisson property for smooth functions F and H,*

$$\{F \circ J\,,\, H \circ J\} = \{F\,,\, H\} \circ J\,. \tag{11.3.25}$$

This relation defines a Lie–Poisson bracket on $su(2)^*$ that inherits the defining properties of a Poisson bracket from the canonical relations

$$\{a_k,\, a_l^*\} = -2i\delta_{kl}\,,$$

for the canonical symplectic form, $\omega = \Im\,(da_j \wedge da_j^*)$.

Remark 11.3.3 The Poisson bracket table in (11.3.23) is the $so(3)^*$ Lie–Poisson bracket table for angular momentum in the *spatial* frame. It differs by an overall sign from the $so(3)^*$ Lie–Poisson bracket table for angular momentum in the *body* frame, see (2.5.13). □

Definition 11.3.3 (Dual pairs) *Let (M, ω) be a symplectic manifold and let P_1, P_2 be two Poisson manifolds. A pair of Poisson mappings*

$$P_1 \xleftarrow{J_1} (M, \omega) \xrightarrow{J_2} P_2$$

*is called a **dual pair** [We1983b] if $\ker TJ_1$ and $\ker TJ_2$ are symplectic orthogonal complements of one another, that is,*

$$(\ker TJ_1)^{\omega} = \ker TJ_2. \tag{11.3.26}$$

A systematic treatment of dual pairs can be found in Chapter 11 of [OrRa2004]. The infinite-dimensional case is treated in [GaVi2010].

Remark 11.3.4 (Summary) In the pair of momentum maps

$$su(2)^* \equiv \mathbb{R}^3 \xleftarrow{\mathbf{Y}} (\mathbb{C}^2, \omega) \xrightarrow{R} \mathbb{R}, \tag{11.3.27}$$

$\mathbf{Y}$ maps the fibres of R, which are three-spheres, into two-spheres, that are coadjoint orbits of $SU(2)$. The restriction of $\mathbf{Y}$ to these three-spheres is the Hopf fibration. Further pursuit of the theory of dual pairs is beyond our present scope. See [HoVi2010] for a recent discussion of dual pairs for resonant oscillators from the present viewpoint. □

Example 11.3.4 (An infinite-dimensional momentum map) *Let $\mathcal{F} : T^*\mathcal{M} \mapsto \mathbb{R}$ be the space of real-valued functions on phase space defined by the cotangent bundle $T^*\mathcal{M}$ with coordinates (q, p) of a manifold $\mathcal{M}$ with coordinates q. The dual of the phase-space functions comprises the phase-space densities, denoted $\mathcal{F}^*$. The pairing $\langle \cdot , \cdot \rangle : \mathcal{F}^* \times \mathcal{F} \mapsto \mathbb{R}$ between a phase-space function $g \in \mathcal{F}$ and a phase-space density $f \, dp \wedge dq \in \mathcal{F}^*$ is defined by the* integral over phase space,

$$\langle f , g \rangle := \int_{T^*\mathcal{M}} f \, g \, dq \wedge dp \, .$$

We identify the space $\mathcal{F}^* \times \mathcal{F} \simeq T^*\mathcal{F}$ as the cotangent bundle of the space of phase-space functions $\mathcal{F}$.

Let $X_h = \{ \cdot , h \}$ be the Hamiltonian vector field defined by applying the canonical Poisson bracket on $T^*\mathcal{M}$ using the phase-space function

$h \in \mathcal{F}$. The momentum map for the action of canonical (symplectic) transformations on phase-space functions is given by

$$J_h(f,g) = \langle f, X_h g\rangle = \langle f, \{g,h\}\rangle = \langle h, \{f,g\}\rangle\,,$$

where we have integrated by parts and invoked homogeneous boundary conditions in the third step. This discussion reveals the following.

Lemma 11.3.1 *The Poisson bracket $\{f,g\}(q,p)$ is the momentum map for the action of canonical transformations on phase-space functions.*

Remark 11.3.5 The preservation of this momentum map (i.e., the Poisson bracket) under canonical transformations on phase-space functions is no surprise; this preservation is part of the definition of a canonical transformation. □

Equivalently, one may write this momentum map as a pairing between the Lie algebra $\mathfrak{g}$ of Hamiltonian vector fields and its dual, $\mathfrak{g}^*$, the one-form densities on phase space. The following lemma provides this representation.

Lemma 11.3.2 *The Poisson one-form density $f dg \otimes dq \wedge dp$ is the momentum map dual to the action of Hamiltonian vector fields on phase-space functions f and g.*

Proof. The momentum map for this action is defined in terms of the pairing,

$$\begin{aligned} J_{X_h}(f,g) &= \langle J(f,g)\,,\, X_h\rangle_{\mathfrak{g}^*\times\mathfrak{g}} = -\langle f\diamond g, X_h\rangle_{\mathfrak{g}^*\times\mathfrak{g}} \\ &= \langle f, X_h g\rangle \\ &= \int_{T^*M} f(g_q h_p - h_q g_p)\, dq\wedge dp \\ &= \int_{T^*M} (X_h \,\lrcorner\, f dg)\, dq\wedge dp \\ &= \langle f\, dg\,,\, X_h\rangle_{\mathfrak{g}^*\times\mathfrak{g}}\,, \end{aligned} \tag{11.3.28}$$

where $X_h \,\lrcorner\, f dg$ denotes substitution of the Hamiltonian vector field $X_h = \{\,\cdot\,, h\}$ into the differential one-form on phase space

$$f\, dg(q,p) = f(g_q dq + g_p dp)\,.$$ ■

Remark 11.3.6 The computation in (11.3.28) shows that the exterior differential (d) on phase space is minus the diamond operation for the action of Hamiltonian vector fields on phase-space functions. □

11.3.2 Overview

Looking back on Section 2.5.5 which discussed the Clebsch variational principle for the rigid body, we notice that the first of Equations (2.5.22) is exactly the momentum map for body angular momentum. Likewise, the momentum map $(q \diamond p)$ emerged in Equation (9.3.8) in the Clebsch variational approach to the Euler–Poincaré equation in Chapter 9.

> This observation affirms the main message of the book: Lie symmetry reduction on the Lagrangian side produces the Euler–Poincaré equation, as discussed in Chapter 9. Its formulation on the Hamiltonian side as a Lie–Poisson equation in Section 9.4 governs the dynamics of the momentum map defined in Equation (11.2.1), which derives from the cotangent lift of the Lie algebra action of the original Lie symmetry on the configuration manifold defined in (11.1.2). This is the relation between the results of reduction by Lie symmetry on the Lagrangian and Hamiltonian sides.
>
> The primary purpose of this book has been to explain that statement, so that it is understood by undergraduate students in mathematics, physics and engineering.

Remark 11.3.7 Looking forward, the reader may wish to pursue these ideas further, and go beyond the cotangent-lift momentum maps discussed here. At that point, the reader will need to consult the books [OrRa2004, MaMiOrPeRa2007] for discussions of momentum maps obtained by singular reduction techniques and reduction by stages. For discussions of the use of momentum maps derived from Hamilton–Pontryagin and Clebsch variational principles, respectively, in designing geometric integrators for discrete dynamical systems, see [BoMa2009] and [CoHo2007]. □

Exercise. The canonical Poisson bracket $\{g, h\}$ between two phase-space functions g and h induces a Lie–Poisson bracket $\{\,\cdot\,,\,\cdot\,\}_{LP}$ between linear functionals of phase-space functions defined by

$$J_{X_h}(f, g) = \big\langle f, \{g, h\}\big\rangle = \big\{\langle f, g\rangle, \langle f, h\rangle\big\}_{LP}.$$

Show that this Lie–Poisson bracket $\{\,\cdot\,,\,\cdot\,\}_{LP}$ satisfies the properties that define a Poisson bracket, including the Jacobi identity. ★

Exercise. Let the components of the angular momentum vector $\mathbf{L} \in \mathbb{R}^3$ be defined by

$$L_i := \int (p \times q)_i\, f(q, p)\, dq \wedge dp = \big\langle f\, (p \times q)_i \big\rangle ,$$

where $i = 1, 2, 3$ and $(\times)$ denotes vector product in $\mathbb{R}^3$. Compute the Lie–Poisson bracket $\big\{L_i,\, L_j\big\}_{LP}$. ★

Exercise. Compute the Lie–Poisson bracket relation

$$\{J_{X_h}(f, g),\, J_{X_H}(f, G)\}_{LP} = J_{X_{\{G,H\}}}(f,\, \{g, h\})$$

for phase-space functions g, h, G, H. ★

12 ROUND, ROLLING RIGID BODIES

Contents

12.1	Introduction	246
	12.1.1 Holonomic versus nonholonomic	246
	12.1.2 The Chaplygin ball	248
12.2	Nonholonomic Hamilton–Pontryagin variational principle	252
	12.2.1 HP principle for the Chaplygin ball	256
	12.2.2 Circular disk rocking in a vertical plane	265
	12.2.3 Euler's rolling and spinning disk	268
12.3	Nonholonomic Euler–Poincaré reduction	275
	12.3.1 Semidirect-product structure	276
	12.3.2 Euler–Poincaré theorem	278
	12.3.3 Constrained reduced Lagrangian	282

This chapter deals with round rigid bodies rolling on a flat surface, by applying the Hamilton–Pontryagin and Euler–Poincaré approaches to Lagrangian reduction for variational principles with symmetries. These approaches were introduced for the heavy top in Chapter 8. To the potential energy and rotational energy of the heavy top, one now adds translational energy for its centre of mass motion and imposes the nonholonomic constraint that the rigid body is rolling. We focus on the constrained rocking, rolling, but not sliding motion of a rigid body on a perfectly rough horizontal plane in two examples. These examples are the Chaplygin ball (a rolling spherical ball with an off-centre mass distribution, as sketched in Figure 12.1) and the Euler disk (modelling a gyrating coin, as sketched in Figure 12.2). The chapter closes by showing that the Hamilton–Pontryagin and Euler–Poincaré approaches to Lagrangian reduction yield the same equations as those obtained from the standard Lagrange–d'Alembert variational approach [Bl2004].

We hope the examples of geometric mechanics problems treated in this chapter will inspire confidence in interested readers. Having mastered the material so far, they should find themselves ready and completely equipped with the tools needed to model nonholonomic dynamics of rolling without sliding.

12.1 Introduction

12.1.1 Holonomic versus nonholonomic

Constraints which restrict the possible configurations for a mechanical system are called ***holonomic***. Examples are restriction of a pendulum to have constant length, or restriction of a particle in three dimensions to move on the surface of a sphere. Constraints on the velocities which cannot be reduced to holonomic constraints are termed ***nonholonomic***.

Nonholonomic systems typically arise when constraints on velocity are imposed, such as the constraint that the bodies roll without slipping on a surface. Cars, bicycles, unicycles – anything with rolling wheels – are all examples of nonholonomic systems.

Nonholonomic mechanics has a rich history that dates back to the time of Euler, Lagrange and d'Alembert. The classical work in this subject is summarised in Routh [Ro1860], Jellett [Je1872] and has been discussed more recently in the context of control theory in Neimark and Fufaev [NeFu1972]. The geometry of nonholonomic systems shares its mathematical foundations with geometric control theory. An introduction to nonholonomic constraints in geometric control applications (such as feedback laws that stabilise or generate locomotion) is given in [Bl2004]. A branch of mathematics known as sub-Riemannian geometry has developed for dealing with nonholonomic geometric control systems [Mo2002]. Control problems and sub-Riemannian geometry are beyond the scope of the present text.

Definition 12.1.1 *A **constrained dynamical system** consists of*

- *a smooth manifold Q, which is the configuration space;*
- *a smooth function $L : TQ \to \mathbb{R}$, which is the Lagrangian (typically taken to be the difference between the kinetic energy and potential energy); and*
- *a smooth distribution $\mathcal{D} \subset TQ$, which determines the constraints.*

Definition 12.1.2 *A **distribution** is a collection of linear subspaces of the tangent spaces of Q such that $\mathcal{D}_q \subset T_qQ$ for $q \in Q$.*

Remark 12.1.1 The $\mathcal{D}_q$ tangent spaces of Q parameterise the allowable directions for the system at a given point q of the configuration space. A curve $q(t) \in Q$ satisfies the constraints, provided $\dot{q}(t) \in \mathcal{D}_{q(t)}$ for all t. (This implies that the constraints are linear in the velocities.) The distribution $\mathcal{D} \subset TQ$ will in general not be the differential of a function on Q. That is, in general, the constraints will be ***nonholonomic***. □

This chapter applies the same ideas of symmetry reduction underlying the Hamilton–Pontryagin and Euler–Poincaré approaches that were explained earlier in studying Hamilton's principle for the heavy top in Chapter 8. In addition, it introduces nonholonomic constraints into the variational principle by using ***Lagrange multipliers***. The approach is illustrated by deriving the dynamical equations for the rocking, rolling, but not sliding motion of two classical nonholonomic problems. The first problem involves a spherical ball whose mass distribution is off-centre, and is rolling without slipping on a horizontal plane in the presence of gravity. This problem is called the ***Chaplygin ball***. The second problem treats a disk that is simultaneously rolling, spinning and falling. This is the problem of the ***Euler disk***. Both of these problems have stimulated many interesting investigations, and they are likely to keep doing so.

12.1.2 The Chaplygin ball

Consider a spherical ball of mass m, radius r and moment of inertia I, whose mass distribution is inhomogeneous, so that its centre of mass lies anywhere in the ball as it rolls without slipping on a horizontal plane in the presence of gravity. This problem was first solved by Chaplygin [Ch1903]. Extensive references for the history of this problem and the associated problem of the tippe top involving both rolling and sliding are given in [GrNi2000]. Modern geometric perspectives of this problem appear in [Cu1998, He1995, Ze1995, Sc2002, BoMaRo2004, GlPeWo2007, CiLa2007, Ki2011].

Definition 12.1.3 (Geometry for the rolling ball) *Let* (E_1, E_2, E_3) *denote the reference system of body coordinates, chosen to coincide with the ball's principal axes, in which the inertia tensor is diagonal,* $I = \mathrm{diag}(I_1, I_2, I_3)$*. The coordinate directions in space are denoted* $(\mathsf{e}_1, \mathsf{e}_2, \mathsf{e}_3)$*. These spatial unit vectors are chosen so that* $(\mathsf{e}_1, \mathsf{e}_2)$ *are horizontal and* e_3 *is vertical. The origin of the body coordinate system coincides with the centre of mass of the ball, which has coordinates* $x(t) = (x_1(t), x_2(t), x_3(t))$ *in space. In the spatial coordinate system, the body frame is the moving frame* $(g(t)E_1, g(t)E_2, g(t)E_3)$*, where* $g(t) \in SO(3)$ *defines the attitude of the body relative to its reference configuration.*

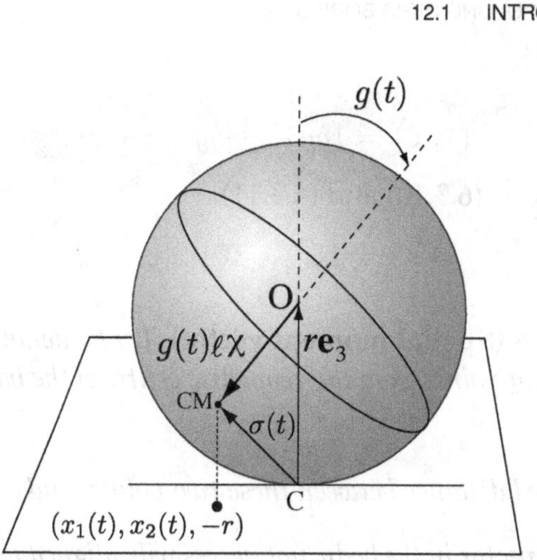

Figure 12.1. The Chaplygin ball. The position $(x_1(t), x_2(t), x_3(t))$ is the position of the centre of mass, not the centre of the sphere. The spatial vector $\sigma(t)$ points from the contact point $\mathbf{C}$ to the centre of mass. The projection of the centre of mass location onto the plane is the point $(x_1(t), x_2(t), -r)$.

Thus, the motion of the ball is given by a curve in the group of special Euclidean transformations

$$(g(t),\, x(t)) \in SE(3) \simeq SO(3) \times \mathbb{R}^3 \,,$$

where $g(t) \in SO(3)$ *and* $x(t) \in \mathbb{R}^3$*. An element of* $SE(3)$ *maps a generic point* P *in the ball's reference coordinates to a point in space,*

$$Q(t) = g(t)P + x(t) \,,$$

at time t.

Exercise. Show that the left-invariant and right-invariant tangent vectors for this motion on $SE(3)$ are, respectively,

$$(\Omega,\, Y) := (g^{-1}\dot{g},\, g^{-1}\dot{x}) \,,$$

and

$$(\omega, \zeta) := (\dot{g}g^{-1}, -\dot{g}g^{-1}x + \dot{x}) .$$

See Equations (6.3.10) and (6.3.11). ★

Definition 12.1.4 (Spatial moment vector) *Let $\ell\chi$ denote the vector in the body frame that points from the geometric centre of the ball to its centre of mass, where*

- *ℓ is the fixed distance between these two points; and*
- *χ is a unit vector in the body, not necessarily aligned with a principal axis.*

At any instant, the vector in space *pointing from the contact point C to the centre of mass at $x(t)$ is the* **spatial moment vector,**

$$\sigma(t) = re_3 + \ell g(t)\chi .$$

Definition 12.1.5 (Rolling constraint) *Physically, rolling of the body means that the velocity of its centre of mass $\dot{x}(t)$* in the spatial frame *at any instant t must be obtained by a rotation with angular frequency*

$$\widehat{\omega} = \dot{g}g^{-1} \in so(3) ,$$

around the fixed contact point. Consequently, the **rolling constraint** *is imposed by requiring*

$$\dot{x}(t) = \dot{g}g^{-1}\sigma(t) = \widehat{\omega}\,\sigma(t) = \omega \times \sigma(t) , \qquad (12.1.1)$$

where we have used the hat map in the last step for the (right-invariant) spatial *angular velocity. Equivalently, one may rewrite this rolling constraint as*

$$\dot{x}(t) = \dot{g}(t)\big(rg^{-1}(t)e_3 + \ell\chi\big) . \qquad (12.1.2)$$

This formula defines the **spatial nonholonomic constraint distribution** *D for the Chaplygin ball.*

Exercise. What is the velocity $\dot{x}(t)$ in the body coordinate system? ★

The Hamilton–Pontryagin approach for the Chaplygin ball

The Hamilton–Pontryagin method proceeds after assembling the following ingredients:

- Potential energy: The potential energy due to gravity for the Chaplygin ball is $m\gamma x_3(t)$, where

$$x_3(t) = \langle e_3,\, g(t)\ell\chi\rangle = \langle g^{-1}(t)e_3,\, \ell\chi\rangle \tag{12.1.3}$$

  is the vertical displacement of the centre of mass relative to the centre of the ball and γ is the constant acceleration of gravity.

- Kinetic energy: The kinetic energy of the Chaplygin ball is the sum of the energy of translation of its centre of mass and the energy of rotation about the centre of mass.

- Lagrangian: The Lagrangian is defined to be the difference between the total kinetic energy and the gravitational potential energy,

$$L = \frac{1}{2}\langle \Omega, I\Omega\rangle + \frac{m}{2}|\dot{x}|^2 - m\gamma\ell\langle g^{-1}e_3, \chi\rangle\,,$$

  where $\Omega = g^{-1}\dot{g}$ is the (left-invariant) body angular velocity.

- Rolling constraint in the body: The spatial distribution (12.1.2) for the rolling constraint transforms into body coordinates upon setting

$$Y = g^{-1}\dot{x}\,,\quad \Gamma = g^{-1}e_3\,,\quad s = g^{-1}\sigma = r\Gamma + \ell\chi\,, \tag{12.1.4}$$

  with vectors Y, Γ, s each in $\mathbb{R}^3$. The rolling constraint in body coordinates then simplifies to

$$Y = \Omega s\,. \tag{12.1.5}$$

The hat map expresses this rolling constraint equivalently in vector form as $Y = \Omega \times s$.

- Reduced Lagrangian in body coordinates: The reduced Lagrangian is defined in body coordinates by using the definitions in (12.1.5), as

$$l(\Omega, Y, \Gamma) = \frac{1}{2}\langle \Omega, I\Omega \rangle + \frac{m}{2}|Y|^2 - m\gamma\ell\langle \Gamma, \chi \rangle . \tag{12.1.6}$$

These considerations have expressed a spatial Lagrangian $L : TG \times TV \times V \to \mathbb{R}$ in its body form as $l : \mathfrak{g} \times TV \times V \to \mathbb{R}$, plus the various constraint relations.

12.2 Nonholonomic Hamilton–Pontryagin variational principle

We shall write Hamilton's principle $\delta S = 0$ for a class of nonholonomically constrained action principles that includes the Chaplygin ball, by simply using Lagrange multipliers to impose the constraints on the reduced Lagrangian, as in the Hamilton–Pontryagin principle for the matrix Euler equations in Section 2.4.1. In particular, we apply the Hamilton–Pontryagin approach to the following class of constrained action integrals,

$$\begin{aligned} S &= \int L(g,\, \dot{g},\, \dot{x},\, e_3)\, dt \\ &= \int \Big\{ l(\Omega, Y, \Gamma) + \Big\langle \Pi,\, g^{-1}\dot{g} - \Omega \Big\rangle \\ &\qquad + \Big\langle \kappa,\, g^{-1}e_3 - \Gamma \Big\rangle + \Big\langle \lambda,\, g^{-1}\dot{x} - Y \Big\rangle \Big\} \, dt , \end{aligned} \tag{12.2.1}$$

where

$$l(\Omega, Y, \Gamma) = L(e,\, g^{-1}\dot{g},\, g^{-1}\dot{x},\, g^{-1}e_3) . \tag{12.2.2}$$

This class of action integrals produces constrained equations of motion determined from the Hamilton–Pontryagin principle.

Theorem 12.2.1 (Hamilton–Pontryagin principle) *The stationarity condition for the nonholonomically constrained Hamilton–Pontryagin principle defined in Equations (12.2.1) and (12.2.2) under the rolling constraint in body coordinates (12.1.5) implies the following equation of motion,*

$$\left(\frac{d}{dt} - \mathrm{ad}^*_\Omega\right)\left(\Pi - \lambda \diamond s\right) = \kappa \diamond \Gamma - \lambda \diamond \dot{s}\,, \tag{12.2.3}$$

where $\Pi = \delta l/\delta\Omega$, $\kappa = \delta l/\delta\Gamma$, $\lambda = \delta l/\delta Y$ *and* $s = s(\Gamma)$ *defines the vector in the body directed from the point of rolling contact to the centre of mass.*

Remark 12.2.1 (The ad^* **and** $\diamond$ **operations)** The ubiquitous coadjoint action $\mathrm{ad}^* : \mathfrak{g}^* \times \mathfrak{g} \to \mathfrak{g}^*$ and the diamond operation $\diamond : V^* \times V \to \mathfrak{g}^*$ in Equation (12.2.3) are defined as the duals of the Lie algebra adjoint action $\mathrm{ad} : \mathfrak{g} \times \mathfrak{g} \to \mathfrak{g}$ and of its (left) action $\mathfrak{g} \times V \to V$ on a vector representation space V, respectively, with respect to the corresponding pairings as

$$\left\langle \mathrm{ad}^*_\Omega \Pi\,,\, \eta \right\rangle_{\mathfrak{g}^* \times \mathfrak{g}} = \left\langle \Pi\,,\, \mathrm{ad}_\Omega \eta \right\rangle_{\mathfrak{g}^* \times \mathfrak{g}}, \tag{12.2.4}$$

with $\Omega,\, \eta \in \mathfrak{g}$ and $\Pi \in \mathfrak{g}^*$ for ad^* and

$$\left\langle \kappa \diamond \Gamma\,,\, \eta \right\rangle_{\mathfrak{g}^* \times \mathfrak{g}} = \left\langle \kappa\,,\, -\eta\Gamma \right\rangle_{V^* \times V}, \tag{12.2.5}$$

with $\eta \in \mathfrak{g}$, $\kappa \in V^*$ and $\Gamma \in V$ for $\diamond$. The ad^* and $\diamond$ operations first appeared in the dual actions of $SE(3)$ in (6.2.9). They arose again in the Clebsch Equation (9.3.8) and in the proof of the Clebsch variational principle for the Euler–Poincaré equation in Proposition 9.3.1. They featured in the definition of the cotangent-lift momentum map (11.2.1) and now they have emerged again in the nonholonomic Hamilton–Pontryagin Equation (12.2.3) □

Proof. One evaluates the variational derivatives in this constrained Hamilton's principle from the definitions of the variables as

$$\begin{aligned}
\delta\Omega = \delta(g^{-1}\dot{g}) &= \frac{d\eta}{dt} + \mathrm{ad}_\Omega \eta\,, \\
\delta\Gamma = \delta(g^{-1}e_3) &= -\,\eta(g^{-1}e_3) = -\,\eta\Gamma\,, \\
\delta Y = \delta(g^{-1}\dot{x}) &= \frac{d}{dt}(\eta s) + (\mathrm{ad}_\Omega \eta)s\,.
\end{aligned} \tag{12.2.6}$$

Here $\eta := g^{-1}\delta g$ and the last equation is computed from

$$\begin{aligned}
\delta(g^{-1}\dot{x}) &= -\eta g^{-1}\dot{x} + g^{-1}\delta\dot{x} \\
&= -\,\eta\Omega s + \frac{d}{dt}(g^{-1}\delta x) + \Omega\eta s\,.
\end{aligned} \tag{12.2.7}$$

Remark 12.2.2 In computing formulas (12.2.6), one must first take variations of the definitions, and only then evaluate the result on the constraint distribution defined by $Y = g^{-1}\dot{x} = \Omega s$ and

$$g^{-1}\delta x = (g^{-1}\delta g)(rg^{-1}e_3 + \ell\chi) = \eta s,$$

which is the analogue of Equation (12.1.2). This type of subtlety in taking variations of Hamilton's principle in deriving Euler–Lagrange equations for nonholonomic systems is also discussed in [Ru2000]. □

Expanding the variations of the Hamilton–Pontryagin action integral (12.2.1) using relations (12.2.6) and then integrating by parts yields

$$\begin{aligned}
\delta S = \int \Big\{ &\Big\langle \frac{\delta l}{\delta\Omega} - \Pi\,,\, \delta\Omega \Big\rangle + \Big\langle \frac{\delta l}{\delta\Gamma} - \kappa\,,\, \delta\Gamma \Big\rangle + \Big\langle \frac{\delta l}{\delta Y} - \lambda\,,\, \delta Y \Big\rangle \\
&- \Big\langle \Big(\frac{d}{dt} - \mathrm{ad}^*_\Omega\Big)(\Pi - \lambda \diamond s) - \kappa \diamond \Gamma + \lambda \diamond \dot{s}\,,\, \eta \Big\rangle \Big\}\, dt \\
&+ \Big\langle \Pi + s \diamond \lambda, \eta \Big\rangle \Big|_a^b .
\end{aligned} \tag{12.2.8}$$

The last entry in the integrand arises from varying in the group element g using formulas (12.2.6) obtained from relation (12.2.7). Stationarity ($\delta S = 0$) for the class of action integrals S in Equations (12.2.1) and (12.2.2) for variations η that vanish at the endpoints now defines the Lagrange multipliers, $\Pi = \delta l/\delta\Omega$, $\kappa = \delta l/\delta\Gamma$, $\lambda = \delta l/\delta Y$, in terms of variational derivatives of the Lagrangian and thereby proves the formula for the constrained equation of motion (12.2.3) in the statement of the theorem. ■

Remark 12.2.3 (Preparation for Noether's theorem) The endpoint term that arises from integration by parts in (12.2.8) will be discussed later in Remark 12.2.13 about the Jellett and Routh integrals for the cylindrically symmetric Chaplygin ball, and in Remark 12.3.5, about Noether's theorem for the nonholonomic Euler–Poincaré equations. As usual, this endpoint term is the source of Noether's theorem for the system. However, in nonholonomic systems the variations must respect the constraints, so Noether symmetries have more than just their geometric meaning. They also have a dynamical meaning. □

Remark 12.2.4 (Vector notation) For $g \in SO(3)$ the equation of constrained motion (12.2.3) arising from stationarity ($\delta S = 0$) of the action in (12.2.8) may be expressed in $\mathbb{R}^3$ vector notation via the hat map as

$$\Big(\frac{d}{dt} + \boldsymbol{\Omega}\times\Big)\big(\boldsymbol{\Pi} - \boldsymbol{\lambda}\times\boldsymbol{s}\big) = \boldsymbol{\kappa}\times\boldsymbol{\Gamma} - \boldsymbol{\lambda}\times\dot{\boldsymbol{s}}\,. \qquad (12.2.9)$$

These equations are completed by the formulas

$$\boldsymbol{\kappa} = \delta l/\delta\boldsymbol{\Gamma}\,, \quad \boldsymbol{\lambda} = \delta l/\delta\boldsymbol{Y}\,, \quad \dot{\boldsymbol{\Gamma}} = -\,\boldsymbol{\Omega}\times\boldsymbol{\Gamma}\,,$$

and $\dot{\boldsymbol{s}}$ with $\boldsymbol{s} = \boldsymbol{s}(\boldsymbol{\Gamma})$. This vector version of Theorem 12.2.1 will yield equations of motion for two classic nonholonomic problems, the Chaplygin ball and the Euler disk. □

Exercise. Set $M = \Pi - \lambda \times s$ and compute $d(M \cdot \Omega)/dt$. How does this simplify when $\ell = 0$, so that the centre of mass coincides with the centre of the sphere? ★

12.2.1 HP principle for the Chaplygin ball

In vector notation, the reduced Lagrangian in Equation (12.1.6) for the Chaplygin ball in body coordinates is

$$l(\boldsymbol{\Omega}, \mathbf{Y}, \boldsymbol{\Gamma}) = \frac{1}{2}\boldsymbol{\Omega} \cdot I\boldsymbol{\Omega} + \frac{m}{2}|\mathbf{Y}|^2 - m\gamma\ell\,\boldsymbol{\Gamma} \cdot \boldsymbol{\chi} . \tag{12.2.10}$$

This is the sum of the kinetic energies due to rotation and translation, minus the potential energy of gravity. One evaluates its vector-valued variational relations as

$$\begin{aligned} \boldsymbol{\Pi} &= \frac{\delta l}{\delta \boldsymbol{\Omega}} = I\boldsymbol{\Omega} , \quad \boldsymbol{\kappa} = \frac{\delta l}{\delta \boldsymbol{\Gamma}} = -\,m\gamma\ell\boldsymbol{\chi} , \\ \boldsymbol{\lambda} &= \frac{\delta l}{\delta \mathbf{Y}} = m\mathbf{Y} = m\,\boldsymbol{\Omega} \times \boldsymbol{s} , \end{aligned} \tag{12.2.11}$$

in which the *last step* is to substitute the rolling constraint.

From their definitions for the Chaplygin ball

$$\boldsymbol{\Gamma}(t) = g^{-1}(t)\mathrm{e}_3 \quad \text{and} \quad \boldsymbol{s} = r\boldsymbol{\Gamma}(t) + \ell\boldsymbol{\chi} , \tag{12.2.12}$$

one also finds the auxiliary equations,

$$\dot{\boldsymbol{\Gamma}} = -\,\boldsymbol{\Omega} \times \boldsymbol{\Gamma} \quad \text{and} \quad \dot{\boldsymbol{s}} = -\,\boldsymbol{\Omega} \times (\boldsymbol{s} - \ell\boldsymbol{\chi}) . \tag{12.2.13}$$

As always, the auxiliary equation for the unit vector $\boldsymbol{\Gamma}$ preserves $|\boldsymbol{\Gamma}|^2 = 1$.

Substituting these two auxiliary equations into the stationarity condition (12.2.9) and rearranging yields the vector motion equation for the Chaplygin ball,

$$\left(\frac{d}{dt}+\boldsymbol{\Omega}\times\right)\boldsymbol{M} = \underbrace{m\gamma\ell\boldsymbol{\Gamma}\times\boldsymbol{\chi}}_{\text{gravity}} - \underbrace{mr\left(\boldsymbol{\Omega}\times\boldsymbol{\Gamma}\right)\times\left(\boldsymbol{\Omega}\times\ell\boldsymbol{\chi}\right)}_{\text{rolling constraint torque}}$$

$$= m\ell\Big[\gamma\boldsymbol{\Gamma}\times\boldsymbol{\chi}+r\,\boldsymbol{\Omega}\left(\boldsymbol{\Omega}\cdot\boldsymbol{\chi}\times\boldsymbol{\Gamma}\right)\Big] \qquad (12.2.14)$$

$$\text{with}\quad \boldsymbol{M} = I\boldsymbol{\Omega}+m\boldsymbol{s}\times(\boldsymbol{\Omega}\times\boldsymbol{s}). \qquad (12.2.15)$$

Exercise. (Energy conservation) Show that the Chaplygin ball in rolling motion without sliding conserves the following energy in body coordinates,

$$E(\boldsymbol{\Omega},\boldsymbol{\Gamma}) = \frac{1}{2}\boldsymbol{\Omega}\cdot I\boldsymbol{\Omega}+\frac{m}{2}\left|\boldsymbol{\Omega}\times\boldsymbol{s}\right|^2 + m\gamma\ell\boldsymbol{\Gamma}\cdot\boldsymbol{\chi}\,. \qquad (12.2.16)$$

This is the sum of the kinetic energies due to rotation and translation, plus the potential energy of gravity. (The rolling constraint does no work.) ★

Exercise. (Constrained reduced Lagrangian) Determine whether the motion Equation (12.2.14) for the Chaplygin ball also results from the constrained reduced Lagrangian,

$$l_c(\boldsymbol{\Omega},\boldsymbol{\Gamma}) = \frac{1}{2}\boldsymbol{\Omega}\cdot I\boldsymbol{\Omega}+\frac{m}{2}|\boldsymbol{\Omega}\times\boldsymbol{s}|^2 - m\gamma\ell\boldsymbol{\Gamma}\cdot\boldsymbol{\chi}\,, \qquad (12.2.17)$$

in which the rolling constraint is applied *before* varying. ★

Remark 12.2.5 The distinction between the motion equations that result from the reduced Lagrangian (12.2.10) when the rolling constraint is applied after the variations and those that result from the *constrained* reduced Lagrangian (12.2.17) on applying the rolling constraint before the variations will be discussed in Section 12.3.3.

□

Remark 12.2.6 (The issue of sliding) The rolling constraint exerts the torque on the body that keeps it rolling. This torque appears in the motion Equation (12.2.14) directed along the angular velocity vector $\boldsymbol{\Omega}$ and proportional to the square of its magnitude. For sufficiently rapid rotation, one could expect that friction might no longer be able to sustain the constraint torque. In this situation, additional modelling steps to include sliding would be required. However, considerations of sliding seriously affect the solution procedure, because they change the nature of the problem.

For pure rolling, one integrates the motion Equation (12.2.14) for angular momentum and solves for the body angular velocity and orientation of the vertical in the body. One then reconstructs the path on the rotation group $g(t)$. Finally, one applies $g(t)$ in Equation (12.1.2) for the rolling constraint to obtain the position of the centre of mass $\mathbf{x}(t)$.

The ball is sliding, to the extent that its ***slip velocity*** is nonzero. The slip velocity of the contact point is the velocity of the contact point on the rigid body relative to the centre of mass. This is

$$\mathbf{v}_S(t) = \dot{\mathbf{x}}(t) - \dot{g}(t)\big(rg^{-1}(t)\mathbf{e}_3 + \ell\boldsymbol{\chi}\big)\,. \qquad (12.2.18)$$

Sliding introduces frictional effects that differ considerably from pure rolling. In particular, sliding introduces an additional friction force in Newton's law for the acceleration of the centre of mass,

$$m\ddot{\mathbf{x}}(t) = \mathbf{F}_S\,.$$

Sometimes, the sliding force $\mathbf{F}_S$ is taken as being proportional to the slip velocity $\mathbf{v}_S(t)$ and of opposite sign. The torque $\mathbf{F}_S \times \mathbf{s}$ associated with the sliding force also enters the angular momentum equation. Moreover, a normal reaction force and its torque must be determined to ensure that the sphere stays in contact with the plane. In some applications, such as the tippe top, sliding is essential. However, we shall forgo the opportunity to discuss sliding in the present text. See [GrNi2000, CiLa2007] for discussions of frictional effects for the Chaplygin ball, in the context of the tippe top phenomenon. □

Remark 12.2.7 (The Chaplygin ball vs the heavy top) The motion Equation (12.2.14) for the Chaplygin ball is reminiscent of the heavy-top motion Equation (8.1.2) and it has the same auxiliary Equation (8.1.3) for $\mathbf{\Gamma}$. However, it also has two important differences. First, the translational kinetic energy associated with rolling of the Chaplygin ball enters effectively as an angular momentum. Second, an additional torque due to the rolling constraint appears on the right-hand side of the motion equation.

Hamilton–Pontryagin principle for the heavy top

The heavy-top equations also arise from the Hamilton–Pontryagin principle (12.2.1) with the following reduced Lagrangian in vector body coordinates, obtained by simply ignoring the kinetic energy of translation in (12.2.10) for the Chaplygin ball:

$$l(\mathbf{\Omega}, \mathbf{\Gamma}) = \frac{1}{2}\,\mathbf{\Omega}\cdot I\mathbf{\Omega} - m\gamma\ell\,\mathbf{\Gamma}\cdot\boldsymbol{\chi}\,. \tag{12.2.19}$$

This is just the difference between the kinetic energy of rotation and the potential energy of gravity. Substituting the vector-valued variational relations in Equation (12.2.11) and the auxiliary equation,

$$\dot{\mathbf{\Gamma}} = -\,\mathbf{\Omega}\times\mathbf{\Gamma}\,, \tag{12.2.20}$$

into the stationarity condition (12.2.9) and rearranging recovers the vector motion Equation (8.1.2) for the heavy top,

$$\left(\frac{d}{dt} + \mathbf{\Omega}\times\right) I\mathbf{\Omega} = m\gamma\ell\,\mathbf{\Gamma}\times\boldsymbol{\chi}\,,$$

which we see now is rather simpler than Equation (12.2.14) for the motion of the Chaplygin ball. For example, one easily sees that the heavy-top equations conserve $|\mathbf{\Gamma}|^2$ and $\mathbf{\Gamma}\cdot I\mathbf{\Omega}$. □

Exercise. Show that the vector Equation (12.2.9) for nonholonomic motion in body coordinates and the auxiliary Equations (12.2.11)–(12.2.13) imply an evolution equation,

$$\frac{d}{dt}(\boldsymbol{M}\cdot\boldsymbol{\Gamma}) = mr\ell(\boldsymbol{\Omega}\cdot\boldsymbol{\Gamma})(\boldsymbol{\Omega}\cdot\boldsymbol{\chi}\times\boldsymbol{\Gamma}), \qquad (12.2.21)$$

for the projection $\boldsymbol{M}\cdot\boldsymbol{\Gamma}$ of the total body angular momentum $\boldsymbol{M}$ onto the vertical direction as seen from the body $\boldsymbol{\Gamma} = g^{-1}(t)\mathbf{e}_3$, with

$$\boldsymbol{M} = \boldsymbol{\Pi} - \boldsymbol{\lambda}\times\boldsymbol{s} \quad \text{with} \quad \boldsymbol{\lambda} = \delta l/\delta\boldsymbol{Y} = m\boldsymbol{Y} = m\,\boldsymbol{\Omega}\times\boldsymbol{s}$$

and the vector $\boldsymbol{\Gamma} = g^{-1}(t)\mathbf{e}_3$, which represents the vertical direction as seen from the body. ★

Answer. The required evolution Equation (12.2.21) follows from a direct calculation,

$$\begin{aligned}\frac{d}{dt}(\boldsymbol{M}\cdot\boldsymbol{\Gamma}) &= -mr(\boldsymbol{\Omega}\times\boldsymbol{\Gamma})\cdot(\boldsymbol{Y}\times\boldsymbol{\Gamma})\\ &= mr(\boldsymbol{\Omega}\cdot\boldsymbol{\Gamma})(\boldsymbol{Y}\cdot\boldsymbol{\Gamma})\\ &= mr(\boldsymbol{\Omega}\cdot\boldsymbol{\Gamma})\big((\boldsymbol{\Omega}\times\boldsymbol{s})\cdot\boldsymbol{\Gamma}\big)\\ &= mr\ell(\boldsymbol{\Omega}\cdot\boldsymbol{\Gamma})(\boldsymbol{\Omega}\cdot\boldsymbol{\chi}\times\boldsymbol{\Gamma}),\end{aligned}$$

as obtained by using Equations (12.2.11)–(12.2.13). ▲

Remark 12.2.8 The corresponding *spatial* quantity is the vertical component of the spatial angular momentum about the point of contact, given by

$$\boldsymbol{M}\cdot\boldsymbol{\Gamma} = \boldsymbol{m}\cdot\mathbf{e}_3 = m_3\,. \qquad (12.2.22)$$

Here the spatial vector $\boldsymbol{m}$ is defined by $\boldsymbol{m} = g(t)\boldsymbol{M}$ for motion along the curve $g(t)\in SO(3)$. Relation (12.2.21) implies that the vertical component of spatial angular momentum m_3 in (12.2.22) is *not* conserved, even for a cylindrical body. □

Remark 12.2.9 The curve $g(t)$ may be reconstructed from the solution of the motion Equation (12.2.14) for the body angular momentum $\mathbf{\Pi}(t) := I\mathbf{\Omega}(t)$ by inverting the Legendre transformation and then solving the linear differential equation $\dot{g}(t) = g(t)\mathbf{\Omega}(t)$ obtained from the definition of body angular velocity. □

Exercise. (Jellett's relation for the Chaplygin ball) Show that the vector Equation (12.2.9) for nonholonomic motion in body coordinates implies the following evolution equation for the projection of the angular momentum $\mathbf{\Pi}$ onto the vector $\boldsymbol{s}$ directed from the point of contact to the centre of mass:

$$\frac{d}{dt}(\mathbf{\Pi} \cdot \boldsymbol{s}) = \Big(- \mathbf{\Omega} \times (\mathbf{\Pi} - \boldsymbol{\lambda} \times \boldsymbol{s}) + \boldsymbol{\kappa} \times \mathbf{\Gamma} \Big) \cdot \boldsymbol{s} + \mathbf{\Pi} \cdot \dot{\boldsymbol{s}} \,. \tag{12.2.23}$$

For the Chaplygin ball, Equations (12.2.11)–(12.2.13) imply *Jellett's relation,*

$$\frac{d}{dt}(\mathbf{\Pi} \cdot \boldsymbol{s}) = \ell \boldsymbol{\chi} \cdot \mathbf{\Pi} \times \mathbf{\Omega} \,. \tag{12.2.24}$$

★

Remark 12.2.10 (Jellett's integral for cylindrical symmetry) The mass distribution of Chaplygin's ball is usually assumed to be cylindrically symmetric about $\boldsymbol{\chi}$. For this case, treated for example in [LyBu2009], the unit vector $\boldsymbol{\chi}$ points along one of the principal axes and the right-hand side of (12.2.24) vanishes. Thus, rotational symmetry about $\boldsymbol{\chi}$ produces an additional constant of motion, $\mathbf{\Pi} \cdot \boldsymbol{s}$, called *Jellett's integral* [Je1872]. See, e.g., [Cu1998, GrNi2000, BoMaRo2004, LyBu2009] for references to the original literature and further discussions of the relation of cylindrical symmetry to the conservation of Jellett's integral.

As emphasised in [LyBu2009], the concept of cylindrical symmetry has two aspects:

(i) a principal axis passes through the geometric centre;

(ii) the moments of inertia about all axes perpendicular to this axis are equal.

Either of these conditions may be satisfied without the other one holding. The rock'n'roller treated in [LyBu2009] satisfies condition (i) but not condition (ii). Chaplygin's ball satisfies both. An example of a body for which (ii) holds but not (i) is a light polystyrene sphere into which an arbitrarily oriented ellipsoid of heavy metal has been embedded off-centre.

□

Exercise. Prove that the right-hand side of (12.2.24) vanishes when both conditions (i) and (ii) of cylindrical symmetry are satisfied for an axis along χ. ★

Answer. Let $\chi = -\mathbf{E}_3$, so that the centre of mass lies along the $\mathbf{E}_3$ principal axis. Then

$$\ell\chi \cdot \mathbf{\Pi} \times \mathbf{\Omega} = -\ell(I_1\Omega_1\Omega_2 - I_2\Omega_2\Omega_1) = -\ell(I_1 - I_2)\Omega_1\Omega_2\,.$$

For cylindrical symmetry about $\mathbf{E}_3$, we have $I_1 = I_2$. Consequently, the right-hand side vanishes and Jellett's integral $\mathbf{\Pi} \cdot \boldsymbol{s}$ is conserved. ▲

Remark 12.2.11 (Equivalent form of Jellett's integral) Jellett's integral $J = \mathbf{\Pi} \cdot \boldsymbol{s}$ may also be written as a linear expression in the total angular momentum, $\boldsymbol{M}$, defined as

$$\boldsymbol{M} = \mathbf{\Pi} + m\,\boldsymbol{s} \times (\mathbf{\Omega} \times \boldsymbol{s})\,.$$

The equivalent form of Jellett's integral is

$$J = \boldsymbol{M} \cdot \boldsymbol{s} = r\boldsymbol{M} \cdot \boldsymbol{\Gamma} + \ell \boldsymbol{M} \cdot \boldsymbol{\chi} = rm_3 - \ell M_3 \tag{12.2.25}$$

for $\chi = -\mathbf{E}_3$. Thus, Jellett's integral is a weighted difference between the geometric spatial and body angular momentum components in the corresponding three directions. Separately, m_3 and M_3 are not conserved. □

Remark 12.2.12 (Routh's integral) An additional integral of motion for the cylindrically symmetric Chaplygin ball was found by Routh [Ro1860],

$$R = \Omega_3 \sqrt{I_1 I_3 + m(\mathbf{s} \cdot I\mathbf{s})}\,.$$

The Routh integral R may also be written equivalently as a linear expression in the total angular momentum [KiPuHo2011]. Namely,

$$R = \frac{\boldsymbol{M} \cdot (I_1 \boldsymbol{\chi} + m(\boldsymbol{s} \cdot \boldsymbol{\chi})\boldsymbol{s})}{\sqrt{I_1 I_3 + m(\mathbf{s} \cdot I\mathbf{s})}}\,. \tag{12.2.26}$$

□

Exercise. (Endpoint term in vector notation) Express the endpoint term in (12.2.8) for the Hamilton–Pontryagin principle for the Chaplygin ball in vector notation. Discuss its relation to Jellett's integral J. ★

Answer. In vector notation, this endpoint term becomes

$$\langle \Pi + s \diamond \lambda, \eta \rangle = (\boldsymbol{\Pi} + m\,\boldsymbol{s} \times (\boldsymbol{\Omega} \times \boldsymbol{s})) \cdot \boldsymbol{\eta} = \boldsymbol{M} \cdot \boldsymbol{\eta}\,.$$

Thus, the endpoint term in (12.2.8) recovers Jellett's integral J when $\boldsymbol{\eta} = \boldsymbol{s}$. ▲

Remark 12.2.13 (Relation of J, R to Noether's theorem) Both Jellett's integral J in (12.2.25) and Routh's integral R in (12.2.26) arise via Noether's theorem from symmetries of the Lagrangian (12.2.10) for the Chaplygin ball.

In vector notation, the reduced Lagrangian in Equation (12.2.10) for the Chaplygin ball in body coordinates has variations

$$\begin{aligned} \delta l(\boldsymbol{\Omega}, \boldsymbol{Y}, \boldsymbol{\Gamma}) &= \delta\, \boldsymbol{\Omega} \cdot \frac{\delta l}{\delta \boldsymbol{\Omega}} + \delta \boldsymbol{Y} \cdot \frac{\delta l}{\delta \boldsymbol{Y}} + \delta\, \boldsymbol{\Gamma} \cdot \frac{\delta l}{\delta \boldsymbol{\Gamma}} \\ &= \delta\, \boldsymbol{\Omega} \cdot I \boldsymbol{\Omega} + \delta \boldsymbol{Y} \cdot m \boldsymbol{Y} + \delta\, \boldsymbol{\Gamma} \cdot \big(- m \gamma \ell\, \boldsymbol{\chi} \big) . \end{aligned} \tag{12.2.27}$$

The vector variations $(\delta\boldsymbol{\Omega}, \delta\boldsymbol{\Gamma}, \delta\boldsymbol{Y})$ here may be expressed in terms of a single vector $\boldsymbol{\eta}$ by rewriting (12.2.6) equivalently as

$$\begin{aligned} \delta\boldsymbol{\Omega} &= \dot{\boldsymbol{\eta}} + \boldsymbol{\Omega} \times \boldsymbol{\eta} , \\ \delta\boldsymbol{\Gamma} &= -\,\boldsymbol{\eta} \times \boldsymbol{\Gamma} , \\ \delta\boldsymbol{Y} &= \frac{d}{dt} (\boldsymbol{\eta} \times \boldsymbol{s}) + (\boldsymbol{\Omega} \times \boldsymbol{\eta}) \times \boldsymbol{s} . \end{aligned} \tag{12.2.28}$$

When $\boldsymbol{\eta} = \boldsymbol{s}$, these expressions become

$$\begin{aligned} \delta\boldsymbol{\Omega} &= \dot{\boldsymbol{s}} + \boldsymbol{\Omega} \times \boldsymbol{s} = \ell \boldsymbol{\Omega} \times \boldsymbol{\chi} , \\ \delta\boldsymbol{\Gamma} &= -\,\boldsymbol{s} \times \boldsymbol{\Gamma} = -\,\ell \boldsymbol{\chi} \times \boldsymbol{\Gamma} , \\ \delta\boldsymbol{Y} &= (\boldsymbol{\Omega} \times \boldsymbol{s}) \times \boldsymbol{s} = \boldsymbol{Y} \times \boldsymbol{s} , \end{aligned} \tag{12.2.29}$$

where we have used $\dot{\boldsymbol{s}} = -\,\boldsymbol{\Omega} \times (\boldsymbol{s} - \ell\boldsymbol{\chi})$ and $\boldsymbol{Y} = \boldsymbol{\Omega} \times \boldsymbol{s}$.

Substituting Equations (12.2.29) for the variations when $\boldsymbol{\eta} = \boldsymbol{s}$ into the variation of the Lagrangian in (12.2.27) yields

$$\delta l(\boldsymbol{\Omega}, \boldsymbol{Y}, \boldsymbol{\Gamma}) = \ell\, \boldsymbol{\Omega} \times \boldsymbol{\chi} \cdot \boldsymbol{\Pi} . \tag{12.2.30}$$

Perhaps not unexpectedly, this variation vanishes and, thus, the reduced Lagrangian is invariant, for $\boldsymbol{\eta} = \boldsymbol{s}$ when cylindrical symmetry $(I_1 = I_2)$ is imposed and the centre of mass lies along the axis of symmetry. Hence, Jellett's integral J in (12.2.25) arises via Noether's theorem from this cylindrical symmetry of the Lagrangian (12.2.10) for the Chaplygin ball.

Routh's integral R in (12.2.26) also arises via Noether's theorem from a symmetry of the Lagrangian (12.2.10) for the Chaplygin ball, but the calculation is more involved [Ki2011]. □

Exercise. (Routh's integral and symmetry) Show that Routh's integral for the Chaplygin ball R in (12.2.26) arises via Noether's theorem from a symmetry of the Lagrangian (12.2.10). ★

Remark 12.2.14 (Chaplygin's concentric sphere, for $\ell \to 0$) When the centre of mass coincides with the centre of symmetry, then $\ell = 0$, the right-hand side of Equation (12.2.14) vanishes and one recovers the equations of motion for ***Chaplygin's concentric sphere***:

$$\frac{d\boldsymbol{\Pi}_{tot}}{dt} + \boldsymbol{\Omega} \times \boldsymbol{\Pi}_{tot} = 0\,, \qquad \frac{d\boldsymbol{\Gamma}}{dt} + \boldsymbol{\Omega} \times \boldsymbol{\Gamma} = 0\,, \tag{12.2.31}$$

$$\boldsymbol{\Pi}_{tot} = I\boldsymbol{\Omega} + mr^2\,\boldsymbol{\Gamma} \times (\boldsymbol{\Omega} \times \boldsymbol{\Gamma})\,. \tag{12.2.32}$$

The equations for Chaplygin's concentric sphere preserve $|\boldsymbol{\Pi}_{tot}|^2$, $|\boldsymbol{\Gamma}|^2$, $I\boldsymbol{\Omega}\cdot\boldsymbol{\Gamma}$ and the corresponding sum of kinetic energies in (12.2.16) when $\ell \to 0$. For the solution behaviour of Chaplygin's concentric sphere, see, e.g., [Ki2001]. □

12.2.2 Circular disk rocking in a vertical plane

Consider an inhomogeneous circular disk that is rocking and rolling in a vertical plane, as its contact point moves backward and forward along a *straight* line. At the initial moment, the centre of mass is assumed to lie a distance ℓ directly beneath the geometric centre of the disk. That is, in the initial configuration,

$$\mathbf{e}_1 = \begin{pmatrix} 1 \\ 0 \end{pmatrix}, \qquad \mathbf{e}_2 = \begin{pmatrix} 0 \\ 1 \end{pmatrix}, \quad \boldsymbol{\chi} = \begin{pmatrix} 0 \\ -1 \end{pmatrix}, \tag{12.2.33}$$

where $\mathbf{e}_1$ and $\mathbf{e}_2$ are spatial unit vectors in the vertical plane. The moment of inertia is $I = I_3$ for rotations about the third axis, $\mathbf{e}_3$, which is *horizontal*. The orientation of the disk is given by

$A(\theta) \in SO(2)$ for rotations about $\mathbf{e}_3$ by an angle variable $\theta(t)$ and its corresponding body angular velocity $\widehat{\Omega} = A^{-1}\dot{A}$ is given by

$$A(\theta) = \begin{bmatrix} \cos\theta & -\sin\theta \\ \sin\theta & \cos\theta \end{bmatrix}, \qquad \widehat{\Omega} = A^{-1}\dot{A} = \begin{bmatrix} 0 & -1 \\ 1 & 0 \end{bmatrix} \dot{\theta},$$

so that in vector notation the body angular velocity is

$$\boldsymbol{\Omega} = \dot{\theta}\,\mathbf{e}_3 .$$

The remaining body vector variables for this problem are

- the vertical unit vector, $\mathbf{e}_2$, as seen from the body,

$$\boldsymbol{\Gamma}(\theta) = A^{-1}(\theta)\mathbf{e}_2 = \begin{pmatrix} \sin\theta \\ \cos\theta \end{pmatrix};$$

- the vector $\boldsymbol{s}(\theta)$ pointing from the contact point C to the centre of mass location in the body,

$$\boldsymbol{s}(\theta) = r\boldsymbol{\Gamma}(\theta) + \ell\boldsymbol{\chi} = \begin{pmatrix} r\sin\theta \\ r\cos\theta - \ell \end{pmatrix};$$

- the rolling constraint in body coordinates (12.1.5),

$$\boldsymbol{Y} = \boldsymbol{\Omega} \times \boldsymbol{s} = \dot{\theta}\,\mathbf{e}_3 \times \boldsymbol{s} = \dot{\theta}\begin{pmatrix} -r\cos\theta + \ell \\ r\sin\theta \end{pmatrix};$$

- the angular momentum in the body,

$$\boldsymbol{\Pi} = I\boldsymbol{\Omega} = I_3\dot{\theta}\,\mathbf{e}_3 ;$$

- the total angular momentum,

$$\begin{aligned} \boldsymbol{M} &= \boldsymbol{\Pi} - m\boldsymbol{Y} \times \boldsymbol{s} \\ &= \left(I_3\dot{\theta} + m|\boldsymbol{s}(\theta)|^2\dot{\theta}\right)\mathbf{e}_3 \end{aligned}$$

  so that $\boldsymbol{\Omega} \times \boldsymbol{M} = 0$.

The constrained motion equation in vector form (12.2.9) with variational derivatives (12.2.11) for $so(2)^*$ in this problem becomes

$$\left(\frac{d}{dt} + \boldsymbol{\Omega}\times\right)\boldsymbol{M} = -m\gamma\ell\boldsymbol{\chi}\times\boldsymbol{\Gamma} - m\boldsymbol{Y}\times\dot{\boldsymbol{s}}\,. \qquad (12.2.34)$$

The problem of the motion of a rolling disk in a vertical plane is considerably simpler than the fully three-dimensional problem, because each term has only a horizontal ($\mathbf{e}_3$) component, which implies that the vector cross product $\boldsymbol{\Omega}\times\boldsymbol{M}$ vanishes. The horizontal $\mathbf{e}_3$-component of each of the remaining terms on the right-hand side of the Euler–Poincaré Equation (12.2.34) may be calculated as

$$\begin{aligned}\boldsymbol{\Gamma}\times m\gamma\ell\boldsymbol{\chi} &= -m\gamma\ell\begin{pmatrix}\sin\theta\\ \cos\theta\end{pmatrix}\times\begin{pmatrix}0\\1\end{pmatrix}\\ &= -m\gamma\ell\sin\theta\,\mathbf{e}_3\,,\end{aligned}$$

$$\begin{aligned}\dot{\boldsymbol{s}}\times m\boldsymbol{Y} &= -(\boldsymbol{\Omega}\times r\,\boldsymbol{\Gamma})\times m\boldsymbol{Y}\\ &= mr\dot{\theta}^2\begin{pmatrix}\cos\theta\\ -\sin\theta\end{pmatrix}\times\begin{pmatrix}-r\cos\theta+\ell\\ r\sin\theta\end{pmatrix}\\ &= mr\ell\dot{\theta}^2\sin\theta\,\mathbf{e}_3\,.\end{aligned}$$

The right-hand side of the motion Equation (12.2.14), or equivalently Equation (12.2.34), for the Chaplygin disk then becomes

$$m\ell\Big[\gamma\boldsymbol{\Gamma}\times\boldsymbol{\chi} + r\,\boldsymbol{\Omega}\,(\,\boldsymbol{\Omega}\cdot\boldsymbol{\chi}\times\boldsymbol{\Gamma})\Big] = \Big[m\ell(\,-\gamma + r\dot{\theta}^2)\sin\theta\Big]\mathbf{e}_3\,.$$

Remark 12.2.15 The Jellett and Routh integrals both vanish for the Chaplygin disk. □

Assembling the terms in the motion Equation (12.2.34) for the Chaplygin disk yields

$$\frac{d}{dt}\Big[\Big(I_3 + m|\boldsymbol{s}(\theta)|^2\Big)\dot{\theta}\Big] = m\ell(\,-\gamma + r\dot{\theta}^2)\sin\theta\,, \qquad (12.2.35)$$

with

$$|\boldsymbol{s}(\theta)|^2 = r^2 + \ell^2 - 2r\ell\cos\theta\,. \qquad (12.2.36)$$

Remark 12.2.16 (Chaplygin disk vs simple pendulum) Compared to the simple pendulum equation,

$$m\ell^2\ddot{\theta} = -m\ell\gamma\sin\theta\,,$$

the rocking and rolling motion Equation (12.2.35) for the Chaplygin disk has two additional terms. These terms represent the rate of change of angular momentum due to translation and the torque arising from the rolling constraint. □

Remark 12.2.17 (Conservation of Chaplygin disk energy) Although the constraint torque is time-dependent, it does no work. Therefore, it preserves the total energy. Energy conservation is seen by expanding the motion Equation (12.2.35), multiplying it by $\dot{\theta}$ and rearranging to find

$$\frac{d}{dt}\left[\frac{1}{2}I_3\dot{\theta}^2 + \frac{1}{2}m|s(\theta)|^2\dot{\theta}^2 + m\gamma\ell(1-\cos\theta)\right] = 0\,, \tag{12.2.37}$$

which expresses the conserved energy for the constrained motion of the Chaplygin disk rocking and rolling, backward and forward, in a vertical plane. The middle term is the translational kinetic energy. □

Remark 12.2.18 (Euler–Lagrange equation) The constrained reduced Lagrangian in Equation (12.2.10) for the Chaplygin ball in body coordinates may be expressed as

$$l_c(\theta,\dot{\theta}) = \frac{1}{2}I_3\dot{\theta}^2 + \frac{1}{2}m|s(\theta)|^2\dot{\theta}^2 + m\gamma\ell\cos\theta\,. \tag{12.2.38}$$

Remarkably, in this case, the equation of motion for the two-dimensional rocking Chaplygin disk is also the Euler–Lagrange equation for the constrained reduced Lagrangian (12.2.38). □

12.2.3 Euler's rolling and spinning disk

One of the best-known examples of a nonholonomically constrained rigid body is the rolling circular disk. Ignoring dissipation, the equations of motion for the rolling disk were formulated by Slesser [Sl1861] and shown to be integrable by Chaplygin [Ch1897], Appel [Ap1900] and Korteweg [Ko1900], by a transformation of variables that takes the dynamics miraculously to Legendre's equation. Nearly a century later these results were used by Cushman *et al.* [CuHeKe1995] and O'Reilly [Or1996] to examine, among other matters, the stability and bifurcations of the steady motions of flat (infinitely thin) circular disks. For additional historical accounts, see [BoMa2002, BoMaKi2002].

Recently, a toy called the ***Euler disk*** appeared, whose inventors [BeShWy1999] optimised the choice of disk weight, finite thickness, shape, material and surface so that the disk will spin and gyrate for well over a minute before coming to rest. In the last few tens of seconds before it comes to rest, it also shows a fascinating increase in its oscillation frequency which has caught the imagination of many people. See, e.g., [McDMcD2000, PeHuGr2002, LeLeGl2005]. Batista [Ba2006] found the transformation which solves the Euler disk in terms of hypergeometric functions.

Here we will use the Hamilton–Pontryagin variational principle to formulate the equations for the dynamics of the rolling and spinning flat circular disk that were originally solved by Chaplygin [Ch1897], Appel [Ap1900] and Korteweg [Ko1900] using the transformation to Legendre's equation.

Consider a flat circular disk with homogeneous mass distribution which rolls without slipping on a horizontal plane, and whose orientation is allowed to tilt away from the vertical plane. Denote its mass and radius by m and r, respectively. Let $I = I_3$ be its moment of inertia about its axis of circular symmetry and $I_1 = J = I_2$ be its moment of inertia about any diameter. Because its mass distribution is homogeneous, the disk's centre of mass coincides with its centre of circular symmetry.

Remark 12.2.19 The case in which the mass distribution of the disk is unbalanced; that is, when its centre of mass does not coincide with the geometric centre of the disk, is investigated in [Ja2011]. This non-integrable, nonholonomic system combines the motion of a symmetric disk with the rolling of an unbalanced spherical ball. □

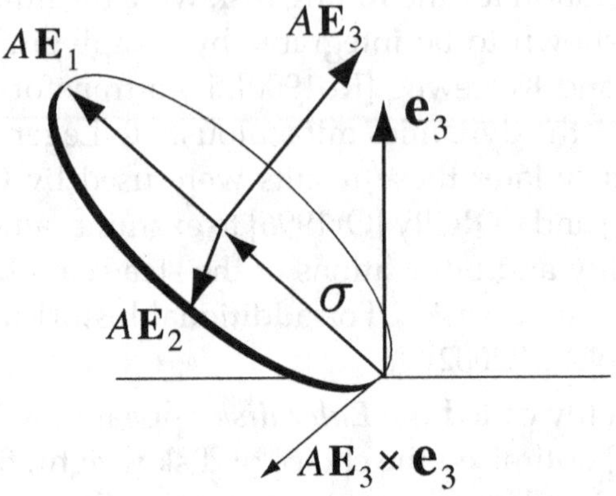

Figure 12.2. The geometry of the Euler disk is essentially the same as that of a rolling coin.

Geometry of the Euler disk

Let (E_1, E_2, E_3) denote an orthonormal frame of unit vectors in the reference system whose origin is attached to the disk at its centre of symmetry, so that the disk lies in the (E_1, E_2) plane in body coordinates (Figure 12.2). Let (e_1, e_2, e_3) be an orthonormal coordinate frame in the inertial (spatial) system. Choose (e_1, e_2) to lie in a fixed horizontal plane and let e_3 be a vertical unit vector. The motion is given by a curve $(A(t), x(t)) \in SE(3)$ where $A(t) \in SO(3)$ and $x(t) \in \mathbb{R}^3$. The origin of the reference system lies at the disk's centre of mass, so that $x(t)$ gives the location of its centre of mass in space.

An element of $SE(3)$ maps a generic point P in the body to a point $A(t)P+x(t)$ in space at time t. The height of the centre of mass at a given time is $z = \langle x(t), e_3 \rangle$. The matrix $A(t) \in SO(3)$ defines the attitude of the body relative to its reference configuration. The vector in the spatial frame, $A(t)E_3$, will be normal to the disk. Consequently, the vector $e_3 \times A(t)E_3$ will be tangent to the edge of the disk at the point of contact with the plane. (One assumes $A(t)E_3 \neq \pm e_3$, so that the disk keeps rolling. The case $\langle e_3, A(t)E_3 \rangle^2 = 1$ will turn out to be a singular situation.)

Definition 12.2.1 (Vector to the centre of mass) *Let $s(t) \in \mathbb{R}^3$ denote the vector in the body frame directed at time t from the point of contact C to the geometric centre of the disk. The corresponding vector in the spatial frame,*

$$\sigma(t) = A(t)s(t)\,,$$

is normal both to $A(t)E_3$ and to the tangent direction $e_3 \times A(t)E_3$. Consequently, $\sigma(t)$ lies along $u(t) := A(t)E_3 \times (e_3 \times A(t)E_3)$ in the spatial frame and because it points from the edge to the centre, its magnitude is equal to the radius of the disk, r. Thus,

$$\sigma(t) = A(t)s(t) = r\frac{u(t)}{|u(t)|} \tag{12.2.39}$$

with

$$u(t) := A(t)E_3 \times (e_3 \times A(t)E_3)\,. \tag{12.2.40}$$

The vector u has squared magnitude

$$|u|^2 = 1 - \langle e_3,\, A(t)E_3\rangle^2 = 1 - \langle A^{-1}(t)e_3,\, E_3\rangle^2 \neq 0\,.$$

The height of the centre of mass at a given time is

$$z(t) = \langle A(t)s(t),\, e_3\rangle = r|u(t)|\,, \tag{12.2.41}$$

so the disk is lying flat ($z = 0$) in the case where $\langle e_3,\, A(t)E_3\rangle^2 = 1$.

Definition 12.2.2 (Rolling constraint) *Physically, rolling of the disk means that the velocity of its centre of mass $\dot{x}(t)$ in the spatial frame at any instant t must be obtained by a rotation of the spatial vector $A(t)s$ at angular frequency $\widehat{\omega}(t) = \dot{A}A^{-1}(t) \in so(3)$ around the fixed contact point C. Consequently, the* ***rolling constraint*** *is imposed by requiring*

$$\dot{x}(t) = \dot{A}A^{-1}\sigma(t) = \widehat{\omega}\sigma(t) = \omega \times r\frac{u(t)}{|u(t)|}\,, \tag{12.2.42}$$

where the hat map $\widehat{\ }: \mathbb{R}^3 \to so(3)$ was used in the last step for the (right-invariant) spatial angular velocity. Equivalently, one may rewrite this rolling constraint using Equation (12.2.39) as

$$\dot{x}(t) = r\dot{A}A^{-1}\frac{u(t)}{|u(t)|}\,. \tag{12.2.43}$$

This formula defines the ***spatial nonholonomic constraint distribution*** *D for this problem.*

The Hamilton–Pontryagin approach for the Euler disk

As before, the Hamilton–Pontryagin method proceeds by first assembling the following ingredients:

- Potential energy: The potential energy due to gravity for the disk is $mgz(t)$, where

$$z(t) = \Big\langle \sigma(t),\, e_3 \Big\rangle = \Big\langle As,\, e_3 \Big\rangle = \Big\langle s(\Gamma),\, \Gamma(t) \Big\rangle \qquad (12.2.44)$$

  is the height of the centre of mass above the horizontal plane and g is the constant acceleration of gravity.

- Kinetic energy: The kinetic energy of the disk is the sum of the energy of translation of its centre of mass and the energy of its rotation about the centre of mass.

- Lagrangian: The Lagrangian is defined as the difference between the total kinetic energy and the gravitational potential energy,

$$L = \frac{1}{2}\langle \Omega, I\Omega \rangle + \frac{m}{2}|\dot{x}|^2 - mg\Big\langle s,\, A^{-1}e_3 \Big\rangle ,$$

  where $\Omega = A^{-1}\dot{A}$ is the (left-invariant) body angular velocity.

- Rolling constraint in the body: The spatial constraint distribution (12.2.43) transforms into body coordinates upon using Equation (12.2.39) and setting

$$Y = A^{-1}\dot{x}\,, \quad \Gamma = A^{-1}e_3\,, \qquad (12.2.45)$$

$$s(\Gamma) = r\,\frac{A^{-1}(t)u(t)}{|u(t)|} = r\,\frac{E_3 \times (\Gamma(t) \times E_3)}{\sqrt{1 - \langle \Gamma(t),\, E_3 \rangle^2}}\,, \qquad (12.2.46)$$

  with vectors Y, Γ, $s(\Gamma)$ each in $\mathbb{R}^3$. Note that

$$s(\Gamma) \cdot \Gamma(t) = r|u(t)| = z(t)$$

  yields the height of the centre of mass and $|s(\Gamma)|^2 = r^2$.

  In this notation, the rolling constraint in body coordinates (12.2.43) simplifies to an algebraic expression,

$$Y = \Omega \times s(\Gamma)\,. \qquad (12.2.47)$$

- Reduced Lagrangian in body coordinates: The reduced Lagrangian is defined in body coordinates by

$$l(\Omega, Y, \Gamma) = \frac{1}{2}\langle \Omega, I\Omega \rangle + \frac{m}{2}|Y|^2 - mg\langle s(\Gamma), \Gamma \rangle \,. \qquad (12.2.48)$$

> These considerations have expressed a spatial Lagrangian
>
> $$L : TG \times TV \times V \to \mathbb{R}$$
>
> in body form as $l : \mathfrak{g} \times TV \times V \to \mathbb{R}$, plus the various constraint relations.

Vector equation for the Euler disk

One may now apply the results of the nonholonomic Hamilton–Pontryagin Theorem 12.2.1 to obtain the motion equation for the Euler disk, in the vector form (12.2.9),

$$\left(\frac{d}{dt} + \boldsymbol{\Omega}\times\right)(\boldsymbol{\Pi} - \boldsymbol{\lambda} \times \boldsymbol{s}) = \boldsymbol{\kappa} \times \boldsymbol{\Gamma} - \boldsymbol{\lambda} \times \dot{\boldsymbol{s}} \,. \qquad (12.2.49)$$

This equation arises from the Lagrangian equivalent to (12.2.48) in terms of vector quantities,

$$L = \frac{1}{2}\,\boldsymbol{\Omega}\cdot I\boldsymbol{\Omega} + \frac{m}{2}|\mathbf{Y}|^2 - mg\,\boldsymbol{s}(\boldsymbol{\Gamma})\cdot\boldsymbol{\Gamma}\,, \qquad (12.2.50)$$

and its vector-valued variational relations,

$$\boldsymbol{\Pi} = \frac{\delta l}{\delta \boldsymbol{\Omega}} = I\boldsymbol{\Omega}\,, \quad \boldsymbol{\kappa} = \frac{\delta l}{\delta \boldsymbol{\Gamma}} = -\,mg\boldsymbol{s}(\boldsymbol{\Gamma})\,, \qquad (12.2.51)$$

$$\boldsymbol{\lambda} = \frac{\delta l}{\delta \boldsymbol{Y}} = m\boldsymbol{Y} = m\boldsymbol{\Omega} \times \boldsymbol{s}(\boldsymbol{\Gamma})\,. \qquad (12.2.52)$$

Recall the definitions,

$$\boldsymbol{\Gamma}(t) = A^{-1}(t)\mathrm{e}_3 \quad \text{and} \quad \boldsymbol{s}(\boldsymbol{\Gamma}) = r\,\frac{\mathbf{E}_3 \times (\boldsymbol{\Gamma} \times \mathbf{E}_3)}{\sqrt{1 - (\boldsymbol{\Gamma}(t)\cdot\mathbf{E}_3)^2}}\,, \qquad (12.2.53)$$

so that

$$\mathbf{\Gamma} \cdot \boldsymbol{s} = \mathbf{e}_3 \cdot \boldsymbol{\sigma} = r\sqrt{1 - (\mathbf{\Gamma}(t) \cdot \mathbf{E}_3)^2} \,. \tag{12.2.54}$$

In vector notation, the two vectors $\mathbf{\Gamma}$ and $\boldsymbol{s}$ satisfy the auxiliary equations,

$$\dot{\mathbf{\Gamma}} = -\,\mathbf{\Omega} \times \mathbf{\Gamma} \tag{12.2.55}$$

and

$$\dot{\boldsymbol{s}} = \frac{r}{\sqrt{1 - (\mathbf{\Gamma} \cdot \mathbf{E}_3)^2}} \Big((\mathbf{E}_3 \times \mathbf{\Gamma}) \cdot (\mathbf{\Omega} \times \mathbf{\Gamma}) \Big) \mathbf{\Gamma} \times \mathbf{E}_3 \,, \tag{12.2.56}$$

so that

$$\mathbf{\Gamma} \cdot \dot{\boldsymbol{s}} = 0 \,.$$

A similar formula holds for the variational derivative, namely

$$\mathbf{\Gamma} \cdot \delta \boldsymbol{s} = 0 \,,$$

which immediately implies the variational derivative

$$\delta l / \delta \mathbf{\Gamma} = -\, mg \boldsymbol{s}(\mathbf{\Gamma})$$

in Equations (12.2.52). The auxiliary equations preserve $|\mathbf{\Gamma}|^2 = 1$ and $|\boldsymbol{s}|^2 = r^2$.

On substituting these vector relations into (12.2.49) one obtains the motion equation for the Euler disk,

$$\begin{aligned} \Big(\frac{d}{dt} + \mathbf{\Omega} \times \Big) \Big(I \mathbf{\Omega} + m \boldsymbol{s}(\mathbf{\Gamma}) \times \big(\mathbf{\Omega} \times \boldsymbol{s}(\mathbf{\Gamma}) \big) \Big) \\ = \underbrace{mg \mathbf{\Gamma} \times \boldsymbol{s}(\mathbf{\Gamma})}_{\text{gravity}} + \underbrace{m \dot{\boldsymbol{s}} \times \big(\mathbf{\Omega} \times \boldsymbol{s}(\mathbf{\Gamma}) \big)}_{\text{rolling constraint torque}} . \end{aligned} \tag{12.2.57}$$

Thus, the motion equation for the Euler disk has the same form as Equation (12.2.14) for the Chaplygin ball. However, the rolling constraint torque is considerably more intricate for the Euler disk, because of the complexity of expression (12.2.56) for $\dot{\boldsymbol{s}}(\mathbf{\Gamma}, \mathbf{\Omega})$.

Exercise. (Energy) Show that the Euler disk in rolling motion without sliding conserves the following energy in body coordinates:

$$E(\mathbf{\Omega},\mathbf{\Gamma}) = \frac{1}{2}\Big\langle \mathbf{\Omega}, I\mathbf{\Omega}\Big\rangle + \frac{m}{2}\big|\mathbf{\Omega}\times \boldsymbol{s}\big|^2 + mg\Big\langle \mathbf{\Gamma}, \boldsymbol{s}(\mathbf{\Gamma})\Big\rangle .$$

This is the sum of the kinetic energies of rotation and translation, plus the potential energy of gravity. (Recall that the rolling constraint does no work.) ★

Exercise. (Jellett's integral for the Euler disk) Show that the Euler disk admits Jellett's integral $(\mathbf{\Pi}\cdot \boldsymbol{s})$ as an additional constant of motion because of the cylindrical symmetry of its mass distribution.

Hint: Compute the analogue for the Euler disk of Equation (12.2.24) for the Chaplygin ball. ★

12.3 Nonholonomic Euler–Poincaré reduction

The treatments of the Chaplygin ball and the Euler disk in the previous section are consistent with an Euler–Poincaré version of the standard ***Lagrange–d'Alembert principle*** [Sc2002, Bl2004]. As in the cases of the Chaplygin ball and the Euler disk, the configuration space will be taken to be a semidirect-product space $\mathfrak{S} = G\circledS V$ such as the special Euclidean group $SE(3) \simeq SO(3)\circledS\mathbb{R}^3$ of orientations and positions, whose various adjoint and coadjoint (AD, Ad, ad, Ad* and ad*) actions needed for Euler–Poincaré theory are discussed in Chapter 7.

12.3.1 Semidirect-product structure

A semidirect-product structure involves a Lie group G and a representation vector space V for the action of G, which we choose to be a left action. Let $\mathfrak{S}$ denote the semidirect product $G \circledS V$. Topologically, $\mathfrak{S}$ is the direct product of spaces $G \times V$. The left group action on $\mathfrak{S}$ is given as in (6.1.1) for the special Euclidean group of rotations and translations, $SE(3)$,

$$(g_1, y_1) \cdot (g_2, y_2) = (g_1 g_2,\, g_1 y_2 + y_1)\,, \tag{12.3.1}$$

where for $g \in G$ and $y \in V$, the left action of G on V is denoted as gy. The formulas for the right action of G on V may be derived by making the appropriate adjustments, as discussed in Section 6.3.3. The induced left action of $\mathfrak{S}$ on its tangent space $\Phi : \mathfrak{S} \times T\mathfrak{S} \to T\mathfrak{S}$ is given by

$$(h, y) \cdot (g, x) = (hg, h\dot{g}, hx + y, h\dot{x})\,, \tag{12.3.2}$$

where dot $(\dot{\ })$ in $(\dot{g}(t), \dot{x}(t))$ denotes the time derivative along a curve $(g(t), x(t))$ in $\mathfrak{S}$. Let $\mathfrak{g}$ denote the Lie algebra of the Lie group G and $\mathfrak{s}$ the Lie algebra of $\mathfrak{S}$. (As a vector space, the Lie algebra $\mathfrak{s}$ is $\mathfrak{g} \times V$.)

The semidirect-product Lie bracket on $\mathfrak{s}$ is of the form in Equation (6.3.1),

$$[(\xi_1, Y_1)\,, (\xi_2, Y_2)] = (\,[\xi_1, \xi_2]\,, \xi_1 Y_2 - \xi_2 Y_1\,)\,, \tag{12.3.3}$$

where the left induced action from $\mathfrak{g} \times V \to V$ is denoted by concatenation from the left, as in $\xi_1 Y_2$. The dual action $V^* \times V \to \mathfrak{g}^*$ for fixed $b \in V^*$ and $v \in V$ is defined for a symmetric real pairing $\langle \cdot\,, \cdot \rangle : V^* \times V \to \mathbb{R}$ by the diamond operation:

$$\langle b,\, -\,\xi v \rangle = \langle b \diamond v,\, \xi \rangle\,. \tag{12.3.4}$$

Definition 12.3.1 *The left- and right-invariant angular velocities* $\xi, \omega \in \mathfrak{g}$ *are defined by*

$$\xi = g^{-1}\dot{g} \quad \textit{and} \quad \omega = \dot{g}g^{-1}\,. \tag{12.3.5}$$

Definition 12.3.2 (Linear constraints in space and body) *Linear constraints in spatial form are expressed as*

$$\dot{x}(t) = \omega(k\,a_0 + g(t)\ell\chi)\,, \tag{12.3.6}$$

and are defined by left Lie algebra and Lie group actions of $\omega \in \mathfrak{g}$ and $g \in G$ on fixed elements a_0 and χ of V, with constant k. In terms of the ***body variables*** *defined by*

$$\Gamma = g^{-1}a_0, \quad Y = g^{-1}\dot{x}\,, \tag{12.3.7}$$

the ***linear constraint in body form*** *corresponding to (12.3.6) may be expressed in its body form as*

$$\begin{aligned} Y &= g^{-1}\dot{x} = g^{-1}\dot{g}g^{-1}(k\,a_0 + g(t)\ell\chi) \\ &= (g^{-1}\dot{g})(k\,g^{-1}a_0 + \ell\chi) = \xi\, s(\Gamma)\,, \end{aligned} \tag{12.3.8}$$

where $s(\Gamma) = k\Gamma + \ell\chi$.

Remark 12.3.1 (Auxiliary equation for Γ) By its definition, the quantity $\Gamma(t) = g^{-1}(t)a_0$ satisfies the auxiliary equation,

$$\frac{d\Gamma}{dt} = -\,\xi(t)\Gamma(t)\,. \tag{12.3.9}$$

Hence, the linear combination $s(\Gamma) = k\Gamma + \ell\chi$ satisfies

$$\dot{s}(\Gamma) = -\,k\xi(t)\Gamma(t)\,. \tag{12.3.10}$$

□

Definition 12.3.3 (Lagrangians on semidirect products) *Denote the Lagrangian by*

$$L_0 : T\mathfrak{G} \to \mathbb{R}\,.$$

Introduce parameter dependence into the Lagrangian $L_0(g, \dot{g}, x, \dot{x})$ by setting it equal to the Lagrangian $L(g, \dot{g}, x, \dot{x}, a_0)$,

$$L : T\mathfrak{G} \times V \to \mathbb{R}\,,$$

with a fixed parameter a_0 for the last factor of V.

Remark 12.3.2 In what follows, we shall assume that Lagrangians L_0 and L are independent of the position x. (This corresponds, for example, to rolling on a horizontal plane.) □

Definition 12.3.4 (Reduced and constrained Lagrangians) *The **reduced Lagrangian** is defined in terms of the x-independent Lagrangian L as*

$$l : \mathfrak{g} \times V, \quad l(\xi, Y, \Gamma) = L(g, g\xi, gY, g\Gamma) = L(g, \dot{g}, \dot{x}, a_0)\,. \qquad (12.3.11)$$

*The **constrained reduced Lagrangian** is defined by evaluating the reduced Lagrangian on the constraint distribution, as*

$$l_c : \mathfrak{g} \times V, \quad l_c(\xi, \xi s(\Gamma), \Gamma) = l(\xi, Y, \Gamma)\big|_{Y=\xi s(\Gamma)}\,. \qquad (12.3.12)$$

Thus, the reduced Lagrangian l is found by transforming from spatial to body variables in the Lagrangian L. Then the constrained reduced Lagrangian l_c is defined by evaluating the reduced Lagrangian l on the constraint distribution.

12.3.2 Euler–Poincaré theorem

The definitions and assumptions for the constraints of the previous section form the semidirect-product framework that leads to the nonholonomic Euler–Poincaré theorem found in [Sc2002].

Theorem 12.3.1 (Nonholonomic Euler–Poincaré theorem) *The following three statements are equivalent.*

- *The curve $(g(t),\, x(t)) \in \mathfrak{S}$ satisfies the Lagrange–d'Alembert principle for the Lagrangian L. That is, $(\dot{g}(t),\, \dot{x}(t)) \in \mathfrak{S}$ satisfies the constraint (12.3.6) and*

$$\delta \int L(g(t), \dot{g}(t), \dot{x}(t), a_0)dt = 0,$$

where δg is an independent variation vanishing at the endpoints, and the variation δx satisfies the constraint condition,

$$\delta x = (\delta g g^{-1})(k a_0 + g\ell\chi)\,. \qquad (12.3.13)$$

- *The reduced Lagrangian l defined on $\mathfrak{s} \times V$ satisfies a constrained variational principle*

$$\delta \int l(\xi(t), Y(t), \Gamma(t)) = 0, \tag{12.3.14}$$

where the variations take the form

$$\begin{aligned} \delta\xi &= \dot{\eta} + ad_\xi \eta\,, \\ \delta\Gamma &= -\eta\Gamma\,, \\ \delta Y &= \dot{\eta} s(\Gamma) + (ad_\xi \eta) s(\Gamma) + \eta \dot{s}(\Gamma)\,, \end{aligned} \tag{12.3.15}$$

and $\eta(t) = g^{-1}\delta g \in \mathfrak{g}$ is an independent variation vanishing at the endpoints.

- *The following nonholonomic Euler–Poincaré equations hold on $\mathfrak{g}^* \times V^*$:*

$$\frac{d}{dt}\frac{\partial l_c}{\partial \xi} - \mathrm{ad}^*_\xi \frac{\partial l_c}{\partial \xi} = \frac{\partial l}{\partial \Gamma} \diamond \Gamma + \dot{s} \diamond \frac{\partial l}{\partial Y}\,, \tag{12.3.16}$$

$$\text{and} \quad \frac{d\Gamma}{dt} + \xi\Gamma = 0\,, \tag{12.3.17}$$

where the constrained Lagrangian l_c is obtained by evaluating the reduced Lagrangian l on the constraint distribution.

Remark 12.3.3 The theorem states that the Lagrange–d'Alembert principle on $T\mathfrak{G}$ and equations of motion on $T^*\mathfrak{G}$ is equivalent to a constrained variational principle on $\mathfrak{s} \times V$ and Euler–Poincaré equations on $\mathfrak{g}^* \times V^*$. □

Definition 12.3.5 (Lagrange–d'Alembert principle) *The **Lagrange–d'Alembert principle** on $T\mathfrak{G}$ for L in Theorem 12.3.1 is the stationary principle,*

$$\delta \int L(g(t), \dot{g}(t), \dot{x}(t))dt = 0\,,$$

where $\delta g(t)$ is an independent variation vanishing at the endpoints, and the variations δx of coordinates $x(t) \in V$ must be consistent with the velocity constraint (12.3.6), rewritten now as a relation between one-forms,

$$dx = (dg\, g^{-1})(ka_0 + g\ell\chi) =: A(g, a_0)\, dg\,.$$

Correspondingly, the variations of L_0 in x must satisfy

$$\delta x(t) = (\delta g g^{-1})(k a_0 + g \ell \chi) =: A(g, a_0)\, \delta g\,. \tag{12.3.18}$$

Stationarity under these variations of $\int L_0\, dt$ produces

$$\frac{d}{dt}\frac{\partial L_{0\,c}}{\partial \dot g} - \frac{\partial L_{0\,c}}{\partial g} = \frac{\partial L_0}{\partial \dot x}\, \mathsf{B}\, \dot g\,,$$

where the constrained Lagrangian $L_{0\,c}$ is obtained by evaluating L_0 on the velocity constraint $\dot x = A(g, a_0)\, \dot g$ as

$$L_{0\,c}(g(t), \dot g(t)) = L_0(g(t), \dot g(t), \dot x(t))\big|_{\dot x = A(g, a_0)\, \dot g(t)}\,. \tag{12.3.19}$$

The quantity B appearing in this equation is viewed geometrically as the curvature associated with the connection A that defines the constraint distribution. The curvature two-form is defined by applying the exterior derivative to the connection one-form as

$$\mathrm{d}(A(g, a_0) dg) = \mathsf{B} \mathrm{d}g \wedge \mathrm{d}g \quad \textit{so that} \quad \mathsf{B} = \mathrm{skew}\left(\frac{\partial A}{\partial g}\right).$$

Of course, the variations of L_0 were taken before rewriting the result using derivatives of $L_{0\,c}$. Details of the computations leading to these formulas and more discussions of their meaning are provided in [BlKrMaMu1996].

Proof. By comparing definitions, one finds that the integrand for the Lagrange–d'Alembert principle for L on $\mathfrak{S}$ in the stationary principle is equal to the integrand l on $\mathfrak{s} \times V$. We must compute what the variations on the group $\mathfrak{S}$ imply on the reduced space $\mathfrak{s} \times V$. Define $\eta = g^{-1}\delta g$. As for the pure Euler–Poincaré theory with left-invariant Lagrangians, the proof of the variational formula expressing $\delta\xi$ in terms of η proceeds by direct computation.

$$\delta\xi = \delta(g^{-1}\dot g) = \dot\eta + \mathrm{ad}_\xi \eta\,.$$

For the $\delta\Gamma$ variation, one calculates

$$\delta\Gamma = \delta(g^{-1}) a_0 = -\, g^{-1}\delta g\, g^{-1} a_0 = -\, \eta\Gamma.$$

The δY variation is computed as follows:

$$\begin{aligned}\delta Y &= \delta(g^{-1})\dot{x} + g^{-1}\delta\dot{x}\\ &= -g^{-1}\,\delta g\,g^{-1}\dot{x} + \frac{d}{dt}(g^{-1}\delta x) - (g^{-1})^{\cdot}\delta x\,.\end{aligned}$$

Upon substituting the constraint $\delta x = \delta g\,g^{-1}(ka_0 + g\ell\chi)$ and expanding terms, one finds $g^{-1}\delta x = \eta s$ and $g^{-1}\dot{x} = \xi s$ so that

$$\begin{aligned}\delta Y &= -\eta\,\xi s + \frac{d}{dt}(\eta\, s) + \xi\eta s\\ &= (\dot{\eta} + ad_{\xi}\eta)s + \eta\dot{s}\,.\end{aligned}$$

After this preparation, the nonholonomic EP equation finally emerges from a *direct* computation of the variation, δS, of the action $S = \int l(\xi, Y, \Gamma)\,dt$:

$$\delta S = \int \left\langle \frac{\partial l}{\partial \xi}, \delta\xi \right\rangle + \left\langle \frac{\partial l}{\partial Y}, \delta Y \right\rangle + \left\langle \frac{\partial l}{\partial \Gamma}, \delta\Gamma \right\rangle dt \qquad (12.3.20)$$

$$\begin{aligned}&= \int \left\langle \frac{\partial l}{\partial \xi}, \dot{\eta} + \mathrm{ad}_{\xi}\eta \right\rangle + \left\langle \frac{\partial l}{\partial Y}, \dot{\eta}s + (\mathrm{ad}_{\xi}\eta)s + \eta\dot{s} \right\rangle\\ &\qquad + \left\langle \frac{\partial l}{\partial \Gamma}, -\eta\Gamma \right\rangle dt\\ &= \int \Big\langle -\frac{d}{dt}\frac{\partial l}{\partial \xi} + \mathrm{ad}^*_{\xi}\frac{\partial l}{\partial \xi} - \frac{d}{dt}\left(s \diamond \frac{\partial l}{\partial Y}\right)\\ &\qquad + \mathrm{ad}^*_{\xi}\left(s \diamond \frac{\partial l}{\partial Y}\right) + \dot{s} \diamond \frac{\partial l}{\partial Y} + \frac{\partial l}{\partial \Gamma} \diamond \Gamma\,,\, \eta \Big\rangle\, dt\\ &\qquad + \left\langle \left(\frac{\partial l}{\partial \xi} + s \diamond \frac{\partial l}{\partial Y}\right), \eta \right\rangle \Big|_a^b .\end{aligned} \qquad (12.3.21)$$

Setting the variation δS equal to zero for any choice of η that vanishes at the endpoints in time, $\eta(a) = 0 = \eta(b)$, yields the nonholonomic Euler–Poincaré equation on $\mathfrak{s} \times V$,

$$\frac{d}{dt}\left(\frac{\partial l}{\partial \xi} + s \diamond \frac{\partial l}{\partial Y}\right) - \mathrm{ad}^*_{\xi}\left(\frac{\partial l}{\partial \xi} + s \diamond \frac{\partial l}{\partial Y}\right) = \dot{s} \diamond \frac{\partial l}{\partial Y} + \frac{\partial l}{\partial \Gamma} \diamond \Gamma\,. \qquad (12.3.22)$$

Remark 12.3.4 (Equivalence theorem) On comparison, one sees that the nonholonomic Euler–Poincaré Equation (12.3.22) coincides with the motion Equation (12.2.3) derived in Theorem 12.2.1 using the Hamilton–Pontryagin method. □

Remark 12.3.5 (Noether's theorem) If the variation of the action δS vanishes because of a Lie symmetry of the Lagrangian, so that its first line in (12.3.20) vanishes,

$$\Big\langle \frac{\partial l}{\partial \xi}, \delta\xi \Big\rangle + \Big\langle \frac{\partial l}{\partial \Gamma}, \delta\Gamma \Big\rangle + \Big\langle \frac{\partial l}{\partial Y}, \delta Y \Big\rangle = 0\,, \tag{12.3.23}$$

with variations $\delta\xi$, $\delta\Gamma$ and the modified variation δY given in (12.3.15) in terms of the symmetry generator η, then according to Noether's theorem the endpoint term in (12.3.21)

$$\Big\langle M, \eta \Big\rangle \quad \text{with} \quad M := \frac{\partial l}{\partial \xi} + s \diamond \frac{\partial l}{\partial Y}$$

must be a constant of the motion governed by the constrained Euler–Poincaré Equation (12.3.22).

Notice that the Noether constant of motion $\langle M, \eta \rangle$ is linear in the total angular momentum, M, just as for the Jellett and Routh integrals (12.2.25) and (12.2.26), respectively, for the cylindrically symmetric Chaplygin ball whose centre of mass lies on the axis of symmetry. The corresponding Lie algebra actions by η for those two constants of motion are symmetries of their Lagrangian (12.2.10). However, as shown in [Ki2011] and [KiPuHo2011], these symmetries should be understood in the sense of infinitesimal invariance of the Lagrangian in (12.3.23) under variations (12.3.15) that respect the nonholonomic rolling constraint (12.3.8) and the dynamical Equations (12.3.9) and (12.3.10) that govern the auxiliary quantities Γ and $s(\Gamma)$. See Remark 12.2.13 for the explicit calculations in the example of Jellett's integral for the Chaplygin ball. □

12.3.3 Constrained reduced Lagrangian

The nonholonomic EP equation may also be written in terms of the constrained reduced Lagrangian. It is defined in (12.3.12) as

$$l_c(\xi, \Gamma) := l(\xi, Y, \Gamma)\big|_{Y = \xi s(\Gamma)}\,.$$

Its differential is given by

$$
\begin{aligned}
dl_c &= \Big\langle \frac{\partial l_c}{\partial \xi}\,,\, d\xi \Big\rangle + \Big\langle \frac{\partial l_c}{\partial \Gamma}\,,\, d\Gamma \Big\rangle \\
&= \Big\langle \frac{\partial l}{\partial \xi}\,,\, d\xi \Big\rangle + \Big\langle \frac{\partial l}{\partial Y}\,,\, d\xi s(\Gamma) + \xi ds(\Gamma) \Big\rangle + \Big\langle \frac{\partial l}{\partial \Gamma}\,,\, d\Gamma \Big\rangle \\
&= \Big\langle \frac{\partial l}{\partial \xi} - \frac{\partial l}{\partial Y} \diamond s(\Gamma)\,,\, d\xi \Big\rangle + \Big\langle \frac{\partial l}{\partial \Gamma} + \Big(\xi \frac{\partial s}{\partial \Gamma}\Big)^T \frac{\partial}{\partial Y}\,,\, d\Gamma \Big\rangle .
\end{aligned}
$$

This calculation implies the following relations among the derivatives of the reduced Lagrangian l evaluated on the constraint, and the derivatives of the constrained reduced Lagrangian l_c:

$$
\frac{\partial_c}{\partial \xi} = \Big(\frac{\partial}{\partial \xi} + s \diamond \frac{\partial}{\partial Y} \Big) \quad \text{and} \quad \frac{\partial_c}{\partial \Gamma} = \frac{\partial l}{\partial \Gamma} + \Big(\xi \frac{\partial s}{\partial \Gamma}\Big)^T \frac{\partial}{\partial Y} . \qquad (12.3.24)
$$

Substituting the first relation into (12.3.22) and recalling the advection equation for Γ yields the expression (12.3.17) in the statement of the theorem for the nonholonomic Euler–Poincaré equations, namely,

$$
\begin{aligned}
\frac{d}{dt}\frac{\partial l_c}{\partial \xi} - \mathrm{ad}^*_\xi \frac{\partial l_c}{\partial \xi} &= \frac{\partial l}{\partial \Gamma} \diamond \Gamma - \frac{\partial l}{\partial Y} \diamond \dot{s}\,, \qquad (12.3.25) \\
\text{and} \quad \frac{d\Gamma}{dt} + \xi\,\Gamma &= 0\,. \qquad (12.3.26)
\end{aligned}
$$

This completes the proof of the nonholonomic Euler–Poincaré Theorem 12.3.1. ∎

Corollary 12.3.1 (Energy conservation) *The nonholonomic Euler–Poincaré motion Equation (12.3.25) conserves the following energy in body coordinates:*

$$
E(\xi, \Gamma) = \Big\langle \frac{\partial l_c}{\partial \xi}\,,\, \xi \Big\rangle - l_c(\xi, \Gamma)\,. \qquad (12.3.27)
$$

This is the Legendre transform of the constrained reduced Lagrangian.

Proof. Conservation of the energy $E(\xi, \Gamma)$ in (12.3.27) may be verified directly by differentiating it with respect to time and following a sequence of steps that use the auxiliary equation $d\Gamma/dt = -\xi\Gamma$, the equation of nonholonomic motion (12.3.25), the relations (12.3.24) and the antisymmetry of the adjoint (ad) and diamond ($\diamond$) operations:

$$\begin{aligned}
\frac{dE}{dt} &= \Big\langle \frac{d}{dt}\frac{\partial l_c}{\partial \xi}, \xi \Big\rangle - \Big\langle \frac{\partial l_c}{\partial \Gamma}, \frac{d\Gamma}{dt} \Big\rangle \\
\text{by (12.3.25)} \quad &= \Big\langle \mathrm{ad}^*_\xi \frac{\partial l_c}{\partial \xi} + \frac{\partial l}{\partial \Gamma} \diamond \Gamma + \dot{s} \diamond \frac{\partial l}{\partial Y}, \xi \Big\rangle + \Big\langle \frac{\partial l_c}{\partial \Gamma}, \xi\Gamma \Big\rangle \\
\text{by ad, ad}^* \text{ and } \diamond \quad &= \Big\langle \dot{s} \diamond \frac{\partial l}{\partial Y}, \xi \Big\rangle + \Big\langle \frac{\partial l_c}{\partial \Gamma} - \frac{\partial l}{\partial \Gamma}, \xi\Gamma \Big\rangle .
\end{aligned}$$

By the second relation in (12.3.24) and the auxiliary equation $d\Gamma/dt = -\xi\Gamma$, one finds

$$\begin{aligned}
\Big\langle \frac{\partial l_c}{\partial \Gamma} - \frac{\partial l}{\partial \Gamma}, \xi\Gamma \Big\rangle &= -\Big\langle \Big(\xi \frac{\partial s}{\partial \Gamma}\Big)^T \frac{\partial}{\partial Y}, \frac{d\Gamma}{dt} \Big\rangle \\
&= -\Big\langle \frac{\partial}{\partial Y}, \xi \frac{\partial s}{\partial \Gamma}\frac{d\Gamma}{dt} \Big\rangle = -\Big\langle \frac{\partial}{\partial Y}, \xi\dot{s}(\Gamma) \Big\rangle .
\end{aligned}$$

Thus,

$$\frac{dE}{dt} = \Big\langle \dot{s} \diamond \frac{\partial l}{\partial Y} + \frac{\partial}{\partial Y} \diamond \dot{s}, \xi \Big\rangle = 0 ,$$

in which the last step invokes the *antisymmetry* of the diamond operation when paired with a vector field, η, which holds because $0 = \pounds_\eta \langle \lambda, s \rangle = \langle \lambda, \eta s \rangle + \langle \eta\lambda, s \rangle = -\langle \lambda \diamond s, \eta \rangle - \langle s \diamond \lambda, \eta \rangle$. As a consequence, $dE/dt = 0$ and the nonholonomic equations of motion (12.3.25) conserve the energy in Equation (12.3.27). ∎

Exercise. Use Noether's theorem to derive conservation of energy (12.3.27) for the nonholonomic Euler–Poincaré Equation (12.3.22). ★

Remark 12.3.6 (More general formulations) It is possible to formulate this nonholonomic reduction procedure more generally in the Euler–Poincaré context. For example, one may allow the constraints and Lagrangian to depend on two or more different parameters, say, $(a_0, b_0) \in V \times V$, and take an arbitrary linear combination for the constraints, as explained, for example, in [Sc2002]. This is useful for nonspherical bodies, whose constraints and potential energy may depend on the other two (nonvertical) directions, as seen from the body. □

Exercise. Show that this generalisation adds other terms and their corresponding auxiliary equations of the same form as those involving Γ to the right-hand side of the nonholonomic Euler–Poincaré Equations (12.3.25). ★

Exercise. What opportunities does this generalisation imply for the rocking and rolling of a nonspherical body? How about the case of a triaxial ellipsoid? ★

Exercise. Derive the Euler–Poincaré equations and discuss the properties of their solutions for the rolling motion of a circular disk whose centre of mass is displaced from its centre of symmetry.

This problem is a combination of the standard Euler disk, in which the centre of mass lies at the centre of symmetry, and the Chaplygin ball, in which it does not. ★

Exercise. Formulate the Clebsch method for dynamical systems with nonholonomic contraints. What role is played by the momentum map in this formulation? ★

A
GEOMETRICAL STRUCTURE OF CLASSICAL MECHANICS

Contents

A.1 Manifolds 288

A.2 Motion: Tangent vectors and flows 296

A.2.1 Vector fields, integral curves and flows 297

A.2.2 Differentials of functions: The cotangent bundle 299

A.3 Tangent and cotangent lifts 300

A.3.1 Summary of derivatives on manifolds 301

A.1 Manifolds

The main text refers to definitions and examples of smooth manifolds that are collected in this appendix for ready reference. We follow [AbMa1978, MaRa1994, Le2003, Ol2000, Wa1983], which should be consulted for further discussions as needed.

Definition A.1.1 *A smooth (i.e., differentiable) manifold M is a set of points together with a finite (or perhaps countable) set of subsets $U_\alpha \subset M$ and one-to-one mappings $\phi_\alpha : U_\alpha \to \mathbb{R}^n$ such that*

- $\bigcup_\alpha U_\alpha = M$;
- *for every nonempty intersection $U_\alpha \cap U_\beta$, the set $\phi_\alpha \left(U_\alpha \cap U_\beta \right)$ is an open subset of $\mathbb{R}^n$ and the one-to-one mapping $\phi_\beta \circ \phi_\alpha^{-1}$ is a smooth function on $\phi_\alpha \left(U_\alpha \cap U_\beta \right)$.*

Remark A.1.1 The sets U_α in the definition are called ***coordinate charts***. The mappings ϕ_α are called ***coordinate functions*** or ***local coordinates***. A collection of charts satisfying both conditions is called an ***atlas***. □

Remark A.1.2 Refinements in the definition of a smooth manifold such as maximality conditions and equivalence classes of charts are ignored. (See [Wa1983, Le2003] for excellent discussions of these matters.) □

Example A.1.1 *Manifolds often arise as level sets*

$$M = \left\{ x \middle| f_i(x) = 0,\ i = 1, \dots, k \right\},$$

for a given set of smooth functions $f_i : \mathbb{R}^n \to \mathbb{R}$, $i = 1, \dots, k$.

If the gradients ∇f_i are linearly independent, or more generally if the rank of $\{\nabla f(x)\}$ is a constant r for all x, then M is a smooth manifold of dimension $n - r$. The proof uses the implicit function theorem to show that an $(n-r)$-dimensional coordinate chart may be defined in a neighbourhood of each point on M. In this situation, the set M is called an ***implicit submanifold*** *of $\mathbb{R}^n$ (see [Le2003, Ol2000]).*

Definition A.1.2 (Submersion) *If $r = k = \dim M$, then the map $\{f_i\} : \mathbb{R}^n \to M$ is a* ***submersion****. (A submersion is a smooth map between smooth manifolds whose derivative is everywhere surjective.)*

Example A.1.2 (Stereographic projection of $S^2 \to \mathbb{R}^2$) *The unit sphere*

$$S^2 = \{(x, y, z) : x^2 + y^2 + z^2 = 1\}$$

is a smooth two-dimensional manifold realised as a submersion in $\mathbb{R}^3$. Let

$$U_N = S^2 \backslash \{0, 0, 1\} \quad \text{and} \quad U_S = S^2 \backslash \{0, 0, -1\}$$

be the subsets obtained by deleting the north and south poles of S^2, respectively. Let

$$\chi_N : U_N \to (\xi_N, \eta_N) \in \mathbb{R}^2 \quad \text{and} \quad \chi_S : U_S \to (\xi_S, \eta_S) \in \mathbb{R}^2$$

be stereographic projections from the north and south poles onto the equatorial plane, $z = 0$.

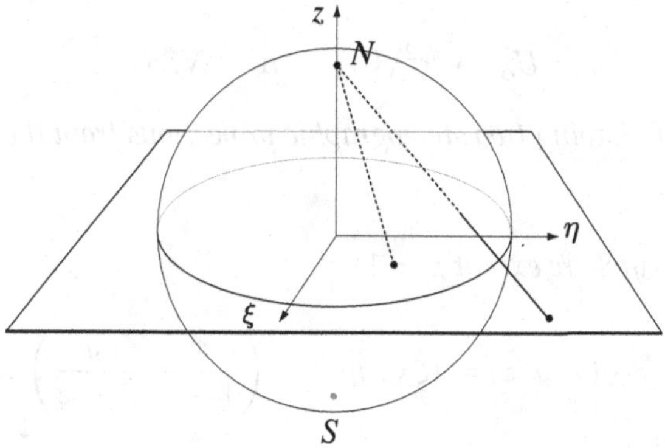

Figure A.1. Two Riemann projections from the north pole of the unit sphere onto the $z = 0$ plane with coordinates (ξ_N, η_N).

Thus, one may place two different coordinate patches in S^2 intersecting everywhere except at the points along the z-axis at $z = 1$ (north pole) and $z = -1$ (south pole).

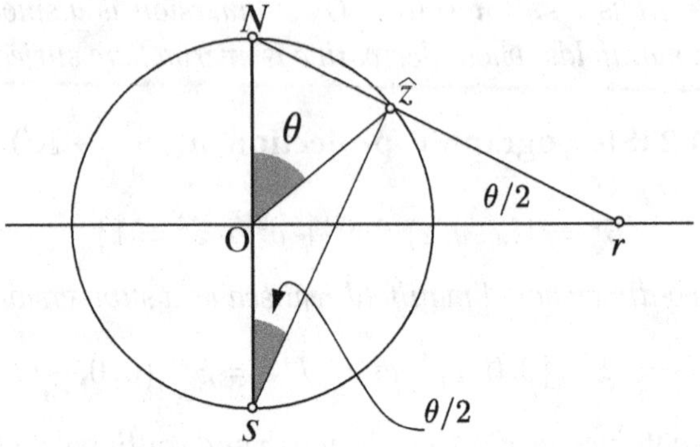

Figure A.2. Riemann projection in the $\xi_N - z$ plane at fixed azimuth $\phi = 0$. The projection through $\hat{z} = (\sin\theta, \cos\theta)$ strikes the ξ_N-axis at distance $r = \cot(\theta/2)$.

In the equatorial plane $z = 0$, one may define two sets of (right-handed) coordinates,

$$\phi_\alpha : U_\alpha \to \mathbb{R}^2\backslash\{0\}, \quad \alpha = N, S,$$

obtained by the following two stereographic projections from the north and south poles:

- *(valid everywhere except $z = 1$)*

$$\phi_N(x, y, z) = (\xi_N, \eta_N) = \left(\frac{x}{1-z}, \frac{y}{1-z}\right),$$

- *(valid everywhere except $z = -1$)*

$$\phi_S(x, y, z) = (\xi_S, \eta_S) = \left(\frac{x}{1+z}, \frac{-y}{1+z}\right).$$

(The two complex planes are identified differently with the plane $z = 0$. An orientation-reversal is necessary to maintain consistent coordinates on the sphere.)

One may check directly that on the overlap $U_N \cap U_S$, the map

$$\phi_N \circ \phi_S^{-1} : \mathbb{R}^2 \backslash \{0\} \to \mathbb{R}^2 \backslash \{0\}$$

is a smooth diffeomorphism, given by the inversion

$$\phi_N \circ \phi_S^{-1}(x, y) = \left(\frac{x}{x^2 + y^2}, \frac{y}{x^2 + y^2} \right).$$

Exercise. Construct the mapping from $(\xi_N, \eta_N) \to (\xi_S, \eta_S)$ and verify that it is a diffeomorphism in $\mathbb{R}^2 \backslash \{0\}$.
Hint: $(1 + z)(1 - z) = 1 - z^2 = x^2 + y^2$. ★

Answer.

$$(\xi_S, -\eta_S) = \frac{1 - z}{1 + z} (\xi_N, \eta_N) = \frac{1}{\xi_N^2 + \eta_N^2} (\xi_N, \eta_N).$$

The map $(\xi_N, \eta_N) \to (\xi_S, \eta_S)$ is smooth and invertible except at $(\xi_N, \eta_N) = (0, 0)$. ▲

Exercise. Show that the circle of points on the sphere with polar angle (colatitude) θ from the north pole project stereographically to another circle given in the complex plane by $\zeta_N = \xi_N + i\eta_N = \cot(\theta/2)e^{i\phi}$. ★

Answer. At a fixed azimuth $\phi = 0$ (in the ξ_N, z-plane as in Figure A.2) a point on the sphere at colatitude θ from the north pole has coordinates

$$\xi_N = \sin\theta, \quad z = \cos\theta.$$

Its projection strikes the (ξ_N, η_N)-plane at distance r and angle ψ, given by

$$\cot\psi = r = \frac{r - \sin\theta}{\cos\theta} .$$

Thus $\psi = \theta/2$, since

$$\cot\psi = r = \frac{\sin\theta}{1 - \cos\theta} = \cot(\theta/2) .$$

The stereographic projection of the sphere from its north pole at polar angle θ thus describes a circle in the complex plane at

$$\zeta_N = \xi_N + i\eta_N = \frac{x + iy}{1 - z} = \cot(\theta/2)e^{i\phi} .$$

The corresponding stereographic projection of the sphere from its south pole describes the circle,

$$\zeta_S = \xi_S + i\eta_S = \frac{x - iy}{1 + z} = \tan(\theta/2)e^{-i\phi} ,$$

so that ζ_S and ζ_N are related by inversion,

$$\zeta_S = \frac{1}{\zeta_N} \quad \text{and} \quad \zeta_N = \frac{1}{\zeta_S} .$$

▲

Exercise. Invert the previous relations to find the spherical polar coordinates $(\theta,\ \phi)$ in terms of the planar variables $\zeta_N = \xi_N + i\eta_N$. ★

Answer. The stereographic formulas give

$$\zeta_N = \xi_N + i\eta_N = \frac{x + iy}{1 - z} = \cot(\theta/2)e^{i\phi} ,$$

with magnitude

$$|\zeta_N|^2 = \xi_N^2 + \eta_N^2 = \frac{1+z}{1-z} = \cot^2(\theta/2)\,,$$

so that

$$1 - z = \frac{2}{|\zeta_N|^2 + 1}\,.$$

Consequently,

$$x + iy = \frac{2\zeta_N}{|\zeta_N|^2 + 1} = \frac{2(\xi_N + i\eta_N)}{|\zeta_N|^2 + 1} = \sin(\theta)e^{i\phi}\,,$$

and

$$z = \frac{|\zeta_N|^2 - 1}{|\zeta_N|^2 + 1} = \cos(\theta)\,.$$

These are the usual invertible polar coordinate relations on the unit sphere. ▲

Example A.1.3 (Torus T^2) *If we start with two identical circles in the xz-plane, of radius r and centred at $x = \pm 2r$, then rotate them round the z-axis in $\mathbb{R}^3$, the result is a torus, written T^2. It is a manifold.*

Exercise. If we begin with a figure eight in the xz-plane, along the x-axis and centred at the origin, and spin it round the z-axis in $\mathbb{R}^3$, we get a pinched surface that looks like a sphere that has been pinched so that the north and south poles touch. Is this a manifold? Prove it. ★

Answer. The origin has a neighbourhood diffeomorphic to a double cone. This is not diffeomorphic to $\mathbb{R}^2$. A proof of this is that if the origin of the cone is removed, two components remain; while if the origin of $\mathbb{R}^2$ is removed, only one component remains. ▲

Exercise. A sphere with unit radius is projected onto a plane, then projected back to a second sphere of different radius. Both projections are stereographic. Show that the polar angle (colatitude) θ of the unit sphere is re-projected to the colatitude θ' of the second sphere by

$$\tan(\theta'/2) = R^{-1}\tan(\theta/2),$$

where R is the radius of the second sphere. ★

Definition A.1.3 (Immersed submanifolds) *An* ***immersed submanifold*** *of a manifold M is a subset S together with a topology and differential structure such that S is a manifold and the inclusion map $i : S \hookrightarrow M$ is an* ***injective immersion****; that is, a smooth map between smooth manifolds whose derivative is everywhere injective. Hence, immersed submanifolds are the images of injective immersions.*

Remark A.1.3 Given any injective immersion $f : N \to M$ the image of N in M uniquely defines an immersed submanifold so that $f : N \to f(N)$ is a ***diffeomorphism***; that is, a smooth invertible map with a smooth inverse. □

Definition A.1.4 (Lie group and Lie subgroup manifolds) *A Lie group is a group that is also a manifold. A Lie subgroup is a submanifold that is invariant under group operations. That is, Lie subgroups are injective immersions.*

Definition A.1.5 (Tangent space to level sets) *Let the submersion defined by*

$$M = \{x \big| f_i(x) = 0,\ i = 1, \dots, k\}$$

be a submanifold of $\mathbb{R}^n$. The ***tangent space*** *at each $x \in M$ is defined by*

$$T_xM = \left\{ v \in \mathbb{R}^n \mid \frac{\partial f_i}{\partial x^a}(x)v^a = 0,\ i = 1, \dots, k \right\}.$$

*Note: We use the **summation convention**, i.e., repeated indices are summed over their range.*

Definition A.1.6 (Tangent vector) *The tangent space T_xM at a point x of a manifold M is a **vector space**. The elements of this space are called **tangent vectors** (Figure A.3).*

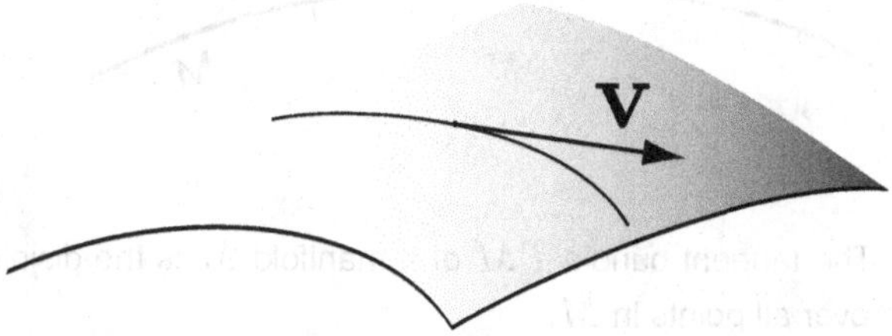

Figure A.3. A tangent vector at a point on a manifold.

Example A.1.4 (Tangent space to the sphere in $\mathbb{R}^3$) *The sphere S^2 is the set of points $(x, y, z) \in \mathbb{R}^3$ solving $x^2 + y^2 + z^2 = 1$. The tangent space to the sphere at such a point (x, y, z) is the plane containing vectors (u, v, w) satisfying $xu + yv + zw = 0$.*

Definition A.1.7 (Tangent bundle) *The **tangent bundle** of a manifold M, denoted by TM, is the smooth manifold whose underlying set is the disjoint union of the tangent spaces to M at the points $x \in M$ (Figure A.4); that is,*

$$TM = \bigcup_{x \in M} T_xM \,.$$

Thus, a single point of TM is (x, v) where $x \in M$ and $v \in T_xM$.

Remark A.1.4 (Dimension of tangent bundle TS^2) Defining TS^2 requires two independent conditions each of codimension 5 in $\mathbb{R}^6$; so $\dim TS^2 = 5 + 5 - 6 = 4$. □

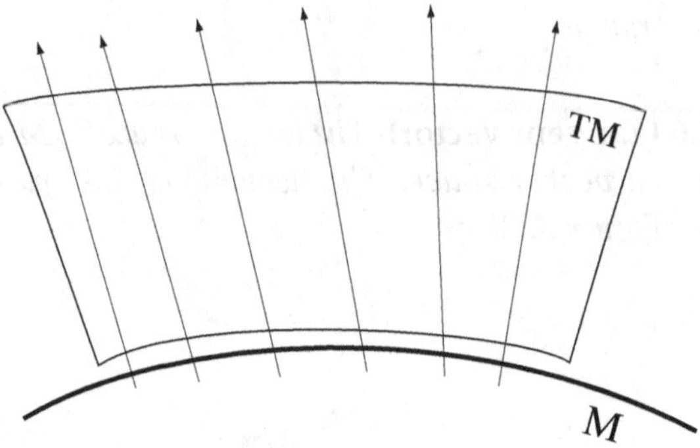

Figure A.4. The tangent bundle TM of a manifold M is the disjoint union of tangent spaces over all points in M.

Exercise. Define the sphere S^{n-1} in $\mathbb{R}^n$. What is the dimension of its tangent space TS^{n-1}? ★

A.2 Motion: Tangent vectors and flows

Envisioning our later considerations of dynamical systems, we shall consider motion along curves $c(t)$ parameterised by time t on a smooth manifold M. Suppose these curves are trajectories of a flow ϕ_t along the tangent vectors of the manifold. We anticipate this means $\phi_t\,(c(0)) = c(t)$ and $\phi_t \circ \phi_s = \phi_{t+s}$ (flow property). The flow will be tangent to M along the curve. To deal with such flows, we will need to know more about tangent vectors.

Recall from Definition A.1.7 that the tangent bundle of M is

$$TM = \bigcup_{x \in M} T_x M\,.$$

We will now add a bit more to that definition. The tangent bundle is an example of a more general structure than a manifold.

Definition A.2.1 (Bundle) *A **bundle** consists of a manifold E, another manifold B called the base space and a projection between them, $\Pi : E \to B$. The inverse images of the projection Π exist and are called the **fibres** of the bundle. Thus, subsets of the bundle E locally have the structure of a Cartesian product space $B \times F$, of the **base space** B with the **fibre space** F. An example is (E, B, Π) consisting of $(\mathbb{R}^2, \mathbb{R}^1, \Pi : \mathbb{R}^2 \to \mathbb{R}^1)$. In this case, $\Pi : (x, y) \in \mathbb{R}^2 \to x \in \mathbb{R}^1$. Likewise, the tangent bundle consists of M, TM and a map $\tau_M : TM \to M$.*

Let $x = (x^1, \ldots, x^n)$ be local coordinates on M, and let $v = (v^1, \ldots, v^n)$ be components of a tangent vector.

$$T_x M = \left\{ v \in \mathbb{R}^n \mid \frac{\partial f_i}{\partial x} \cdot v = 0, i = 1, \ldots, m \right\},$$

for

$$M = \{ x \in \mathbb{R}^n \mid f_i(x) = 0, i = 1, \ldots, m \}.$$

These $2n$ numbers (x, v) give local coordinates on TM, whose dimension is $\dim TM = 2 \dim M$. The ***tangent bundle projection*** is a map $\tau_M : TM \to M$ which takes a tangent vector v to a point $x \in M$ where the tangent vector v is attached (that is, $v \in T_x M$). The inverse of this projection $\tau_M^{-1}(x)$ is called the fibre over x in the tangent bundle.

A.2.1 Vector fields, integral curves and flows

Definition A.2.2 *A **vector field** on a manifold M is a map $X : M \to TM$ that assigns a vector $X(x)$ at each point $x \in M$. This implies that $\tau_M \circ X = Id$.*

Definition A.2.3 *An **integral curve** of X with initial conditions x_0 at $t = 0$ is a differentiable map $c :]a, b[\to M$, where $]a, b[$ is an open interval containing 0, such that $c(0) = x_0$ and $c'(t) = X(c(t))$ for all $t \in]a, b[$.*

Remark A.2.1 A standard result from the theory of ordinary differential equations states that X being Lipschitz is sufficient for its integral curves to be unique and C^1 [CoLe1984]. The integral curves $c(t)$ are differentiable for smooth X. □

Definition A.2.4 *The* ***flow*** *of X is the collection of maps*

$$\phi_t : M \to M\,,$$

where $t \to \phi_t(x)$ *is the integral curve of X with initial condition* x *(Figure A.5).*

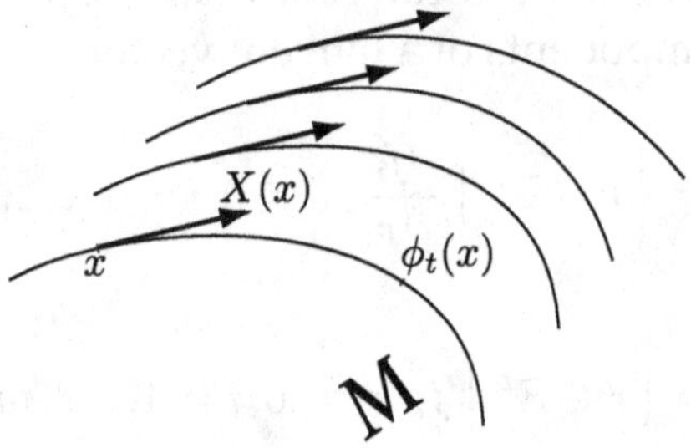

Figure A.5. A collection of maps $\phi_t : M \to M$ shows the flow of a vector field X acting on a manifold M.

Remark A.2.2

- Existence and uniqueness results for solutions of $c'(t) = X(c(t))$ guarantee that flow ϕ of X is smooth in (x, t), for smooth X.
- Uniqueness implies the flow property

$$\phi_{t+s} = \phi_t \circ \phi_s, \tag{FP}$$

for initial condition $\phi_0 = Id$.

- The flow property (FP) generalises to the nonlinear case the familiar linear situation where M is a vector space, $X(x) = Ax$ is a linear vector field for a bounded linear operator A, and $\phi_t(x) = e^{At}x$ (exponentiation).

□

Definition A.2.5 (Equivariance) *A map between two spaces is* ***equivariant*** *when it respects group actions on these spaces.*

A.2.2 Differentials of functions: The cotangent bundle

We are now ready to define differentials of smooth functions and the cotangent bundle.

Let $f : M \to \mathbb{R}$ be a smooth function. We differentiate f at $x \in M$ to obtain $T_x f : T_x M \to T_{f(x)}\mathbb{R}$. As is standard, we identify $T_{f(x)}\mathbb{R}$ with $\mathbb{R}$ itself, thereby obtaining a linear map $df(x) : T_x M \to \mathbb{R}$. The result $df(x)$ is an element of the cotangent space $T_x^* M$, the dual space of the tangent space $T_x M$. The natural pairing between elements of the tangent space and the cotangent space is denoted as $\langle \cdot\,,\, \cdot \rangle : T_x^* M \times T_x M \mapsto \mathbb{R}$.

In coordinates, the linear map $df(x) : T_x M \to \mathbb{R}$ may be written as the directional derivative,

$$\langle df(x)\,,\, v \rangle = df(x) \cdot v = \frac{\partial f}{\partial x^i} \cdot v^i\,,$$

for all $v \in T_x M$.

(Reminder: The summation convention is applied over repeated indices.) Hence, elements $df(x) \in T_x^* M$ are dual to vectors $v \in T_x M$ with respect to the pairing $\langle \cdot\,,\, \cdot \rangle$.

Definition A.2.6 *df is the* ***differential*** *of the function f.*

Definition A.2.7 *The dual space of the tangent bundle TM is the* ***cotangent bundle*** *T^*M. That is,*

$$(T_x M)^* = T_x^* M \quad and \quad T^* M = \bigcup_x T_x^* M\,.$$

Thus, replacing $v \in T_xM$ with $df \in T_x^*M$, for all $x \in M$ and for all smooth functions $f : M \to \mathbb{R}$, yields the cotangent bundle T^*M.

Differential bases

When the basis of vector fields is denoted as $\frac{\partial}{\partial x^i}$ for $i = 1, \ldots, n$, its dual basis is often denoted as dx^i. In this notation, the differential of a function at a point $x \in M$ is expressed as

$$df(x) = \frac{\partial f}{\partial x^i} dx^i .$$

The corresponding pairing $\langle \cdot\,,\cdot \rangle$ of bases is written in this notation as

$$\left\langle dx^j, \frac{\partial}{\partial x^i} \right\rangle = \delta_i^j .$$

Here δ_i^j is the Kronecker delta, which equals unity for $i = j$ and vanishes otherwise. That is, defining T^*M requires a pairing $\langle \cdot\,,\cdot \rangle : T^*M \times TM \to \mathbb{R}$.

A.3 Tangent and cotangent lifts

We next define derivatives of differentiable maps between manifolds (tangent lifts).

We expect that a smooth map $f : U \to V$ from a chart $U \subset M$ to a chart $V \subset N$ will lift to a map between the tangent bundles TM and TN so as to make sense from the viewpoint of ordinary calculus,

$$U \times \mathbb{R}^m \subset TM \longrightarrow V \times \mathbb{R}^n \subset TN$$
$$(q^1, \ldots, q^m; X^1, \ldots, X^m) \longmapsto (Q^1, \ldots, Q^n; Y^1, \ldots, Y^n) .$$

Namely, the relations between the vector field components should be obtained from the differential of the map $f : U \to V$. Perhaps not unexpectedly, these vector field components will be related by

$$Y^i \frac{\partial}{\partial Q^i} = X^j \frac{\partial}{\partial q^j} , \quad \text{so} \quad Y^i = \frac{\partial Q^i}{\partial q^j} X^j ,$$

in which the quantity called the ***tangent lift***

$$Tf = \frac{\partial Q}{\partial q} \tag{A.3.1}$$

of the function f arises from the chain rule and is equal to the Jacobian for the transformation $Tf : TM \mapsto TN$.

The dual of the tangent lift is the cotangent lift, which plays a major role in the definition of momentum maps in Chapter 11. Roughly speaking, the ***cotangent lift*** of the function f,

$$T^*f = \frac{\partial q}{\partial Q}, \tag{A.3.2}$$

arises from

$$\beta_i dQ^i = \alpha_j dq^j\,, \quad \text{so} \quad \beta_i = \alpha_j \frac{\partial q^j}{\partial Q^i}\,,$$

and $T^*f : T^*N \mapsto T^*M$. Note the directions of these maps.

$$\begin{aligned} Tf \quad &: \quad q\,, X \in TM \mapsto Q, Y \in TN\,, \\ f \quad &: \quad q \in M \mapsto Q \in N\,, \\ T^*f \quad &: \quad Q\,, \beta \in T^*N \mapsto q\,, \alpha \in T^*M\,. \end{aligned}$$

A.3.1 Summary of derivatives on manifolds

Definition A.3.1 (Differentiable map) *A map $f : M \to N$ from manifold M to manifold N is said to be **differentiable** (resp. C^k) if it is represented in local coordinates on M and N by differentiable (resp. C^k) functions.*

Definition A.3.2 (Derivative of a differentiable map) *The **derivative** of a differentiable map*

$$f : M \to N$$

at a point $x \in M$ is defined to be the linear map

$$T_x f : T_x M \to T_x N\,,$$

constructed as follows. For $v \in T_xM$, choose a curve that maps an open interval $(-\epsilon, \epsilon)$ around the point $t = 0$ to the manifold M,

$$c : (-\epsilon, \epsilon) \longrightarrow M\,,$$
$$\textit{with} \quad c(0) = x\,,$$
$$\textit{and velocity vector} \quad \frac{dc}{dt}\Big|_{t=0} = v\,.$$

Then $T_xf \cdot v$ is the velocity vector at $t = 0$ of the curve $f \circ c : \mathbb{R} \to N$. That is,

$$T_xf \cdot v = \frac{d}{dt} f(c(t))\Big|_{t=0}\,. \tag{A.3.3}$$

Definition A.3.3 *The union $Tf = \bigcup_x T_xf$ of the derivatives $T_xf : T_xM \to T_xN$ over points $x \in M$ is called the **tangent lift** of the map $f : M \to N$.*

Remark A.3.1 The chain-rule definition of the derivative T_xf of a differentiable map at a point x depends on the function f and the vector v. Other degrees of differentiability are possible. For example, if M and N are manifolds and $f : M \to N$ is of class C^{k+1}, then the tangent lift (Jacobian) $T_xf : T_xM \to T_xN$ is C^k. □

Exercise. Let $\phi_t : S^2 \to S^2$ rotate points on S^2 about a fixed axis through an angle $\psi(t)$. Show that ϕ_t is the flow of a certain vector field on S^2. ★

Exercise. Let $f : S^2 \to \mathbb{R}$ be defined by $f(x, y, z) = z$. Compute df using spherical coordinates (θ, ϕ). ★

Exercise. Compute the tangent lifts for the two stereographic projections of $S^2 \to \mathbb{R}^2$ in Example A.1.2. That is, assuming (x, y, z) depend smoothly on t, find

- How $(\dot{\xi}_N, \dot{\eta}_N)$ depend on $(\dot{x}, \dot{y}, \dot{z})$. Likewise, for $(\dot{\xi}_S, \dot{\eta}_S)$.
- How $(\dot{\xi}_N, \dot{\eta}_N)$ depend on $(\dot{\xi}_S, \dot{\eta}_S)$.

Hint: Recall $(1 + z)(1 - z) = 1 - z^2 = x^2 + y^2$ and use $x\dot{x} + y\dot{y} + z\dot{z} = 0$ when $(\dot{x}, \dot{y}, \dot{z})$ is tangent to S^2 at (x, y, z).

★

B

LIE GROUPS AND LIE ALGEBRAS

Contents

B.1	Matrix Lie groups	306
B.2	Defining matrix Lie algebras	310
B.3	Examples of matrix Lie groups	312
B.4	Lie group actions	314
	B.4.1 Left and right translations on a Lie group	316
B.5	Tangent and cotangent lift actions	317
B.6	Jacobi–Lie bracket	320
B.7	Lie derivative and Jacobi–Lie bracket	323
	B.7.1 Lie derivative of a vector field	323
	B.7.2 Vector fields in ideal fluid dynamics	325

B.1 Matrix Lie groups

The main text refers to definitions and examples of Lie group properties that are collected in this appendix for ready reference. We follow [AbMa1978, MaRa1994, Ol2000], which should be consulted for further discussions as needed by the reader. See also [Ho2008] for another introductory discussion along similar lines.

Definition B.1.1 (Group) *A **group** G is a set of elements possessing*

- *A binary product (multiplication), $G \times G \to G$, such that the following properties hold:*
  - *The product of g and h is written gh.*
  - *The product is associative, $(gh)k = g(hk)$.*
- *Identity element e: $eg = g$ and $ge = g$, for all $g \in G$.*
- *Inverse operation $G \to G$, so that $gg^{-1} = g^{-1}g = e$.*

Definition B.1.2 (Lie group) *A **Lie group** is a smooth manifold G which is also a group and for which the group operations of multiplication, $(g, h) \to gh$ for $g, h \in G$, and inversion, $g \to g^{-1}$ with $gg^{-1} = g^{-1}g = e$, are smooth functions.*

Definition B.1.3 *A **matrix Lie group** is a set of invertible $n \times n$ matrices which is closed under matrix multiplication and which is a submanifold of $\mathbb{R}^{n\times n}$.*

Remark B.1.1 The conditions showing that a matrix Lie group is a Lie group are easily checked:

- A matrix Lie group is a manifold, because it is a submanifold of $\mathbb{R}^{n\times n}$.
- Its group operations are smooth, since they are algebraic operations on the matrix entries.

□

Example B.1.1 (The general linear group $GL(n, \mathbb{R})$) *The matrix Lie group $GL(n, \mathbb{R})$ is the group of linear isomorphisms of $\mathbb{R}^n$ to itself. The dimension of the $n \times n$ matrices in $GL(n, \mathbb{R})$ is n^2, the number of independent elements.*

Remark B.1.2 Definition A.1.4 of Lie groups in terms of manifolds also defines Lie subgroups in terms of invariant submanifolds. This definition suggests a strategy for characterising Lie subgroups of $GL(n, \mathbb{R})$ by imposing submanifold invariance conditions. □

Proposition B.1.1 (Isotropy subgroup) *Let $K \in GL(n, \mathbb{R})$ be a symmetric matrix, $K^T = K$. The mapping*

$$S = \{U \in GL(n, \mathbb{R}) | \, U^T K U = K\}$$

defines the subgroup S of $GL(n, \mathbb{R})$. Moreover, this subgroup is a submanifold of $\mathbb{R}^{n\times n}$ of dimension $n(n-1)/2$.

Proof. Is the mapping S in Proposition B.1.1 a subgroup? We check the following three defining properties:

- Identity: $I \in S$ because $I^T K I = K$.
- Inverse: $U \in S \Longrightarrow U^{-1} \in S$ because

$$K = U^{-T}(U^T K U)U^{-1} = U^{-T}(K)U^{-1}.$$

- Closed under multiplication: $U, V \in S \Longrightarrow UV \in S$ because

$$(UV)^T K UV = V^T(U^T K U)V = V^T(K)V = K.$$

Hence, S is a subgroup of $GL(n, \mathbb{R})$.

Is S a submanifold of $\mathbb{R}^{n\times n}$ of dimension $n(n-1)/2$? Indeed, S is the zero locus of the mapping $UKU^T - K$. Its non-redundant entries, i.e., those on or above the diagonal, define a submersion and thus make S a submanifold.

For a submersion, the dimension of the level set is the dimension of the domain minus the dimension of the range space. In this case, this dimension is $n^2 - n(n+1)/2 = n(n-1)/2$. ■

Exercise. Explain why one can conclude that the zero locus map for S is a submersion. In particular, pay close attention to establishing the constant rank condition for the linearisation of this map. ★

Answer. Here is the strategy for proving that S is a submanifold of $R^{n\times n}$.

(i) The mapping S is the zero locus of

$$U \to U^T K U - K \qquad \text{(locus map)}\,.$$

(ii) Let $U \in S$, and let δU be an arbitrary element of $R^{n\times n}$. Linearise the locus map to find

$$\begin{aligned}&(U+\delta U)^T K(U+\delta U) - K \\ &= U^T K U - K + \delta U^T K U + U^T K \delta U + O(\delta U)^2\,.\end{aligned} \tag{B.1.1}$$

(iii) We may conclude that S is a submanifold of $R^{n\times n}$ provided we can show that the linearisation of the locus map, namely the linear mapping defined by

$$L \equiv \delta U \to \delta U^T K U + U^T K \delta U\,, \quad R^{n\times n} \mapsto R^{n\times n}\,, \tag{B.1.2}$$

has constant rank for all $U \in S$.

▲

To apply this strategy, one needs a lemma.

Lemma B.1.1 *The linearisation map L in (B.1.2) is onto the space of $n \times n$ symmetric matrices. Hence the original map is a submersion.*

Proof. Both the original locus map and the image of L lie in the subspace of $n \times n$ symmetric matrices. Indeed, given U and any symmetric matrix A we can find δU such that

$$\delta U^T K U + U^T K \delta U = A\,.$$

Namely

$$\delta U = K^{-1} U^{-T} A/2\,.$$

Thus, the linearisation map L is onto the space of $n \times n$ symmetric matrices and the original locus map $U \to UKU^T - K$ to the space of symmetric matrices is a submersion. ■

Remark B.1.3 The subgroup S leaves invariant a certain symmetric quadratic form under linear transformations, $S \times \mathbb{R}^n \to \mathbb{R}^n$ given by $\mathbf{x} \to U\mathbf{x}$, since

$$\mathbf{x}^T K \mathbf{x} = \mathbf{x}^T U^T K U \mathbf{x}\,.$$

So the matrices $U \in S$ change the basis for this quadratic form, but they leave its value unchanged. This means S is the ***isotropy subgroup*** of the quadratic form associated with K. □

Corollary B.1.1 (S is a matrix Lie group) *S is both a subgroup and a submanifold of the general linear group $GL(n, \mathbb{R})$. Thus, by Definition B.1.3, the subgroup S is a matrix Lie group.*

Exercise. What is the tangent space to S at the identity, $T_I S$? ★

Proposition B.1.2 *The linear space of matrices A satisfying*

$$A^T K + KA = 0$$

defines $T_I S$, the tangent space at the identity of the matrix Lie group S defined in Proposition B.1.1.

Proof. Near the identity the defining condition for S expands to

$$(I + \epsilon A^T + O(\epsilon^2))K(I + \epsilon A + O(\epsilon^2)) = K\,, \quad \text{for} \quad \epsilon \ll 1\,.$$

At linear order $O(\epsilon)$ one finds

$$A^T K + KA = 0\,.$$

This relation defines the linear space of matrices $A \in T_I S$. ■

Exercise. If $A, B \in T_I S$, does it follow that $[A, B] \in T_I S$? This is closure. ★

Proposition B.1.3 (Closure) *For any pair of matrices $A, B \in T_I S$, the matrix commutator $[A, B] \equiv AB - BA \in T_I S$.*

Proof. Using $[A, B]^T = [B^T, A^T]$, we may check closure by a direct computation,

$$\begin{aligned}
&[B^T, A^T]K + K[A, B] \\
&\quad = B^T A^T K - A^T B^T K + KAB - KBA \\
&\quad = B^T A^T K - A^T B^T K - A^T KB + B^T KA = 0\,.
\end{aligned}$$

Hence, the tangent space of S at the identity $T_I S$ is closed under the matrix commutator $[\cdot, \cdot]$. ■

Remark B.1.4 In a moment, we will show that the matrix commutator for $T_I S$ also satisfies the Jacobi identity. This will imply that the condition $A^T K + KA = 0$ defines a matrix Lie algebra. □

B.2 Defining matrix Lie algebras

We are ready to prove the following, in preparation for defining matrix Lie algebras.

Proposition B.2.1 *Let S be a matrix Lie group, and let $A, B \in T_I S$ (the tangent space to S at the identity element). Then $AB - BA \in T_I S$.*

The proof makes use of a lemma.

Lemma B.2.1 *Let R be an arbitrary element of a matrix Lie group S, and let $B \in T_I S$. Then $RBR^{-1} \in T_I S$.*

Proof. Let $R_B(t)$ be a curve in S such that $R_B(0) = I$ and $R'(0) = B$. Define $S(t) = RR_B(t)R^{-1} \in S$ for all t. Then $S(0) = I$ and $S'(0) = RBR^{-1}$. Hence, $S'(0) \in T_I S$, thereby proving the lemma. ∎

Proof of Proposition B.2.1. Let $R_A(s)$ be a curve in S such that $R_A(0) = I$ and $R'_A(0) = A$. Define $S(t) = R_A(t)BR_A(t)^{-1} \in T_I S$. Then the lemma implies that $S(t) \in T_I S$ for every t. Hence, $S'(t) \in T_I S$, and in particular, $S'(0) = AB - BA \in T_I S$. ∎

Definition B.2.1 (Matrix commutator) *For any pair of $n \times n$ matrices A, B, the* ***matrix commutator*** *is defined as*

$$[A, B] = AB - BA\,.$$

Proposition B.2.2 (Properties of the matrix commutator) *The matrix commutator has the following two properties:*

- *Any two $n \times n$ matrices A and B satisfy*

  $$[B, A] = -[A, B]\,.$$

  (This is the property of skew-symmetry.)

- *Any three $n \times n$ matrices A, B and C satisfy*

  $$[[A, B], C] + [[B, C], A] + [[C, A], B] = 0\,.$$

  (This is known as the ***Jacobi identity.****)*

Definition B.2.2 (Matrix Lie algebra) *A matrix Lie algebra $\mathfrak{g}$ is a set of $n \times n$ matrices which is a vector space with respect to the usual operations of matrix addition and multiplication by real numbers (scalars) and which is closed under the matrix commutator $[\cdot, \cdot]$.*

Proposition B.2.3 *For any matrix Lie group S, the tangent space at the identity $T_I S$ is a matrix Lie algebra.*

Proof. This follows by Proposition B.2.1 and because $T_I S$ is a vector space. ∎

B.3 Examples of matrix Lie groups

Example B.3.1 (The orthogonal group $O(n)$) *The mapping condition $U^T K U = K$ in Proposition B.1.1 specialises for $K = I$ to $U^T U = I$, which defines the orthogonal group. Thus, in this case, S specialises to $O(n)$, the group of $n \times n$ orthogonal matrices. The orthogonal group is of special interest in mechanics.*

Corollary B.3.1 ($O(n)$ is a matrix Lie group) *By Proposition B.1.1 the orthogonal group $O(n)$ is both a subgroup and a submanifold of the general linear group $GL(n, \mathbb{R})$. Thus, by Definition B.1.3, the orthogonal group $O(n)$ is a matrix Lie group.*

Example B.3.2 (The special linear group $SL(n, \mathbb{R})$) *The subgroup of $GL(n, \mathbb{R})$ with $\det(U) = 1$ is called $SL(n, \mathbb{R})$.*

Example B.3.3 (The special orthogonal group $SO(n)$) *The special case of S with $\det(U) = 1$ and $K = I$ is called $SO(n)$. In this case, the mapping condition $U^T K U = K$ specialises to $U^T U = I$ with the extra condition $\det(U) = 1$.*

Example B.3.4 (Tangent space of $SO(n)$ at the identity) *The special case with $K = I$ of $T_I SO(n)$ yields*

$$A^T + A = 0\,.$$

These are antisymmetric matrices. Lying in the tangent space at the identity of a matrix Lie group, this linear vector space forms the matrix Lie algebra $so(n)$.

Example B.3.5 (The special unitary group $SU(n)$) *The Lie group $SU(n)$ comprises complex $n \times n$ unitary matrices U with $U^\dagger U = I$ and* *unit determinant* $\det U = 1$. *An element A in its tangent space at the identity satisfies*

$$A + A^\dagger = 0 \quad for \quad A \in T_I SU(n)\,,$$

so that $A \in su(n)$ is an $n \times n$ traceless skew-Hermitian matrix.

Example B.3.6 (The symplectic group $Sp(l)$) *Suppose $n = 2l$ (that is, let n be even) and consider the nonsingular skew-symmetric matrix*

$$J = \begin{bmatrix} 0 & I \\ -I & 0 \end{bmatrix},$$

where I is the $l \times l$ identity matrix. One may verify that

$$Sp(l) = \{U \in GL(2l, \mathbb{R}) |\, U^T JU = J\}$$

is a group. This is called the symplectic group. Reasoning as before, the matrix algebra $T_I Sp(l)$ is defined as the set of $n \times n$ matrices A satisfying

$$JA^T + AJ = 0\,.$$

This matrix Lie algebra is denoted as $sp(l)$.

Example B.3.7 (The special Euclidean group $SE(3)$) *Consider the Lie group of 4×4 matrices of the form*

$$E(R, v) = \begin{bmatrix} R & v \\ 0 & 1 \end{bmatrix},$$

where $R \in SO(3)$ and $v \in \mathbb{R}^3$. This is the special Euclidean group, denoted $SE(3)$. The special Euclidean group is of central interest in mechanics since it describes the set of rigid motions and coordinate transformations of three-dimensional space.

Exercise. A point P in $\mathbb{R}^3$ undergoes a rigid motion associated with $E(R_1, v_1)$ followed by a rigid motion associated with $E(R_2, v_2)$. What matrix element of $SE(3)$ is associated with the composition of these motions in the given order?

★

Exercise. Multiply the special Euclidean matrices of $SE(3)$. Investigate their matrix commutators in their tangent space at the identity. (This is an example of a semidirect-product Lie group.) ★

Exercise. (Tripos question) When does a stone at the equator of the Earth weigh the most?

Two hints: (i) Assume the Earth's orbit is a circle around the Sun and ignore the declination of the Earth's axis of rotation. (ii) This is an exercise in using $SE(2)$. ★

Exercise. Suppose the $n \times n$ matrices A and M satisfy

$$AM + MA^T = 0\,.$$

Show that $\exp(At)M\exp(A^T t) = M$ for all t.

Hint: $A^n M = M(-A^T)^n$. This direct calculation shows that for $A \in so(n)$ or $A \in sp(l)$, we have $\exp(At) \in SO(n)$ or $\exp(At) \in Sp(l)$, respectively. ★

B.4 Lie group actions

The action of a Lie group G on a manifold M is a group of transformations of M associated with elements of the group G, whose composition acting on M corresponds to group multiplication in G.

Definition B.4.1 (Left action of a Lie group) *Let M be a manifold and let G be a Lie group. A **left action** of a Lie group G on M is a smooth mapping $\Phi : G \times M \to M$ such that*

(i) $\Phi(e, x) = x$ *for all* $x \in M$;

(ii) $\Phi(g, \Phi(h, x)) = \Phi(gh, x)$ *for all* $g, h \in G$ *and* $x \in M$; *and*

(iii) $\Phi(g, \cdot)$ *is a diffeomorphism on M for each* $g \in G$.

We often use the convenient notation gx for $\Phi(g, x)$ and think of the group element g acting on the point $x \in M$. The associativity condition (ii) then simply reads $(gh)x = g(hx)$.

Similarly, one can define a ***right action***, which is a map $\Psi : M \times G \to M$ satisfying $\Psi(x, e) = x$ and $\Psi(\Psi(x, g), h) = \Psi(x, gh)$. The convenient notation for right action is xg for $\Psi(x, g)$, the right action of a group element g on the point $x \in M$. Associativity $\Psi(\Psi(x, g), h) = \Psi(x, gh)$ may then be expressed conveniently as $(xg)h = x(gh)$.

Example B.4.1 (Properties of Lie group actions) *The action $\Phi : G \times M \to M$ of a group G on a manifold M is said to be*

- ***transitive**, if for every $x, y \in M$ there exists a $g \in G$, such that $gx = y$;*
- ***free**, if it has no fixed points, that is, $\Phi_g(x) = x$ implies $g = e$; and*
- ***proper**, if whenever a convergent subsequence $\{x_n\}$ in M exists, and the mapping $g_n x_n$ converges in M, then $\{g_n\}$ has a convergent subsequence in G.*

Definition B.4.2 (Group orbits) *Given a Lie group action of G on M, for a given point $x \in M$, the subset*

$$\operatorname{Orb} x = \{gx |\, g \in G\} \subset M$$

*is called the **group orbit** through x.*

Remark B.4.1 In finite dimensions, it can be shown that group orbits are always smooth (possibly immersed) manifolds. Group orbits generalise the notion of orbits of a dynamical system. □

Exercise. The flow of a vector field on M can be thought of as an action of $\mathbb{R}$ on M. Show that in this case the general notion of group orbit reduces to the familiar notion of orbit used in dynamical systems. ★

Theorem B.4.1 *Orbits of proper Lie group actions are embedded submanifolds.*

This theorem is stated in Chapter 9 of [MaRa1994], who refer to [AbMa1978] for the proof.

Example B.4.2 (Orbits of $SO(3)$) *A simple example of a group orbit is the action of $SO(3)$ on $\mathbb{R}^3$ given by matrix multiplication: The action of $A \in SO(3)$ on a point $\mathbf{x} \in \mathbb{R}^3$ is simply the product $A\mathbf{x}$. In this case, the orbit of the origin is a single point (the origin itself), while the orbit of any other point is the sphere through that point.*

Example B.4.3 (Orbits of a Lie group acting on itself) *The action of a group G on itself from either the left or the right also produces group orbits. This action sets the stage for discussing the tangent-lifted action of a Lie group on its tangent bundle.*

B.4.1 Left and right translations on a Lie group

Left and right translations on the group are denoted L_g and R_g, respectively. For example, $L_g : G \to G$ is the map given by $h \to gh$, while $R_g : G \to G$ is the map given by $h \to hg$, for $g, h \in G$.

- ***Left translation** $L_g : G \to G$; $h \to gh$ defines a transitive and free action of G on itself. Right multiplication $R_g : G \to G$; $h \to hg$*

defines a right action, while $h \to hg^{-1}$ defines a left action of G on itself.

- *G acts on G by conjugation, $g \to I_g = R_{g^{-1}} \circ L_g$. The map $I_g : G \to G$ given by $h \to ghg^{-1}$ is the* ***inner automorphism*** *associated with g. Orbits of this action are called* ***conjugacy classes.***

- *Differentiating conjugation at the identity e gives the* ***adjoint action*** *of G on $\mathfrak{g}$:*

$$\mathrm{Ad}_g := T_e I_g \,:\, T_e G = \mathfrak{g} \to T_e G = \mathfrak{g}\,.$$

Explicitly, the adjoint action of G on $\mathfrak{g}$ is given by

$$\mathrm{Ad} : G \times \mathfrak{g} \to \mathfrak{g}\,, \quad \mathrm{Ad}_g(\xi) = T_e(R_{g^{-1}} \circ L_g)\xi\,.$$

We have already seen an example of adjoint action for matrix Lie groups acting on matrix Lie algebras, when we defined $S(t) = R_A(t)BR_A(t)^{-1} \in T_I S$ as a key step in the proof of Proposition B.2.1.

- *The* ***coadjoint action*** *of G on $\mathfrak{g}^*$, the dual of the Lie algebra $\mathfrak{g}$ of G, is defined as follows. Let $\mathrm{Ad}_g^* : \mathfrak{g}^* \to \mathfrak{g}^*$ be the dual of Ad_g, defined by*

$$\langle \mathrm{Ad}_g^* \alpha, \xi \rangle = \langle \alpha, \mathrm{Ad}_g \xi \rangle$$

for $\alpha \in \mathfrak{g}^*$, $\xi \in \mathfrak{g}$ and pairing $\langle \cdot\,, \cdot \rangle : \mathfrak{g}^* \times \mathfrak{g} \to \mathbb{R}$. Then the map

$$\Phi^* \,:\, G \times \mathfrak{g}^* \to \mathfrak{g}^* \quad \textit{given by} \quad (g, \alpha) \mapsto \mathrm{Ad}_{g^{-1}}^* \alpha$$

is the coadjoint action of G on $\mathfrak{g}^*$.

B.5 Tangent and cotangent lift actions

Definition B.5.1 (Tangent lift action) *Let $\Phi : G \times M \to M$ be a left action, and write $\Phi_g(m) = \Phi(g, m)$. The* ***tangent lift*** *action of G on the tangent bundle TM is defined by*

$$gv = T_x \Phi_g(v)\,, \tag{B.5.1}$$

for every $v \in T_x M$.

Definition B.5.2 (Cotangent lift action) *The cotangent lift action of G on T^*M is defined by*

$$g\alpha = (T_m\Phi_{g^{-1}})(\alpha), \tag{B.5.2}$$

for every $\alpha \in T^*M$.

Remark B.5.1 In standard calculus notation, the expression for tangent lift may be written as

$$T_x\Phi \cdot v = \frac{d}{dt}\Phi(c(t))\Big|_{t=0} = \frac{\partial\Phi}{\partial c}c'(t)\Big|_{t=0} =: D\Phi(x) \cdot v, \tag{B.5.3}$$

with $c(0) = x$ and $c'(0) = v$. □

Definition B.5.3 *If X is a vector field on M and ϕ is a differentiable map from M to itself, then the **push-forward** of X by ϕ is the vector field ϕ_*X defined by*

$$(\phi_*X)(\phi(x)) = T_x\phi(X(x)) = D\phi(x) \cdot X(x).$$

That is, the following diagram commutes:

$$
\begin{array}{ccc}
 & T\phi & \\
TM & \longrightarrow & TM \\
X \uparrow & & \uparrow \phi_*X \\
 & \phi & \\
M & \longrightarrow & M
\end{array}
$$

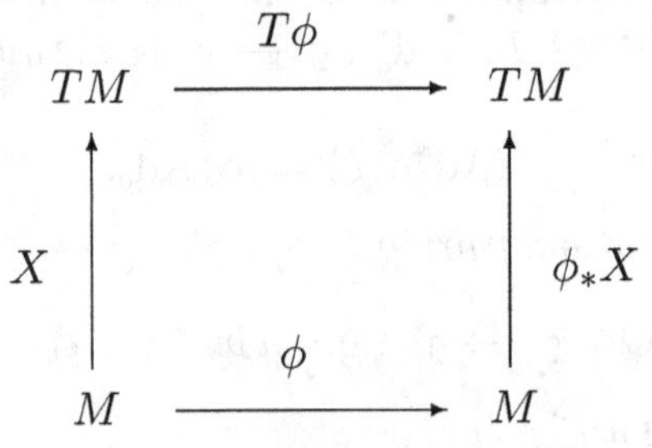

*If ϕ is a diffeomorphism then the **pull-back** ϕ^*X is also defined:*

$$(\phi^*X)(x) = T_{\phi(x)}\phi^{-1}(X(\phi(x))) = D\phi^{-1}(\phi(x)) \cdot X(\phi(x)).$$

This denotes the standard calculus operation of computing the Jacobian of the inverse map ϕ^{-1} acting on a vector field.

Definition B.5.4 (Group action on vector fields) *Let $\Phi : G \times M \to M$ be a left action, and write $\Phi_g(m) = \Phi(g, m)$. Then G has a left action on $\mathfrak{X}(M)$ (the set of vector fields on M) by push-forwards: $gX = (\Phi_g)_* X$. (The definition for right actions is the same, but written Xg instead of gX.)*

Definition B.5.5 (Group-invariant vector fields) *Let G act on M on the left. A vector field X on M is **invariant** with respect to this action (we often say "G-invariant" if the action is understood) if $gX = X$ for all $g \in G$; equivalently (using all of the above definitions!) $g\,(X(x)) = X(gx)$ for all $g \in G$ and all $x \in X$. (Similarly for right actions.)*

Definition B.5.6 (Left-invariant vector fields) *Consider the left action of G on itself by left multiplication,*

$$\Phi_g(h) = L_g(h) = gh\,.$$

*A vector field on G that is invariant with respect to this action is called **left-invariant**. From Definition B.5.5, we see that X is left-invariant if and only if $g\,(X(h)) = X(gh)$, which in less compact notation means $T_hL_gX(h) = X(gh)$. The set of all such vector fields is written $\mathfrak{X}^L(G)$.*

Theorem B.5.1 *Given a $\xi \in T_eG$, define $X_\xi^L(g) = g\xi$ (recall $g\xi \equiv T_eL_g\xi$). Then X_ξ^L is the unique left-invariant vector field such that $X_\xi^L(e) = \xi$.*

Proof. To show that X_ξ^L is left-invariant, we need to show that $g\left(X_\xi^L(h)\right) = X_\xi^L(gh)$ for every $g, h \in G$. This follows from the definition of X_ξ^L and the associativity property of Lie group actions,

$$g\left(X_\xi^L(h)\right) = g\,(h\xi) = (gh)\,\xi = X_\xi^L(gh)\,.$$

We repeat the last line in less compact notation,

$$T_hL_g\left(X_\xi^L(h)\right) = T_hL_g\,(h\xi) = T_eL_{gh}\xi = X_\xi^L(gh)\,.$$

For uniqueness, suppose X is left-invariant and $X(e) = \xi$. Then for any $g \in G$, we have $X(g) = g\,(X(e)) = g\xi = X_\xi^L(g)$. ■

The proof also yields the following.

Corollary B.5.1 *The map $\xi \mapsto X_\xi^L$ is a **vector space isomorphism** from T_eG to $\mathfrak{X}^L(G)$.*

Remark B.5.2 Because of this vector space isomorphism, the left-invariant vector fields could have been used to define the Lie algebra of a Lie group. See [Ol2000] for an example of this approach using the corresponding isomorphism for the right-invariant vector fields. □

All of the above definitions and theorems have analogues for right actions. The definitions of right-invariant, $\mathfrak{X}^R(G)$ and X_ξ^R, use the right action of G on itself defined by $\Psi(g,h) = R_g(h) = hg$. For example, the tangent lift for right action of G on the tangent bundle TM is defined by $vg = T_x\Psi_g(v)$ for every $v \in T_xM$.

Exercise. Show that $\Phi_g(h) = hg^{-1}$ defines a left action of G on itself. ★

The map $\xi \mapsto X_\xi^L$ may be used to relate the Lie bracket on $\mathfrak{g}$, defined as $[\xi, \eta] = \mathrm{ad}_\xi \eta$, with the Jacobi–Lie bracket on vector fields.

B.6 Jacobi–Lie bracket

Definition B.6.1 (Jacobi–Lie bracket) *The **Jacobi–Lie bracket** on* $\mathfrak{X}$ (M) *is defined in local coordinates by*

$$[X,Y]_{J-L} \equiv (DX)\cdot Y - (DY)\cdot X\,,$$

which, in finite dimensions, is equivalent to

$$[X,Y]_{J-L} \equiv -(X\cdot\nabla)Y + (Y\cdot\nabla)X \equiv -[X,Y]\,.$$

Theorem B.6.1 (Properties of the Jacobi–Lie bracket)

- *The Jacobi–Lie bracket satisfies*

$$[X,Y]_{J-L} = \pounds_X Y \equiv \frac{d}{dt}\bigg|_{t=0} \Phi_t^* Y,$$

where Φ_t is the flow of X. (This is coordinate-free, and can be used as an alternative definition.)

- *This bracket makes $\mathfrak{X}^L(M)$ a Lie algebra with $[X,Y]_{J-L} = -[X,Y]$, where $[X,Y]$ is the Lie algebra bracket on $\mathfrak{X}(M)$.*
- *$\phi_*[X,Y] = [\phi_* X, \phi_* Y]$ for any differentiable $\phi : M \to M$.*

Remark B.6.1 The first property of the Jacobi–Lie bracket is proved for matrices in Section B.7. The other two properties are proved below for the case that M is the Lie group G. □

Theorem B.6.2 *$\mathfrak{X}^L(G)$ is a subalgebra of $\mathfrak{X}(G)$.*

Proof. Let $X, Y \in \mathfrak{X}^L(G)$. Using the last item of the previous theorem, and then the G-invariance of X and Y, gives the push-forward relations

$$(L_g)_* [X,Y]_{J-L} = [(L_g)_* X, (L_g)_* Y]_{J-L},$$

for all $g \in G$. Hence $[X,Y]_{J-L} \in \mathfrak{X}^L(G)$. This is the second property in Theorem B.6.1. ■

Theorem B.6.3

$$[X^L_\xi, X^L_\eta]_{J-L}(e) = [\xi, \eta],$$

for every $\xi, \eta \in \mathfrak{g}$, where the bracket on the right is the Jacobi–Lie bracket. (That is, the Lie bracket on $\mathfrak{g}$ is the pull-back of the Jacobi–Lie bracket by the map $\xi \mapsto X^L_\xi$.)

Proof. The proof of this theorem for matrix Lie algebras is relatively easy: we have already seen that $\mathrm{ad}_A B = AB - BA$. On the other hand, since $X^L_A(C) = CA$ for all C, and this is linear in C, we have $DX^L_B(I) \cdot A = AB$, so

$$\begin{aligned}
[A, B] &= [X^L_A, X^L_B]_{J-L}(I) \\
&= DX^L_B(I) \cdot X^L_A(I) - DX^L_A(I) \cdot X^L_B(I) \\
&= DX^L_B(I) \cdot A - DX^L_A(I) \cdot B \\
&= AB - BA.
\end{aligned}$$

This is the third property of the Jacobi–Lie bracket listed in Theorem B.6.1. For the general proof, see [MaRa1994], Proposition 9.14. ■

Remark B.6.2 This theorem, together with the second property in Theorem B.6.1, proves that the Jacobi–Lie bracket makes $\mathfrak{g}$ into a Lie algebra. □

Remark B.6.3 By Theorem B.6.2, the vector field $[X_\xi^L, X_\eta^L]$ is left-invariant. Since $[X_\xi^L, X_\eta^L]_{J-L}(e) = [\xi, \eta]$, it follows that

$$[X_\xi^L, X_\eta^L] = X_{[\xi,\eta]}^L .$$

□

Definition B.6.2 (Infinitesimal generator) *Let $\Phi : G \times M \to M$ be a left action, and let $\xi \in \mathfrak{g}$. Let $g(t)$ be a path in G such that $g(0) = e$ and $g'(0) = \xi$. Then the* ***infinitesimal generator*** *of the action in the ξ direction is the vector field ξ_M on M defined by*

$$\xi_M(x) = \frac{d}{dt}\bigg|_{t=0} \Phi_{g(t)}(x) .$$

Remark B.6.4 Note: This definition does not depend on the choice of $g(t)$. For example, the choice in [MaRa1994] is $\exp(t\xi)$, where exp denotes the exponentiation on Lie groups (not defined here). □

Exercise. Consider the action of $SO(3)$ on the unit sphere S^2 around the origin, and let $\xi = (0, 0, 1)$. Sketch the vector field ξ_M.

Hint: The vectors all point eastward. ★

Theorem B.6.4 *For any left action of G, the Jacobi–Lie bracket of infinitesimal generators is related to the Lie bracket on $\mathfrak{g}$ as follows (note the minus sign):*

$$[\xi_M, \eta_M] = -\,[\xi, \eta]_M \,.$$

For a proof, see [MaRa1994], Proposition 9.3.6.

Exercise. Express the statements and formulas of this appendix for the case of $SO(3)$ action on its Lie algebra $so(3)$. (Hint: Look at the previous section.) Wherever possible, translate these formulas to $\mathbb{R}^3$ by using the hat map $\widehat{\ }: so(3) \to \mathbb{R}^3$.

Write the Lie algebra for $so(3)$ using the Jacobi–Lie bracket in terms of linear vector fields on $\mathbb{R}^3$. What are the characteristic curves of these linear vector fields? ★

B.7 Lie derivative and Jacobi–Lie bracket

B.7.1 Lie derivative of a vector field

Definition B.7.1 (Lie derivative) *Let X and Y be two vector fields on the same manifold M. The* ***Lie derivative*** *of Y with respect to X is*

$$\mathcal{L}_X Y \equiv \frac{d}{dt}\Phi_t^* Y\bigg|_{t=0} ,$$

where Φ is the flow of X.

Remark B.7.1 The Lie derivative $\mathcal{L}_X Y$ is the derivative of Y in the direction given by X. Its definition is coordinate-independent. In contrast, $DY \cdot X$ is also the derivative of Y in the X direction, but its value depends on the coordinate system, and in particular does not usually equal $\mathcal{L}_X Y$ in a given coordinate system. □

Theorem B.7.1

$$\mathcal{L}_X Y = [X, Y],$$

where the bracket on the right is the Jacobi–Lie bracket.

Proof. Assume that M is finite-dimensional and work in local coordinates. That is, consider everything as matrices, which allows the use of the product rule and the identities $(M^{-1})' = -M^{-1}M'M^{-1}$ and $\frac{d}{dt}(D\Phi_t(x)) = D\left(\frac{d}{dt}\Phi_t\right)(x)$.

$$\begin{aligned}
\mathcal{L}_X Y(x) &= \left.\frac{d}{dt}\Phi_t^* Y(x)\right|_{t=0} \\
&= \left.\frac{d}{dt}\left(D\Phi_t(x)\right)^{-1} Y\left(\Phi_t(x)\right)\right|_{t=0} \\
&= \left[\left(\frac{d}{dt}\left(D\Phi_t(x)\right)^{-1}\right) Y\left(\Phi_t(x)\right)\right. \\
&\qquad\qquad \left.+ \left(D\Phi_t(x)\right)^{-1}\frac{d}{dt}Y\left(\Phi_t(x)\right)\right]_{t=0} \\
&= \left[-\left(D\Phi_t(x)\right)^{-1}\left(\frac{d}{dt}D\Phi_t(x)\right)\left(D\Phi_t(x)\right)^{-1} Y\left(\Phi_t(x)\right)\right. \\
&\qquad\qquad \left.+ \left(D\Phi_t(x)\right)^{-1}\frac{d}{dt}Y\left(\Phi_t(x)\right)\right]_{t=0}.
\end{aligned}$$

Regrouping yields

$$\begin{aligned}
\mathcal{L}_X Y(x) &= \left[-\left(\frac{d}{dt}D\Phi_t(x)\right) Y(x) + \frac{d}{dt}Y\left(\Phi_t(x)\right)\right]_{t=0} \\
&= -D\left(\left.\frac{d}{dt}\Phi_t(x)\right|_{t=0}\right) Y(x) + DY(x)\left(\left.\frac{d}{dt}\Phi_t(x)\right|_{t=0}\right) \\
&= -DX(x)\cdot Y(x) + DY(x)\cdot X(x) \\
&= [X, Y]_{J-L}(x).
\end{aligned}$$

Therefore $\mathcal{L}_X Y = [X, Y]_{J-L}$. ■

B.7.2 Vector fields in ideal fluid dynamics

The Lie derivative formula for vector fields also applies in infinite dimensions, although the proof is more elaborate. For example, the equation for the vorticity dynamics of an Euler fluid with velocity $\mathbf{u}$ (with $\operatorname{div}\mathbf{u} = 0$) and vorticity $\omega = \operatorname{curl}\mathbf{u}$ may be written as a Lie derivative formula,

$$\begin{aligned}
\partial_t \omega &= -\mathbf{u}\cdot\nabla\omega + \omega\cdot\nabla\mathbf{u} \\
&= -\left[u,\omega\right]_{J-L} \\
&= \operatorname{ad}_u \omega \\
&= -\mathcal{L}_u \omega \,.
\end{aligned}$$

These equations each express the invariance of the vorticity vector field ω under the flow of its corresponding divergenceless velocity vector field u. This invariance is also encapsulated in the language of fluid dynamics as

$$\frac{d}{dt}\left(\omega\cdot\frac{\partial}{\partial\mathbf{x}}\right) = 0\,, \quad \text{along} \quad \frac{d\mathbf{x}}{dt} = \mathbf{u}(\mathbf{x},t) = \operatorname{curl}^{-1}\omega\,.$$

Here, the curl-inverse operator is defined by the ***Biot–Savart law,***

$$\mathbf{u} = \operatorname{curl}^{-1}\omega = \operatorname{curl}(-\Delta)^{-1}\omega\,,$$

which follows from the identity

$$\operatorname{curl}\operatorname{curl}\mathbf{u} = -\,\Delta\mathbf{u} + \nabla\operatorname{div}\mathbf{u}\,.$$

Thus, in coordinates,

$$\begin{aligned}
\frac{d\mathbf{x}}{dt} = \mathbf{u}(\mathbf{x},t) \quad &\Longrightarrow \quad \mathbf{x}(t,\mathbf{x}_0) \quad \text{with} \quad \mathbf{x}(0,\mathbf{x}_0) = \mathbf{x}_0 \text{ at } t=0\,, \\
\text{and} \quad \omega^j(\mathbf{x}(t,\mathbf{x}_0),t)\frac{\partial}{\partial x^j(t,\mathbf{x}_0)} \quad &= \quad \omega^A(\mathbf{x}_0)\frac{\partial}{\partial x_0^A}\,,
\end{aligned}$$

in which one sums over repeated indices. Consequently, one may write

$$\omega^j(\mathbf{x}(t,\mathbf{x}_0),t) = \omega^A(\mathbf{x}_0)\frac{\partial x^j(t,\mathbf{x}_0)}{\partial x_0^A}\,. \tag{B.7.1}$$

This is Cauchy's (1859) solution of ***Euler's equation for vorticity,***

$$\partial_t \omega = -\left[\mathrm{curl}^{-1}\omega, \omega\right]_{J-L} = \mathrm{ad}_u \omega \quad \text{with} \quad \mathbf{u} = \mathrm{curl}^{-1}\omega \,.$$

This type of Lie derivative equation often appears in the text. In it, the vorticity ω evolves by the ad action of the right-invariant vector field $\mathbf{u} = \mathrm{curl}^{-1}\omega$. The Cauchy solution is the tangent lift of this flow.

Exercise. Discuss the cotangent lift of this flow and how it might be related to the Kelvin circulation theorem,

$$\frac{d}{dt}\int_{c(\mathbf{u})} \mathbf{u}\cdot d\mathbf{x} = 0 \,,$$

for a circuit $c(\mathbf{u})$ moving with the flow. ★

C ENHANCED COURSEWORK

Contents

C.1 Variations on rigid-body dynamics 328

C.1.1 Two times 328

C.1.2 Rotations in complex space 334

C.1.3 Rotations in four dimensions: $SO(4)$ 338

C.2 $\mathbb{C}^3$ oscillators 343

C.3 Momentum maps for $GL(n,\mathbb{R})$ 348

C.4 Motion on the symplectic Lie group $Sp(2)$ 354

C.5 Two coupled rigid bodies 359

C.1 Variations on rigid-body dynamics

C.1.1 Two times

Scenario C.1.1 *The Bichrons are an alien life form who use two-dimensional time $u = (s,t) \in \mathbb{R}^2$ for time travel. To decide whether we are an intelligent life form, they require us to define spatial and body angular velocity for free rigid rotation in their two time dimensions. What should we tell them?*

Answer. (Bichrons) Following the approach to rotating motion taken in Section 2, let's define a trajectory of a moving point $\mathbf{x} = \mathbf{r}(u) \in \mathbb{R}^3$ as a smooth invertible map $\mathbf{r} : \mathbb{R}^2 \to \mathbb{R}^3$ with "time" $u \in \mathbb{R}^2$, so that $u = (s,t)$. Suppose the components of the trajectory are given in terms of a fixed and a moving orthonormal frame by

$$\begin{aligned} \mathbf{r}(u) &= r_0^A(u)\mathbf{e}_A(0) \quad \text{fixed frame,} \\ &= r^a \mathbf{e}_a(u) \quad \text{moving frame,} \end{aligned}$$

with moving orthonormal frame defined by $O : \mathbb{R}^2 \to SO(3)$, so that

$$\mathbf{e}_a(u) = O(u)\mathbf{e}_a(0)\,,$$

where $O(u)$ is a *surface* in $SO(3)$ parameterised by the two times $u = (s,t)$. The exterior derivative[1] of the moving frame relation above yields the infinitesimal spatial displacement,

$$d\mathbf{r}(u) = r^a d\mathbf{e}_a(u) = r^a dOO^{-1}(u)\mathbf{e}_a(u) = \widehat{\omega}(u)\mathbf{r}\,,$$

in which $\widehat{\omega}(u) = dOO^{-1}(u) \in so(3)$ is the one-form for spatial angular displacement. One denotes

$$dO = O'\,ds + \dot{O}\,dt\,,$$

[1]This subsection uses the notation of differential forms and wedge products. Readers unfamiliar with it may regard this subsection as cultural background.

so that the spatial angular displacement is the right-invariant $so(3)$-valued one-form

$$\widehat{\omega}(u) = dOO^{-1} = (O'\,ds + \dot{O}\,dt)O^{-1}\,.$$

Likewise, the body angular displacement is the left-invariant $so(3)$-valued one-form

$$\begin{aligned}\widehat{\Omega}(u) &= O^{-1}\widehat{\omega}(u)O = O^{-1}dO \qquad \text{(C.1.1)}\\ &= O^{-1}(O'\,ds + \dot{O}\,dt) =: \widehat{\Omega}_s ds + \widehat{\Omega}_t dt\,,\end{aligned}$$

and $\widehat{\Omega}_s$ and $\widehat{\Omega}_t$ are its *two body angular velocities*.
This is the answer the Bichrons wanted: For them, free rotation takes place on a space-time surface in $SO(3)$ and it has two body angular velocities because such a surface has two independent tangent vectors. ▲

Scenario C.1.2 *What would the Bichrons do with this information?*

Answer. To give an idea of what the Bichrons might do with our answer, let us define the ***coframe*** at position $\mathbf{x} = \mathbf{r}(u)$ as the infinitesimal displacement in *body coordinates*,

$$\Xi = O^{-1}d\mathbf{r}\,. \qquad \text{(C.1.2)}$$

Taking its exterior derivative gives the two-form,

$$d\Xi = -O^{-1}dO \wedge O^{-1}d\mathbf{r} = -\widehat{\Omega} \wedge \Xi\,, \qquad \text{(C.1.3)}$$

in which the left-invariant $so(3)$-valued one-form $\widehat{\Omega} = O^{-1}dO$ encodes the exterior derivative of the coframe as a rotation by the body angular displacement. In differential geometry, $\widehat{\Omega}$ is called the ***connection form*** and Equation (C.1.3) is called ***Cartan's first structure equation*** for a moving orthonormal frame [Fl1963, Da1994]. Taking another exterior derivative gives zero (because $d^2 = 0$) in the form of

$$0 = d^2\Xi = -d\widehat{\Omega} \wedge \Xi - \widehat{\Omega} \wedge d\Xi = -(d\widehat{\Omega} + \widehat{\Omega} \wedge \widehat{\Omega}) \wedge \Xi\,.$$

Hence we have ***Cartan's second structure equation,***

$$d\widehat{\Omega} + \widehat{\Omega} \wedge \widehat{\Omega} = 0\,. \qquad \text{(C.1.4)}$$

The left-hand side of this equation is called the ***curvature two-form*** associated with the connection form $\widehat{\Omega}$. The interpretation of (C.1.4) is that the connection form $\widehat{\Omega} = O^{-1}dO$ has zero curvature. This makes sense because the rotating motion takes place in Euclidean space, $\mathbb{R}^3$, which is flat.

Of course, one may also prove the ***zero curvature relation*** (C.1.4) directly from the definition $\widehat{\Omega} = O^{-1}dO$ by computing

$$d\widehat{\Omega} = d(O^{-1}dO) = -O^{-1}dO \wedge O^{-1}dO = -\widehat{\Omega} \wedge \widehat{\Omega}\,.$$

Expanding this out using the two angular velocities $\widehat{\Omega}_s = O^{-1}O'$ and $\widehat{\Omega}_t = O^{-1}\dot{O}$ gives (by using antisymmetry of the wedge product, $ds \wedge dt = -dt \wedge ds$)

$$\begin{aligned}
d\widehat{\Omega}(u) &= d(\widehat{\Omega}_s ds + \widehat{\Omega}_t dt) \\
&= -\left(\widehat{\Omega}_s ds + \widehat{\Omega}_t dt\right) \wedge \left(\widehat{\Omega}_s ds + \widehat{\Omega}_t dt\right) \\
&= -\widehat{\Omega} \wedge \widehat{\Omega} \\
&= \frac{\partial \widehat{\Omega}_s}{\partial t} dt \wedge ds + \frac{\partial \widehat{\Omega}_t}{\partial s} ds \wedge dt \\
&= -\widehat{\Omega}_s \widehat{\Omega}_t \, ds \wedge dt - \widehat{\Omega}_t \widehat{\Omega}_s \, dt \wedge ds \\
&= \left(\frac{\partial \widehat{\Omega}_t}{\partial s} - \frac{\partial \widehat{\Omega}_s}{\partial t}\right) ds \wedge dt \\
&= (\widehat{\Omega}_t \widehat{\Omega}_s - \widehat{\Omega}_s \widehat{\Omega}_t)\, ds \wedge dt \\
&=: \left[\widehat{\Omega}_t\,,\, \widehat{\Omega}_s\right] ds \wedge dt\,.
\end{aligned}$$

Since $ds \wedge dt \neq 0$, this equality implies that the coefficients are equal. In other words, this calculation proves the following.

Proposition C.1.1 *The zero curvature relation (C.1.4) may be expressed equivalently as*

$$\frac{\partial \widehat{\Omega}_t}{\partial s} - \frac{\partial \widehat{\Omega}_s}{\partial t} = \widehat{\Omega}_t \widehat{\Omega}_s - \widehat{\Omega}_s \widehat{\Omega}_t = \left[\widehat{\Omega}_t \,,\, \widehat{\Omega}_s \right], \qquad \text{(C.1.5)}$$

in terms of the two angular velocities, $\widehat{\Omega}_s = O^{-1}O'$ *and* $\widehat{\Omega}_t = O^{-1}\dot{O}$. ▲

Remark C.1.1 The component form (C.1.5) of the zero curvature relation (C.1.4) arises in Equation (10.3.10) of Chapter 11 in the derivation of the Euler–Poincaré equations for the dynamics of the classical spin chain. It also arises in many places in the theory of soliton solutions of completely integrable Hamiltonian partial differential equations, see, e.g., [FaTa1987]. □

Exercise. Why is $\widehat{\Omega}$ called a connection form? ★

Answer. Consider the one-form Equation (C.1.2) written in components as

$$\Xi^j = \Xi^j_\alpha(r)dr^\alpha \,, \qquad \text{(C.1.6)}$$

in which the matrix $\Xi^j_\alpha(r)$ depends on spatial location, and it need not be orthogonal. In the basis $\Xi^j(r)$, a one-form v may be expanded in components as

$$v = v_j \Xi^j \,. \qquad \text{(C.1.7)}$$

Its differential is computed in this basis as

$$\begin{aligned} dv &= d(v_j \Xi^j) \\ &= dv_j \wedge \Xi^j + v_j d\Xi^j \,. \end{aligned}$$

Substituting Equation (C.1.3) in components as

$$d\,\Xi^j = -\,\widehat{\Omega}^j_k \wedge \Xi^k \qquad \text{(C.1.8)}$$

then yields the differential two-form,

$$
\begin{aligned}
dv &= (dv_k - v_j\widehat{\Omega}^j_k) \wedge \Xi^k \\
&=: (dv_k - v_j\Gamma^j_{kl}\Xi^l) \wedge \Xi^k \\
&=: Dv_k \wedge \Xi^k \,. \qquad \text{(C.1.9)}
\end{aligned}
$$

The last equation defines the covariant exterior derivative operation D in the basis of one-form displacements $\Xi(r)$. The previous equation introduces the quantities Γ^j_{kl} defined as

$$\widehat{\Omega}^j_k = \Gamma^j_{kl}\Xi^l \,. \qquad \text{(C.1.10)}$$

Γ^j_{kl} are the Christoffel coefficients in the ***local coframe*** given by Equation (C.1.6). These are the standard connection coefficients for curvilinear geometry. ▲

Exercise. Prove from their definition in formula (C.1.9) that the Christoffel coefficients are symmetric under the exchange of indices, $\Gamma^j_{kl} = \Gamma^j_{lk}$. ★

Definition C.1.1 (Body covariant derivative) *The relation in Equation (C.1.9)*

$$Dv_k := dv_k - v_j\widehat{\Omega}^j_k = dv_k - v_j\Gamma^j_{kl}\Xi^l \qquad \text{(C.1.11)}$$

defines the components of the ***covariant derivative*** *of the one-form v in the body frame; that is, in the Ξ-basis. Thus, $\widehat{\Omega}$ is a connection form in the standard sense of differential geometry [Fl1963, Da1994].*

Remark C.1.2 (Metric tensors) The metric tensors in the two bases of infinitesimal displacements dr and Ξ are related by requiring that the element of length measured in either basis must be the same. That is,

$$ds^2 = g_{\alpha\beta}\, dr^\alpha \otimes dr^\beta = \delta_{jk}\, \Xi^j \otimes \Xi^k \,, \qquad \text{(C.1.12)}$$

where $\otimes$ is the symmetric tensor product. This implies a relation between the metrics,

$$g_{\alpha\beta} = \delta_{jk}\, \Xi^j_\alpha \Xi^k_\beta \,, \tag{C.1.13}$$

which, in turn, implies

$$\Gamma^\nu_{\beta\mu}(r) = \frac{1}{2} g^{\nu\alpha}\left[\frac{\partial g_{\alpha\mu}(r)}{\partial r^\beta} + \frac{\partial g_{\alpha\beta}(r)}{\partial r^\mu} - \frac{\partial g_{\beta\mu}(r)}{\partial r^\alpha}\right]. \tag{C.1.14}$$

This equation identifies $\Gamma^\nu_{\beta\mu}(r)$ as the Christoffel coefficients in the ***spatial basis***. Note that the spatial Christoffel coefficients are symmetric under the exchange of indices, $\Gamma^\nu_{\beta\mu}(r) = \Gamma^\nu_{\mu\beta}(r)$. □

Definition C.1.2 (Spatial covariant derivative) *For the spatial metric $g_{\alpha\mu}$, the **covariant derivative** of the one-form $v = v_\beta dr^\beta$ in the spatial coordinate basis dr^β is defined by the standard formula, cf. Equation (C.1.11),*

$$Dv_\beta = dv_\beta - v_\nu \Gamma^\nu_{\alpha\beta} dr^\alpha \,,$$

or, in components,

$$\nabla_\alpha v_\beta = \partial_\alpha v_\beta - v_\nu \Gamma^\nu_{\alpha\beta} \,.$$

Remark C.1.3 Thus, in differential geometry, the connection one-form (C.1.10) in the local coframe encodes the Riemannian Christoffel coefficients for the spatial coordinates, via the equivalence of metric length (C.1.13) as measured in either set of coordinates. The left-invariant $so(3)$-valued one-form $\widehat{\Omega} = O^{-1}dO$ that the Bichrons need for keeping track of the higher-dimensional time components of their rotations in body coordinates in (C.1.1) plays the same role for their time surfaces in $SO(3)$ as the connection one-form does for taking covariant derivatives in a local coframe. For more discussion of connection one-forms and their role in differential geometry, see, e.g., [Fl1963, Da1994]. □

Exercise. Write the two-time version of the Euler–Poincaré equation for a left-invariant Lagrangian defined on $so(3)$.

Hint: Take a look at Chapter 10. ★

C.1.2 Rotations in complex space

Scenario C.1.3 *The Bers are another alien life form who use one-dimensional time $t \in \mathbb{R}$ (thankfully), but their spatial coordinates are complex $\mathbf{z} \in \mathbb{C}^3$, while ours are real $\mathbf{x} \in \mathbb{R}^3$. They test us to determine whether we are an intelligent life form by requiring us to write the equations for rigid-body motion for body angular momentum coordinates $\mathbf{L} \in \mathbb{C}^3$.*

Their definition of a rigid body requires its moment of inertia $\mathbb{I}$, rotational kinetic energy $\frac{1}{2}\mathbf{L}\cdot\mathbb{I}^{-1}\mathbf{L}$ and magnitude of body angular momentum $\sqrt{\mathbf{L}\cdot\mathbf{L}}$ all to be real. They also tell us these rigid-body equations must be invariant under the operations of parity $\mathcal{P}\,\mathbf{z} \to -\mathbf{z}^*$ and time reversal $\mathcal{T} : t \to -t$. What equations should we give them? Are these equations the same as ours in real body angular momentum coordinates? Keep your approach general for as long as you like, but if you wish to simplify, work out your results with the simple example in which $\mathbb{I} = \mathrm{diag}(1, 2, 3)$.

Answer. Euler's equations for free rotational motion of a rigid body about its centre of mass may be expressed in real vector coordinates $\mathbf{L} \in \mathbb{R}^3$ ($\mathbf{L}$ is the body angular momentum vector) as

$$\dot{\mathbf{L}} = \frac{\partial C}{\partial \mathbf{L}} \times \frac{\partial E}{\partial \mathbf{L}}\,, \qquad \text{(C.1.15)}$$

where C and E are conserved quadratic functions defined by

$$C(\mathbf{L}) = \frac{1}{2}\mathbf{L}\cdot\mathbf{L}\,, \quad E(\mathbf{L}) = \frac{1}{2}\mathbf{L}\cdot\mathbb{I}^{-1}\mathbf{L}\,. \qquad \text{(C.1.16)}$$

Here, $\mathbb{I}^{-1} = \mathrm{diag}\,(I_1^{-1}, I_2^{-1}, I_3^{-1})$ is the inverse of the (real) moment of inertia tensor in principal axis coordinates. These equations are $\mathcal{PT}$-symmetric; they are invariant under spatial reflections of the angular momentum components in the body $P : \mathbf{L} \to \mathbf{L}$ composed with time reversal $T : \mathbf{L} \to -\mathbf{L}$. The simplifying choice $\mathbb{I}^{-1} = \mathrm{diag}(1, 2, 3)$ reduces the dynamics (C.1.15) to

$$\dot{L}_1 = L_2 L_3\,, \quad \dot{L}_2 = -\,2L_1L_3\,, \quad \dot{L}_3 = L_1L_2\,, \qquad \text{(C.1.17)}$$

which may also be written equivalently as

$$\dot{\mathbf{L}} = \mathbf{L} \times \mathsf{K}\mathbf{L}\,, \tag{C.1.18}$$

with $\mathsf{K} = \mathrm{diag}(-1, 0, 1)$.

Since $\mathbf{L}$ is complex, we set $\mathbf{L} = \mathbf{x} + i\mathbf{y}$ and obtain *four* conservation laws, namely the real and imaginary parts of $C(\mathbf{L}) = \frac{1}{2}\mathbf{L}\cdot\mathbf{L}$ and $H(\mathbf{L}) = \frac{1}{2}\mathbf{L}\cdot\mathsf{K}\mathbf{L}$, expressed as

$$C(\mathbf{L}) = \frac{1}{2}\mathbf{x}\cdot\mathbf{x} - \frac{1}{2}\mathbf{y}\cdot\mathbf{y} + i\mathbf{x}\cdot\mathbf{y}\,, \tag{C.1.19}$$

$$H(\mathbf{L}) = \frac{1}{2}\mathbf{x}\cdot\mathsf{K}\mathbf{x} - \frac{1}{2}\mathbf{y}\cdot\mathsf{K}\mathbf{y} + i\mathbf{x}\cdot\mathsf{K}\mathbf{y}\,. \tag{C.1.20}$$

The solutions to Euler's equations that have been studied in the past are the *real* solutions to (C.1.17), that is, the solutions for which $\mathbf{y} = 0$. For this case the phase space is three-dimensional and the two conserved quantities are

$$C = \frac{1}{2}\left(x_1^2 + x_2^2 + x_3^2\right), \quad H = -\frac{1}{2}x_1^2 + \frac{1}{2}x_3^2\,. \tag{C.1.21}$$

If we take $C = \frac{1}{2}$, then the phase-space trajectories are constrained to a sphere of radius 1. There are six critical points located at $(\pm1, 0, 0)$, $(0, \pm1, 0)$ and $(0, 0, \pm1)$. These are the conventional trajectories that are discussed in standard textbooks on dynamical systems [MaRa1994].

Exercise. When $H = 0$, show that the resulting equation is a first integral of the simple pendulum problem. ★

Let us now examine the complex $\mathcal{PT}$-symmetric solutions to Euler's equations. The equation set (C.1.16) is six-dimensional. However, a reduction in dimension occurs because the requirement of $\mathcal{PT}$ symmetry requires the constants of motion C and H in (C.1.20) to be real. The vanishing of the imaginary parts of C and H gives the two equations

$$\mathbf{x} \cdot \mathbf{y} = 0, \quad \mathbf{x} \cdot \mathsf{K}\mathbf{y} = 0 . \tag{C.1.22}$$

These two bilinear constraints may be used to eliminate the $\mathbf{y}$ terms in the complex Equations (C.1.17). When this elimination is performed using the definition $\mathsf{K} = \mathrm{diag}(-1, 0, 1)$, one obtains the following real equations for $\mathbf{x}$ on the $\mathcal{PT}$ constraint manifolds (C.1.22):

$$\dot{\mathbf{x}} = \mathbf{x} \times \mathsf{K}\mathbf{x} + M(\mathbf{x})\,\mathbf{x} . \tag{C.1.23}$$

Here, the scalar function $M = PN/D$, where the functions P, N and D are given by

$$P(\mathbf{x}) = 2x_1x_2x_3 \,, \; N(\mathbf{x}) = x_1^2 + x_2^2 + x_3^2 - 1 \,, \tag{C.1.24}$$

$$D(\mathbf{x}) = \left| \mathrm{Re}\left(\frac{\partial C}{\partial \mathbf{L}} \times \frac{\partial H}{\partial \mathbf{L}} \right) \right|^2 = x_1^2x_2^2 + x_2^2x_3^2 + 4x_1^2x_3^2 .$$

The system (C.1.23) has nonzero divergence, so it cannot be Hamiltonian even though it arises from constraining a Hamiltonian system. Nonetheless, the system has two *additional* real conservation laws, and it reduces to the integrable form

$$\dot{x}_1 = x_2x_3\left(1 + 2x_1^2N/D\right) , \tag{C.1.25}$$
$$\dot{x}_2 = -2x_1x_3\left(1 - x_2^2N/D\right) ,$$
$$\dot{x}_3 = x_1x_2\left(1 + 2x_3^2N/D\right) , \tag{C.1.26}$$

on level sets of two conserved quantities:

$$A = \frac{(N+1)^2N}{D} , \tag{C.1.27}$$

$$B = \frac{x_1^2 - x_3^2}{D}\left(2x_2^2x_3^2 + 4x_1^2x_3^2 + x_2^4 + 2x_1^2x_2^2 - x_2^2\right) .$$

Hence, the motion takes place in $\mathbb{R}^3$ on the intersection of the level sets of these two conserved quantities. These quantities vanish when either $N = 0$ (the unit sphere) or $x_3^2 - x_1^2 = 0$ (the degenerate hyperbolic cylinder). On these level sets of the conserved quantities the motion Equations (C.1.23) restrict to Equations (C.1.17) for the original real rigid body. ▲

Remark C.1.4 We are dealing with rotations of the group of complex 3×3 orthogonal matrices with unit determinant acting on complex three-vectors. These are the linear maps, $SO(3,\mathbb{C})\times\mathbb{C}^3\mapsto\mathbb{C}^3$.

Euler's Equations (C.1.15) for complex body angular momentum describe geodesic motion on $SO(3,\mathbb{C})$ with respect to the metric given by the trace norm $g(\Omega,\Omega) = \frac{1}{2}\mathrm{trace}\,(\Omega^T\mathbb{I}\Omega)$ for the real symmetric moment of inertia tensor $\mathbb{I}$ and left-invariant Lie algebra element $\Omega(t) = g^{-1}(t)\dot{g}(t) \in so(3,\mathbb{C})$. Because $SO(3,\mathbb{C})$ is orthogonal, $\Omega \in so(3,\mathbb{C})$ is a 3×3 complex skew-symmetric matrix, which may be identified with complex vectors $\widehat{\boldsymbol{\Omega}} \in \mathbb{C}^3$ by $(\Omega)_{jk} = -\widehat{\Omega}^i\epsilon_{ijk}$. Euler's Equations (C.1.15) follow from Hamilton's principle in Euler–Poincaré or Lie–Poisson form:

$$\dot{\mu} = \mathrm{ad}^*_\Omega\,\mu = \{\mu,\, H\}\,, \qquad \text{(C.1.28)}$$

where

$$\frac{\delta l}{\delta\Omega} = \mu\,, \quad g^{-1}\dot{g} = \Omega\,, \quad \Omega = \frac{\partial H}{\partial\mu}\,. \qquad \text{(C.1.29)}$$

These are Hamiltonian with the standard Lie–Poisson bracket defined on the dual Lie algebra $so(3,\mathbb{C}^3)^*$. Because of the properties of the trace norm, we may take $\mu = \mathrm{skew}\,\mathbb{I}\Omega$. (Alternatively, we may set the preserved symmetric part of μ initially to zero.) Hence, μ may be taken as a skew-symmetric complex matrix, which again may be identified with the components of a complex three-vector $\mathbf{z}$ as $(\mu)_{jk} = -z^i\epsilon_{ijk}$. On making this identification, Euler's Equations (C.1.15) emerge for $\mathbf{z} \in \mathbb{C}^3$, with real $\mathbb{I}$. The $\mathcal{PT}$-symmetric initial conditions on the real level sets of the preserved complex quantities C and H form an invariant manifold of this system of three complex ordinary differential equations. On this invariant manifold, the complex angular motion is completely integrable. By following the approach established by Manakov [Man1976] this reasoning may also extend to the rigid body on $SO(n,\mathbb{C})$. □

Exercise. The Bers left behind a toy monopole. This is a rigid body that rotates by complex angles and whose three moments of inertia are the complex cube roots of unity. What are the equations of motion for this toy monopole? For a hint, take a look at [Iv2006]. ★

C.1.3 Rotations in four dimensions: $SO(4)$

Scenario C.1.4 *The Tets are yet another alien life form who also use one-dimensional time $t \in \mathbb{R}$ (we sigh with relief), but their spatial coordinates are $\mathbf{X} \in \mathbb{R}^4$, while ours are $\mathbf{x} \in \mathbb{R}^3$. They test us to determine whether we are an intelligent life form by requiring us to write the equations for rigid-body motion for four-dimensional rotations.*

Hint: The angular velocity of rotation $\widehat{\Psi} = O^{-1}\dot{O}(t)$ for rotations $O(t) \in SO(4)$ in four dimensions will be represented by a 4×4 skew-symmetric matrix. Write a basis for the 4×4 skew-symmetric matrices by adding a row and column to the 3×3 basis.

Answer. Any 4×4 skew-symmetric matrix may be represented as a linear combination of 4×4 basis matrices with three-dimensional vector coefficients $\Omega, \Lambda \in \mathbb{R}^3$ in the form

$$\begin{aligned}
\widehat{\Psi} &= \begin{pmatrix} 0 & -\Omega_3 & \Omega_2 & -\Lambda_1 \\ \Omega_3 & 0 & -\Omega_1 & -\Lambda_2 \\ -\Omega_2 & \Omega_1 & 0 & -\Lambda_3 \\ \Lambda_1 & \Lambda_2 & \Lambda_3 & 0 \end{pmatrix} \\
&= \Omega \cdot \widehat{J} + \Lambda \cdot \widehat{K} \\
&= \Omega_a \widehat{J}_a + \Lambda_b \widehat{K}_b \,.
\end{aligned}$$

This is the formula for the angular velocity of rotation in four dimensions.

The 4×4 basis set $\widehat{J} = (J_1, J_2, J_3)^T$ and $\widehat{K} = (K_1, K_2, K_3)^T$ consists of the following six linearly independent 4×4 skew-symmetric matrices, $\widehat{J}_a$, $\widehat{K}_b$ with $a, b = 1, 2, 3$:

$$\widehat{J}_1 = \begin{pmatrix} 0 & 0 & 0 & 0 \\ 0 & 0 & -1 & 0 \\ 0 & 1 & 0 & 0 \\ 0 & 0 & 0 & 0 \end{pmatrix}, \quad \widehat{K}_1 = \begin{pmatrix} 0 & 0 & 0 & -1 \\ 0 & 0 & 0 & 0 \\ 0 & 0 & 0 & 0 \\ 1 & 0 & 0 & 0 \end{pmatrix},$$

$$\widehat{J}_2 = \begin{pmatrix} 0 & 0 & 1 & 0 \\ 0 & 0 & 0 & 0 \\ -1 & 0 & 0 & 0 \\ 0 & 0 & 0 & 0 \end{pmatrix}, \quad \widehat{K}_2 = \begin{pmatrix} 0 & 0 & 0 & 0 \\ 0 & 0 & 0 & -1 \\ 0 & 0 & 0 & 0 \\ 0 & 1 & 0 & 0 \end{pmatrix},$$

$$\widehat{J}_3 = \begin{pmatrix} 0 & -1 & 0 & 0 \\ 1 & 0 & 0 & 0 \\ 0 & 0 & 0 & 0 \\ 0 & 0 & 0 & 0 \end{pmatrix}, \quad \widehat{K}_3 = \begin{pmatrix} 0 & 0 & 0 & 0 \\ 0 & 0 & 0 & 0 \\ 0 & 0 & 0 & -1 \\ 0 & 0 & 1 & 0 \end{pmatrix}.$$

The matrices $\widehat{J}_a$ with $a = 1, 2, 3$ embed the basis for 3×3 skew-symmetric matrices into the 4×4 matrices by adding a row and column of zeros. The skew matrices $\widehat{K}_a$ with $a = 1, 2, 3$ then extend the 3×3 basis to 4×4.

Commutation relations

The skew matrix basis $\widehat{J}_a$, $\widehat{K}_b$ with $a, b = 1, 2, 3$ satisfies the commutation relations,

$$\begin{aligned} \left[\widehat{J}_a,\, \widehat{J}_b\right] &= \widehat{J}_a\widehat{J}_b - \widehat{J}_b\widehat{J}_a = \epsilon_{abc}\widehat{J}_c\,, \\ \left[\widehat{J}_a,\, \widehat{K}_b\right] &= \widehat{J}_a\widehat{K}_b - \widehat{K}_b\widehat{J}_a = \epsilon_{abc}\widehat{K}_c\,, \\ \left[\widehat{K}_a,\, \widehat{K}_b\right] &= \widehat{K}_a\widehat{K}_b - \widehat{K}_b\widehat{K}_a = \epsilon_{abc}\widehat{J}_c\,. \end{aligned}$$

These commutation relations may be verified by a series of direct calculations, as $[\,\widehat{J}_1,\ \widehat{J}_2\,] = \widehat{J}_3$, etc.

Hat map for* 4×4 *skew matrices

The map above for the 4×4 skew matrix $\widehat{\Psi}$ may be written as

$$\widehat{\Psi} = \Omega \cdot \widehat{J} + \Lambda \cdot \widehat{K} = \Omega_a \widehat{J}_a + \Lambda_b \widehat{K}_b \,, \text{ sum on } a, b = 1, 2, 3 \,.$$

This map provides the 4×4 version of the hat map, written now as $(\,\cdot\,)\widehat{\ }: \mathbb{R}^3 \times \mathbb{R}^3 \mapsto so(4)$. Here $so(4)$ is the Lie algebra of the 4×4 special orthogonal matrices, which consists of the 4×4 skew matrices represented in the six-dimensional basis of $\widehat{J}$'s and $\widehat{K}$'s.

Commutator as intertwined vector product

The commutator of 4×4 skew matrices corresponds to an intertwined vector product, as follows. For any vectors Ω, Λ, ω, $\lambda \in \mathbb{R}^3$, one has

$$\begin{aligned}&\Big[\Omega \cdot \widehat{J} + \Lambda \cdot \widehat{K} \,,\, \omega \cdot \widehat{J} + \lambda \cdot \widehat{K} \Big] \\ &\qquad = \Big(\Omega \times \omega + \Lambda \times \lambda \Big) \cdot \widehat{J} + \Big(\Omega \times \lambda - \Lambda \times \omega \Big) \cdot \widehat{K} \,.\end{aligned}$$

Likewise, the matrix pairing $\langle\, A, B \,\rangle = \mathrm{tr}(A^T B)$ is related to the vector dot-product pairing in $\mathbb{R}^3$ by

$$\Big\langle \Omega \cdot \widehat{J} + \Lambda \cdot \widehat{K} \,,\, \omega \cdot \widehat{J} + \lambda \cdot \widehat{K} \Big\rangle = \Omega \cdot \omega + \Lambda \cdot \lambda \,.$$

That is,

$$\Big\langle\, \widehat{J}_a \,,\, \widehat{J}_b \,\Big\rangle = \delta_{ab} = \Big\langle\, \widehat{K}_a \,,\, \widehat{K}_b \,\Big\rangle \quad \text{and} \quad \Big\langle\, \widehat{J}_a \,,\, \widehat{K}_b \,\Big\rangle = 0 \,.$$

***Euler–Poincaré equation on* $so(4)^*$**

For

$$\Phi = O^{-1} \delta O(t) = \xi \cdot \widehat{J} + \eta \cdot \widehat{K} \in so(4) \,,$$

Hamilton's principle $\delta S = 0$ for $S = \int_a^b \ell(\Psi)\, dt$ with

$$\Psi = O^{-1}\dot{O}(t) = \Omega \cdot \widehat{J} + \Lambda \cdot \widehat{K} \in so(4)$$

leads to

$$\delta S = \int_a^b \Big\langle \frac{\delta \ell}{\delta \Psi}\,,\, \delta\Psi \Big\rangle\, dt = \int_a^b \Big\langle \frac{\delta \ell}{\delta \Psi}\,,\, \dot{\Phi} + \mathrm{ad}_\Psi \Phi \Big\rangle\, dt\,,$$

where

$$\begin{aligned}
\mathrm{ad}_\Psi \Phi &= [\Psi,\, \Phi] = \big[\Omega \cdot \widehat{J} + \Lambda \cdot \widehat{K},\, \xi \cdot \widehat{J} + \eta \cdot \widehat{K}\big] \\
&= \Big(\Omega \times \xi + \Lambda \times \eta\Big) \cdot \widehat{J} + \Big(\Omega \times \eta - \Lambda \times \xi\Big) \cdot \widehat{K}\,.
\end{aligned}$$

Thus,

$$\begin{aligned}
\delta S &= \int_a^b \Big\langle -\frac{d}{dt}\frac{\delta \ell}{\delta \Psi}\,,\, \Phi \Big\rangle + \Big\langle \frac{\delta \ell}{\delta \Psi}\,,\, \mathrm{ad}_\Psi \Phi \Big\rangle\, dt \\
&= \int_a^b \Big\langle -\frac{d}{dt}\frac{\delta \ell}{\delta \Omega} \cdot \widehat{J} - \frac{d}{dt}\frac{\delta \ell}{\delta \Lambda} \cdot \widehat{K}\,,\, \xi \cdot \widehat{J} + \eta \cdot \widehat{K} \Big\rangle\, dt \\
&\quad + \int_a^b \Big\langle \frac{\delta \ell}{\delta \Omega} \cdot \widehat{J} + \frac{\delta \ell}{\delta \Lambda} \cdot \widehat{K}\,, \\
&\qquad \Big(\Omega \times \xi + \Lambda \times \eta\Big) \cdot \widehat{J} + \Big(\Omega \times \eta - \Lambda \times \xi\Big) \cdot \widehat{K} \Big\rangle\, dt \\
&= \int_a^b \Big(-\frac{d}{dt}\frac{\delta \ell}{\delta \Omega} + \frac{\delta \ell}{\delta \Omega} \times \Omega - \frac{\delta \ell}{\delta \Lambda} \times \Lambda \Big) \cdot \xi \\
&\quad + \Big(-\frac{d}{dt}\frac{\delta \ell}{\delta \Lambda} + \frac{\delta \ell}{\delta \Lambda} \times \Omega + \frac{\delta \ell}{\delta \Omega} \times \Lambda \Big) \cdot \eta\, dt\,.
\end{aligned}$$

Hence, $\delta S = 0$ yields

$$\begin{aligned}
\frac{d}{dt}\frac{\delta \ell}{\delta \Omega} &= \frac{\delta \ell}{\delta \Omega} \times \Omega - \frac{\delta \ell}{\delta \Lambda} \times \Lambda \\
\text{and} \quad \frac{d}{dt}\frac{\delta \ell}{\delta \Lambda} &= \frac{\delta \ell}{\delta \Lambda} \times \Omega + \frac{\delta \ell}{\delta \Omega} \times \Lambda\,. \qquad \text{(C.1.30)}
\end{aligned}$$

These are the $\widehat{J}$, $\widehat{K}$ basis components of the Euler–Poincaré equation on $so(4)^*$,

$$\frac{d}{dt}\frac{\delta \ell}{\delta \Psi} = \mathrm{ad}^*_\Psi \frac{\delta \ell}{\delta \Psi},$$

written with $\Psi = \Omega \cdot \widehat{J} + \Lambda \cdot \widehat{K}$ in this basis.

Hamiltonian form on $so(4)^*$

Legendre-transforming yields the pairs

$$\Pi = \frac{\delta \ell}{\delta \Omega}, \qquad \Omega = \frac{\delta h}{\delta \Pi}, \quad \text{and} \quad \Xi = \frac{\delta \ell}{\delta \Lambda}, \qquad \Lambda = \frac{\delta h}{\delta \Xi}.$$

Hence, these equations may be expressed in Hamiltonian form as

$$\frac{d}{dt}\begin{bmatrix} \Pi \\ \Xi \end{bmatrix} = \begin{bmatrix} \Pi\times & \Xi\times \\ \Xi\times & \Pi\times \end{bmatrix} \begin{bmatrix} \delta h/\delta \Pi \\ \delta h/\delta \Xi \end{bmatrix}. \tag{C.1.31}$$

The corresponding Lie–Poisson bracket is given by

$$\begin{aligned} \{f,\, h\} &= -\Pi \cdot \left(\frac{\delta f}{\delta \Pi} \times \frac{\delta h}{\delta \Pi} + \frac{\delta f}{\delta \Xi} \times \frac{\delta h}{\delta \Xi} \right) \\ &\quad - \Xi \cdot \left(\frac{\delta f}{\delta \Pi} \times \frac{\delta h}{\delta \Xi} - \frac{\delta h}{\delta \Pi} \times \frac{\delta f}{\delta \Xi} \right). \end{aligned}$$

This Lie–Poisson bracket has an extra term proportional to Π, relative to the $se(3)^*$ bracket (7.1.10) for the heavy top. Its Hamiltonian matrix has two null eigenvectors for the variational derivatives of $C_1 = |\Pi|^2 + |\Xi|^2$ and $C_2 = \Pi \cdot \Xi$. The functions C_1, C_2 are the Casimirs of the $so(4)$ Lie–Poisson bracket. That is, $\{C_1,\, H\} = 0 = \{C_2,\, H\}$ for every Hamiltonian $H(\Pi,\, \Xi)$.

The Hamiltonian matrix in Equation (C.1.31) is similar to that for the Lie–Poisson formulation of heavy-top dynamics, except for the one extra term $\{\Xi,\, \Xi\} \neq 0$. ▲

C.2 $\mathbb{C}^3$ oscillators

Scenario C.2.1 *For* $\mathbf{a} \in \mathbb{C}^3$ *one may write the* 3×3 *Hermitian matrix* $Q = \mathbf{a} \otimes \mathbf{a}^*$ *as the sum* $Q = S + iA$ *of a* 3×3 *real symmetric matrix* S *plus* i *times a* 3×3 *real antisymmetric matrix* A*:*

$$\begin{aligned} Q &= \begin{bmatrix} M_1 & N_3 - iL_3 & N_2 + iL_2 \\ N_3 + iL_3 & M_2 & N_1 - iL_1 \\ N_2 - iL_2 & N_1 + iL_1 & M_3 \end{bmatrix} \\ &= \begin{bmatrix} M_1 & N_3 & N_2 \\ N_3 & M_2 & N_1 \\ N_2 & N_1 & M_3 \end{bmatrix} + i \begin{bmatrix} 0 & -L_3 & L_2 \\ L_3 & 0 & -L_1 \\ -L_2 & L_1 & 0 \end{bmatrix} . \end{aligned}$$

(i) *Compute the Poisson brackets of the L's, M's and N's among themselves, given that* $\{a_j, a_k^*\} = -2i\delta_{jk}$ *for* $j, k = 1, 2, 3$.

(ii) *Transform into a rotating frame in which the real symmetric part of* Q *is diagonal. Write the Hamiltonian equations for the L's, M's and N's in that rotating frame for a rotationally invariant Hamiltonian.*

Answer. (Oscillator variables in three dimensions) The nine elements of Q are the S^1-invariants

$$Q_{jk} = a_j a_k^* = S_{jk} + iA_{jk} , \quad j, k = 1, 2, 3.$$

The Poisson brackets among these variables are evaluated from the canonical relation,

$$\{a_j , a_k^*\} = -2i\,\delta_{jk} ,$$

by using the Leibniz property (product rule) for Poisson brackets to find

$$\{Q_{jk} , Q_{lm}\} = 2i\,(\delta_{kl}Q_{jm} - \delta_{jm}Q_{kl}) , \quad j, k, l, m = 1, 2, 3 .$$

Remark C.2.1 The quadratic S^1-invariant quantities in $\mathbb{C}^3$ Poisson commute among themselves. This property of ***closure*** is to be expected for a simple reason. The Poisson bracket between two homogeneous polynomials of weights w_1 and w_2 produces a homogeneous polynomial of weight $w = w_1 + w_2 - 2$ and $2+2-2 = 2$; so the quadratic homogeneous polynomials Poisson-commute among themselves.

The result is also a simple example of Poisson reduction by symmetry, obtained by ***transforming to quantities that are invariant*** under the action of a Lie group. The action in this case is the (diagonal) S^1 phase shift $a_j \to a_j e^{i\phi}$ for $j = 1, 2, 3$. □

(i) One defines $L_a := -\frac{1}{2}\epsilon_{ajk}A_{jk} = (\mathbf{p}\times\mathbf{q})_a$ and finds the Poisson bracket relations,

$$\begin{aligned}\{L_a\,,\,L_b\} &= A_{ab} - A_{ba} = \epsilon_{abc}L_c\,,\\ \{L_a\,,\,Q_{jk}\} &= \frac{1}{2}\left[\epsilon_{ajc}Q_{ck} - \epsilon_{akc}Q_{jc}\right].\end{aligned}$$

Thus, perhaps not unexpectedly, the Poisson bracket for quadratic S^1-invariant quantities in $\mathbb{C}^3$ contains the angular momentum Poisson bracket among the variables L_a with $a = 1, 2, 3$. This could be expected, because the 3×3 form of Q contains the 2×2 form, which we know admits the Hopf fibration into quantities which satisfy Poisson bracket relations dual to the Lie algebra $so(3) \simeq su(2)$. Moreover, the imaginary part $\operatorname{Im} Q = \mathbf{L}\cdot\widehat{\mathbf{J}} = L_a\widehat{J}_a$, where $\widehat{J}_a$ with $a = 1, 2, 3$ is a basis set for $so(3)$ as represented by the 3×3 skew-symmetric real matrices.

Another interesting set of Poisson bracket relations among the M's, N's and L's may be found. These relations are

$$\begin{aligned}\{N_a - iL_a\,,\, N_b - iL_b\} &= 2i\,\epsilon_{abc}(N_c + iL_c)\,,\\ \{M_a\,,\, M_b\} &= 0\,,\\ \{M_a\,,\, N_b - iL_b\} &= 2i\,\mathrm{sgn}(b-a)(-1)^{a+b}\\ &\qquad (N_b - iL_b)\,,\end{aligned}$$

where $\mathrm{sgn}(b-a)$ is the sign of the difference $(b-a)$, which vanishes when $b = a$.

Additional Poisson bracket relations may also be read off from the Poisson commutators of the real and imaginary components of $Q = S + iA$ among themselves as

$$\begin{aligned}\{S_{jk}\,,\, S_{lm}\} &= \delta_{jl}A_{mk} + \delta_{kl}A_{mj} - \delta_{jm}A_{kl} - \delta_{km}A_{jl}\,,\\ \{S_{jk}\,,\, A_{lm}\} &= \delta_{jl}S_{mk} + \delta_{kl}S_{mj} - \delta_{jm}S_{kl} - \delta_{km}S_{jl}\,,\\ \{A_{jk}\,,\, A_{lm}\} &= \delta_{jl}A_{mk} - \delta_{kl}A_{mj} + \delta_{jm}A_{kl} - \delta_{km}A_{jl}\,.\end{aligned} \tag{C.2.1}$$

These relations produce the following five tables of Poisson brackets in addition to $\{M_a\,,\, M_b\} = 0$:

$\{\cdot\,,\cdot\}$	L_1	L_2	L_3
L_1	0	L_3	$-L_2$
L_2	$-L_3$	0	L_1
L_3	L_2	$-L_1$	0

$\{\cdot\,,\cdot\}$	N_1	N_2	N_3
N_1	0	$-L_3$	L_2
N_2	L_3	0	$-L_1$
N_3	$-L_2$	L_1	0

$\{\cdot\,,\cdot\}$	L_1	L_2	L_3
M_1	0	$2N_2$	$-2N_3$
M_2	$-2N_1$	0	$2N_3$
M_3	$2N_1$	$-2N_2$	0

$\{\cdot,\cdot\}$	N_1	N_2	N_3
M_1	0	$-2L_2$	$2L_3$
M_2	$2L_1$	0	$-2L_3$
M_3	$-2L_1$	$2L_2$	0

$\{\cdot,\cdot\}$	L_1	L_2	L_3
N_1	$M_2 - M_3$	$-N_3$	N_2
N_2	N_3	$M_3 - M_1$	$-N_1$
N_3	$-N_2$	N_1	$M_1 - M_2$

As expected, the system is closed and it has the angular momentum Poisson bracket table as a closed subset. Next, we will come to understand that this is because the Lie algebra $su(2)$ is a subalgebra of $su(3)$.

(ii) The rotation group $SO(3)$ is a subgroup of $SU(3)$. An element $Q \in su(3)^*$ transforms under $SO(3)$ by the coAdjoint action

$$\mathrm{Ad}^*_R\, Q = R^{-1}QR = R^{-1}SR + iR^{-1}AR\,.$$

Choose $R \in SO(3)$ so that $R^{-1}SR = D = \mathrm{diag}\,(d_1, d_2, d_3)$ is diagonal. (That is, rotate into principal axis coordinates for S.) The eigenvalues are unique up to their order, which one may fix as, say, $d_1 \geq d_2 \geq d_3$. While it diagonalises the symmetric part of Q, the rotation R takes the antisymmetric part from the spatial frame to the body frame, where S is diagonal. At the same time the spatial angular momentum matrix A is transformed to $B = R^{-1}AR$, which is the body angular momentum. Thus,

$$\mathrm{Ad}^*_R\, Q = R^{-1}SR + iR^{-1}AR =: D + iB\,.$$

Define the body angular velocity $\Omega = R^{-1}\dot{R} \in so(3)$, which is left-invariant. The Hamiltonian dynamical system obeys

$$\dot{Q} = \{Q,\, H(Q)\}\,.$$

For $B = R^{-1}AR$, this implies

$$\dot{B} + [\Omega,\, B] = R^{-1}\dot{A}R = R^{-1}\{A,\, H(Q)\}R\,.$$

However, $H(Q)$ being rotationally symmetric means the spatial angular momentum A will be time-independent $\dot{A} = \{A,\, H(Q)\} = 0$. Hence,

$$\dot{B} + [\Omega, B] = 0\,.$$

Thus, the equation for the body angular momentum B is formally identical to Euler's equations for rigid-body motion. Physically, this represents conservation of spatial angular momentum, because of the rotational symmetry of the Hamiltonian. Likewise, for $D = R^{-1}SR$, one finds

$$\dot{D} + [\Omega,\, D] = R^{-1}\dot{S}R = R^{-1}\{S,\, H(Q)\}R \neq 0\,.$$

The body angular momentum B satisfies Euler's rigid-body equations, but this body is not rigid! While the rotational degrees of freedom satisfy spatial angular momentum conservation, the shape of the body depends on the value of the Poisson bracket $R^{-1}\{S,\, H(Q)\}R$ which is likely to be highly nontrivial! For example, the Hamiltonian $H(Q)$ may be chosen to be a function of the following three rotationally invariant quantities:

$$\begin{aligned} \mathrm{tr}(A^TA) &= \mathrm{tr}(B^TB)\,, \\ \mathrm{tr}(A^TSA) &= \mathrm{tr}(B^TDB)\,, \\ \mathrm{tr}(A^TS^2A) &= \mathrm{tr}(B^TD^2B)\,. \end{aligned}$$

Dependence of the Hamiltonian on these quantities will bring the complications of the Poisson bracket relations in (C.2.1) into the dynamics of the triaxial ellipsoidal shape represented by D. ▲

Remark C.2.2 The quantity

$$\widetilde{Q} = \mathbf{a} \otimes \mathbf{a}^* - \frac{1}{3}\,\mathrm{Id}|\mathbf{a}|^2 : \mathbb{C}^3 \mapsto su(3)^*$$

corresponds for the action of $SU(3)$ on $\mathbb{C}^3$ to the momentum map $J : \mathbb{C}^2 \mapsto su(2)^*$ in Example 11.3.1 for the action of $SU(2)$ on $\mathbb{C}^2$. □

C.3 Momentum maps for $GL(n, \mathbb{R})$

Scenario C.3.1 ($GL(n, \mathbb{R})$ invariance) *Begin with the Lagrangian*

$$L = \frac{1}{2}\,\mathrm{tr}\Big(\dot{S}S^{-1}\dot{S}S^{-1}\Big) + \frac{1}{2}\,\dot{\mathbf{q}}^T S^{-1}\dot{\mathbf{q}}\,,$$

where $S = S^T$ is an $n \times n$ symmetric matrix and $\mathbf{q} \in \mathbb{R}^n$ is an n-component column vector.

(i) Legendre-transform to construct the corresponding Hamiltonian and canonical equations.

(ii) Show that the system is invariant under the group action

$$\mathbf{q} \to A\mathbf{q} \quad \text{and} \quad S \to ASA^T\,,$$

for any constant invertible $n \times n$ matrix, A.

(iii) Compute the infinitesimal generator for this group action and construct its corresponding momentum map. Is this momentum map equivariant?

(iv) Verify directly that this momentum map is a conserved $n \times n$ matrix quantity by using the equations of motion.

(v) Is this system completely integrable for any value of $n > 2$?

Answer. ($GL(n,\mathbb{R})$ invariance)

(i) Legendre-transform

$$P = \frac{\partial L}{\partial \dot{S}} = S^{-1}\dot{S}S^{-1} \quad \text{and} \quad \mathbf{p} = \frac{\partial L}{\partial \dot{\mathbf{q}}} = S^{-1}\dot{\mathbf{q}}\,.$$

Thus, $P = P^T$ is also a symmetric matrix. The Hamiltonian $H(Q,P)$ and its canonical equations are

$$H(\mathbf{q},\mathbf{p},S,P) = \frac{1}{2}\mathrm{tr}\big(PS\cdot PS\big) + \frac{1}{2}\mathbf{p}\cdot S\mathbf{p}\,,$$

$$\dot{S} = \frac{\partial H}{\partial P} = SPS\,,\ \dot{P} = -\frac{\partial H}{\partial S} = -\Big(PSP + \frac{1}{2}\mathbf{p}\otimes\mathbf{p}\Big)\,,$$

$$\dot{\mathbf{q}} = \frac{\partial H}{\partial \mathbf{p}} = S\mathbf{p}\,,\quad \dot{\mathbf{p}} = \frac{\partial H}{\partial \mathbf{q}} = 0\,.$$

(ii) Under the group action $\mathbf{q} \to G\mathbf{q}$ and $S \to GSG^T$ for any constant invertible $n\times n$ matrix, G, one finds

$$\dot{S}S^{-1} \to G\dot{S}S^{-1}G^{-1} \quad \text{and} \quad \dot{\mathbf{q}}\cdot S^{-1}\dot{\mathbf{q}} \to \dot{\mathbf{q}}\cdot S^{-1}\dot{\mathbf{q}}.$$

Hence, $L \to L$. Likewise, $P \to G^{-T}PG^{-1}$ so $PS \to G^{-T}PSG^T$ and $\mathbf{p} \to G^{-T}\mathbf{p}$ so that $S\mathbf{p} \to GS\mathbf{p}$. Hence, $H \to H$, as well; so both L and H for the system are invariant.

(iii) The infinitesimal actions for $G(\epsilon) = Id + \epsilon A + O(\epsilon^2)$, where $A \in gl(n)$ are

$$X_A\mathbf{q} = \frac{d}{d\epsilon}\Big|_{\epsilon=0} G(\epsilon)\mathbf{q} = A\mathbf{q} \quad \text{and}$$

$$X_A S = \frac{d}{d\epsilon}\Big|_{\epsilon=0}\Big(G(\epsilon)SG(\epsilon)^T\Big) = AS + SA^T\,.$$

The defining relation for the corresponding momentum map yields

$$\begin{aligned}\langle J, A\rangle &= \langle (Q,P), X_A\rangle \\ &= \mathrm{tr}(P^T X_A S) + \mathbf{p}\cdot X_A\mathbf{q} \\ &= \mathrm{tr}\big(P^T(AS + SA^T)\big) + \mathbf{p}\cdot A\mathbf{q}\,.\end{aligned}$$

Hence,

$$\langle J, A\rangle := \operatorname{tr}\big(JA^T\big) = \operatorname{tr}\Big(\big(S(P+P^T) + \mathbf{q}\otimes\mathbf{p}\big)A\Big),$$

which implies

$$J = (P^T + P)S + \mathbf{p}\otimes\mathbf{q} = 2PS + \mathbf{p}\otimes\mathbf{q}\,.$$

This momentum map is a cotangent lift, so it is equivariant.

(iv) Conservation of the momentum map is verified directly by

$$\begin{aligned}\dot{J} &= (2\dot{P}S + 2P\dot{S} + \mathbf{p}\otimes\dot{\mathbf{q}}) \\ &= -2PSPS - (\mathbf{p}\otimes\mathbf{p})S + 2PSPS + \mathbf{p}\otimes S\mathbf{p} = 0\,.\end{aligned}$$

(v) Integrability would be a good question for further study in this problem. ▲

Scenario C.3.2 (Ellipsoidal motions on $GL(3,\mathbb{R})$) *Choose the Lagrangian in three dimensions,*

$$L = \frac{1}{2}\operatorname{tr}\Big(\dot{Q}^T\dot{Q}\Big) - V\Big(\operatorname{tr}(Q^TQ), \det(Q)\Big),$$

where $Q(t) \in GL(3,\mathbb{R})$ is a 3×3 matrix function of time and the potential energy V is an arbitrary function of $\operatorname{tr}(Q^TQ)$ and $\det(Q)$.

(i) Legendre-transform this Lagrangian. That is, find the momenta P_{ij} canonically conjugate to Q_{ij}. Then construct the Hamiltonian $H(Q,P)$ and write Hamilton's canonical equations of motion for this problem.

(ii) Show that the Hamiltonian is invariant under left action $Q \to UQ$ where $U \in SO(3)$. Construct the cotangent lift of this action on P. Hence, construct the momentum map of left action.

(iii) Show that the Hamiltonian is also invariant under right action $Q \to QU$ where $U \in SO(3)$. Construct the cotangent lift of this action on P. That is, construct the momentum map of right action.

(iv) Find the Poisson bracket relations among the various Hamiltonians for the two types of momentum maps. Explain your results in terms of the Lie algebra homomorphism that preserves bracket relations for equivariant momentum maps.

Answer. (Ellipsoidal motions on $GL(3,\mathbb{R})$)

(i) Legendre-transform

$$P_{ij} = \frac{\partial L}{\partial \dot{Q}^T_{ij}} = \dot{Q}_{ij}\,.$$

Thus, the Hamiltonian $H(Q,P)$ and its canonical equations are

$$\begin{aligned} H(Q,P) &= \frac{1}{2}\mathrm{tr}(P^TP) + V\Big(\mathrm{tr}(Q^TQ), \det(Q)\Big)\,, \\ \dot{Q}_{ij} &= \frac{\partial H}{\partial P^T_{ij}} = P_{ij}\,, \\ \dot{P}_{ij} &= -\frac{\partial H}{\partial Q^T_{ij}} \\ &= -\Bigg(\frac{\partial V}{\partial \mathrm{tr}(Q^TQ)} 2Q_{ij} \\ &\qquad + \frac{\partial V}{\partial \det(Q)}(\det Q)(Q^{-1})_{ij}\Bigg) \end{aligned}$$

where one uses the identity $d(\det Q) = (\det Q)\mathrm{tr}\,(Q^{-1}dQ^T)$. The corresponding canonical Poisson bracket is

$$\begin{aligned} \{F,H\} &= \frac{\partial F}{\partial Q_{ij}}\frac{\partial H}{\partial P_{ij}} - \frac{\partial F}{\partial P_{ij}}\frac{\partial H}{\partial Q_{ij}} \\ &= \mathrm{tr}\left(\frac{\partial F}{\partial Q}\frac{\partial H}{\partial P^T} - \frac{\partial F}{\partial P}\frac{\partial H}{\partial Q^T}\right). \qquad \text{(C.3.1)} \end{aligned}$$

(ii) Under $Q \to UQ$, $\dot{Q} \to U\dot{Q}$, one finds $P \to UP$. For $U \in SO(3)$, this means

$$\begin{aligned} \det Q &\to \det UQ = \det U \det Q = \det Q\,, \\ Q^TQ &\to Q^TU^TUQ = Q^TQ\,, \quad \text{and} \\ P^TP &\to P^TU^TUP = P^TP\,. \end{aligned}$$

Hence, $H(Q,P)$ is left-invariant under $Q \to UQ$ for $U \in SO(3)$.
The infinitesimal generator of left action X_A^L is found from

$$X_A^L Q = \frac{d}{d\epsilon}\Big|_{\epsilon=0} U(\epsilon)Q$$
$$= \frac{d}{d\epsilon}\Big|_{\epsilon=0} (Id + \epsilon A + O(\epsilon^2))Q = AQ\,,$$

with $A \in so(3)$ a 3×3 antisymmetric matrix; that is, $A^T = -A$.
Likewise, $P \to UP$ with $U \in SO(3)$ has infinitesimal action $X_A^L P = AP$, $A \in so(3)$. The action

$$X_A^L Q = AQ = \frac{\partial \langle J_L, A\rangle}{\partial P^T} \quad \text{and}$$

$$X_A^L P = AP = -\frac{\partial \langle J_L, A\rangle}{\partial Q^T}$$

is generated canonically by the Hamiltonian

$$\langle J_L, A\rangle = -\operatorname{tr}(J_L^T A) = \operatorname{tr}(P^T AQ)$$
$$= \operatorname{tr}(QP^T A) = -\operatorname{tr}(PQ^T A)\,.$$

Hence, the momentum map for left action is

$$J_L = \left[QP^T\right] = -\left[PQ^T\right],$$

where the square brackets mean taking the antisymmetric part.

(iii) Likewise, $H(Q,P)$ is right-invariant under $(Q,P) \to (QU, PU)$ for $U \in SO(3)$. The corresponding infinitesimal generator is found from

$$X_A^R Q = QA = \frac{\partial \langle J_R, A\rangle}{\partial P^T} \quad \text{and}$$

$$X_A^R P = PA = -\frac{\partial \langle J_R, A\rangle}{\partial Q^T}$$

which is generated by the Hamiltonian

$$\begin{aligned}\langle J_R, A\rangle &= -\operatorname{tr}(J_R^T A) = \operatorname{tr}(J_R A)\\ &= \operatorname{tr}(P^T Q A) = -\operatorname{tr}(Q^T P A)\,.\end{aligned}$$

Hence, the momentum map for right action is

$$J_R = \left[P^T Q\right] = -\left[Q^T P\right]\,.$$

Note that $J_R = Q^{-1} J_L Q = \mathrm{Ad}_{Q^{-1}} J_L$.

(iv) Define Hamiltonians for the momentum maps as the functions

$$J_L^A(Q,P) := \langle J_L, A\rangle\,, \quad J_R^B(Q,P) := \langle J_R, B\rangle\,, \quad \text{etc.}$$

Then by using the canonical Poisson bracket (C.3.1) on $SO(3)$ and antisymmetry of the matrices $A, B \in so(3)$ we find

$$\begin{aligned}\{J_L^A, J_L^B\} &= J_L^{[A,B]}\,,\\ \{J_R^A, J_R^B\} &= -\,J_R^{[A,B]}\,,\\ \{J_L^A, J_R^B\} &= 0\,.\end{aligned}$$

The Poisson brackets for the left (resp., right) actions give plus (resp., minus) representations of the Lie algebra, which is $so(3)$ in this case. This result is expected, because cotangent-lift momentum maps are equivariant, and thus are infinitesimally equivariant, which in turn implies the Lie algebra homomorphism (11.2.4) that preserves bracket relations. Likewise, since left and right actions commute, the two momentum maps commute and their Hamiltonians Poisson-commute.

▲

C.4 Motion on the symplectic Lie group $Sp(2)$

Let the set of 2×2 matrices M_i with $i = 1, 2, 3$ satisfy the defining relation for the symplectic Lie group $Sp(2)$,

$$M_i J M_i^T = J \quad \text{with} \quad J = \begin{pmatrix} 0 & -1 \\ 1 & 0 \end{pmatrix}.$$

The corresponding elements of its Lie algebra $m_i = \dot{M}_i M_i^{-1} \in sp(2)$ satisfy $(Jm_i)^T = Jm_i$ for each $i = 1, 2, 3$. Thus, $\mathsf{X}_i = Jm_i$ satisfying $\mathsf{X}_i^T = \mathsf{X}_i$ is a set of three symmetric 2×2 matrices. For definiteness, we may choose a basis given by

$$\mathsf{X}_1 = Jm_1 = \begin{pmatrix} 2 & 0 \\ 0 & 0 \end{pmatrix}, \ \mathsf{X}_2 = Jm_2 = \begin{pmatrix} 0 & 0 \\ 0 & 2 \end{pmatrix}, \ \mathsf{X}_3 = Jm_3 = \begin{pmatrix} 0 & 1 \\ 1 & 0 \end{pmatrix}.$$

This basis corresponds to the momentum map $\mathbb{R}^6 \to sp(2)^*$ of quadratic phase-space functions $\mathbf{X} = (|\mathbf{q}|^2, |\mathbf{p}|^2, \mathbf{q} \cdot \mathbf{p})^T$. One sees this by using the symmetric matrices $\mathsf{X}_1, \mathsf{X}_2, \mathsf{X}_3$ above to compute the following three quadratic forms defined using $\mathbf{z} = (\mathbf{q}, \mathbf{p})^T$:

$$\frac{1}{2}\mathbf{z}^T \mathsf{X}_1 \mathbf{z} = |\mathbf{q}|^2 = X_1, \ \frac{1}{2}\mathbf{z}^T \mathsf{X}_2 \mathbf{z} = |\mathbf{p}|^2 = X_2, \ \frac{1}{2}\mathbf{z}^T \mathsf{X}_3 \mathbf{z} = \mathbf{q} \cdot \mathbf{p} = X_3.$$

Exercise. (The Lie bracket) For $\mathsf{X} = Jm$ and $\mathsf{Y} = Jn \in sym(2)$ with $m, n \in sp(2)$, prove

$$[\mathsf{X}, \mathsf{Y}]_J := \mathsf{X}J\mathsf{Y} - \mathsf{Y}J\mathsf{X} = -J(mn - nm) = -J[m, n].$$

Use this equality to show that the J-bracket $[\mathsf{X}, \mathsf{Y}]_J$ satisfies the Jacobi identity. ★

Answer. The first part is a straightforward calculation using $J^2 = -\mathrm{Id}_{2\times 2}$ with the definitions of X and Y.

The second part follows from the Jacobi identity for the symplectic Lie algebra and linearity in the definitions of X, Y $\in sym(2)$ in terms of $m, n \in sp(2)$. ▲

Exercise. (A variational identity) If $\mathsf{X} = J\dot{M}M^{-1}$ for derivative $\dot{M} = \partial M(s,\sigma)/\partial s|_{\sigma=0}$ and $\mathsf{Y} = JM'M^{-1}$ for variational derivative $\delta M = M' = \partial M(s,\sigma)/\partial\sigma|_{\sigma=0}$, show that equality of cross derivatives in s and σ implies the relation

$$\delta\mathsf{X} = \mathsf{X}' = \dot{\mathsf{Y}} + [\mathsf{X},\mathsf{Y}]_J\,.$$

★

Answer. This relation follows from an important standard calculation in geometric mechanics, performed earlier in deriving Equation (9.1.4). It begins by computing the time derivative of $MM^{-1} = Id$ along the curve $M(s)$ to find $(MM^{-1})^{\cdot} = 0$, so that

$$(M^{-1})^{\cdot} = -\,M^{-1}\dot{M}M^{-1}\,.$$

Next, one defines $m = \dot{M}M^{-1}$ and $n = M'M^{-1}$. Then the previous relation yields

$$\begin{aligned} m' &= \dot{M}'M^{-1} - \dot{M}M^{-1}M'M^{-1} \\ \dot{n} &= \dot{M}'M^{-1} - M'M^{-1}\dot{M}M^{-1} \end{aligned}$$

so that subtraction yields the relation

$$m' - \dot{n} = nm - mn =: -[m,n].$$

Then, upon substituting the definitions of X and Y, one finds

$$\begin{aligned} \mathsf{X}' = Jm' &= J\dot{n} - J[m,n] \\ &= \dot{\mathsf{Y}} + [\mathsf{X},\mathsf{Y}]_J = \dot{\mathsf{Y}} + 2\mathrm{sym}(\mathsf{X}J\mathsf{Y}). \end{aligned}$$

▲

Exercise. (Hamilton's principle for $sp(2)$) Use the previous relation to compute the Euler–Poincaré equation for evolution resulting from Hamilton's principle,

$$0 = \delta S = \delta \int \ell(\mathsf{X}(s))\, ds = \int \mathrm{tr}\left(\frac{\partial \ell}{\partial \mathsf{X}}\, \delta\mathsf{X}\right) ds.$$

★

Answer. Integrate by parts and rearrange as follows:

$$\begin{aligned} 0 = \delta S &= \int \mathrm{tr}\left(\frac{\partial \ell}{\partial \mathsf{X}}\mathsf{X}'\right) ds \\ &= \int \mathrm{tr}\left(\frac{\partial \ell}{\partial \mathsf{X}}(\dot{\mathsf{Y}} - \mathsf{Y}J\mathsf{X} + \mathsf{X}J\mathsf{Y})\right) ds \\ &= \int \mathrm{tr}\left(\Big(-\frac{d}{ds}\frac{\partial \ell}{\partial \mathsf{X}} - J\mathsf{X}\frac{\partial \ell}{\partial \mathsf{X}} + \frac{\partial \ell}{\partial \mathsf{X}}\mathsf{X}J\Big)\mathsf{Y}\right) ds \\ &= \int \mathrm{tr}\left(\Big(-\frac{d}{ds}\frac{\partial \ell}{\partial \mathsf{X}} - 2\mathrm{sym}\Big(J\mathsf{X}\frac{\partial \ell}{\partial \mathsf{X}}\Big)\Big)\mathsf{Y}\right) ds\,, \end{aligned}$$

upon setting the boundary term $\mathrm{tr}(\frac{\partial \ell}{\partial \mathsf{X}}Y)|_{s_0}^{s_1}$ equal to zero. This results in the Euler–Poincaré equation,

$$\frac{d}{ds}\frac{\partial \ell}{\partial \mathsf{X}} = -2\mathrm{sym}\Big(J\mathsf{X}\frac{\partial \ell}{\partial \mathsf{X}}\Big) = 2\mathrm{sym}\Big(\frac{\partial \ell}{\partial \mathsf{X}}\mathsf{X}J\Big). \quad \text{(C.4.1)}$$

▲

Exercise. (Geodesic motion on $sp(2)^*$) Specialise this evolution equation to the case that $\ell(\mathsf{X}) = \frac{1}{2}\mathrm{tr}(\mathsf{X}^2)$, where tr denotes the trace of a matrix. (This is geodesic motion on the matrix Lie group $Sp(2)$ with respect to the trace norm of matrices.)

★

Answer. When $\ell(\mathsf{X}) = \frac{1}{2}\mathrm{tr}(\mathsf{X}^2)$ we have $\partial\ell/\partial\mathsf{X} = \mathsf{X}$, so the Euler–Poincaré Equation (C.4.1) becomes

$$\dot{\mathsf{X}} = -2\mathrm{sym}\big(J\mathsf{X}^2\big) = \mathsf{X}^2 J - J\mathsf{X}^2 = [\mathsf{X}^2, J]\,. \quad \text{(C.4.2)}$$

This is called a ***Bloch–Iserles equation*** [BlIs2006]. ▲

Exercise. (Lie–Poisson Hamiltonian formulation) Write the Hamiltonian form of the Euler–Poincaré equation on $SP(2)$ and identify the associated Lie–Poisson bracket. ★

Answer. The Hamiltonian form of the Euler–Poincaré Equation (C.4.1) is found from the Legendre transform via the dual relations

$$\mu = \frac{\partial\ell}{\partial\mathsf{X}} \quad \text{and} \quad \mathsf{X} = \frac{\partial h}{\partial\mu} \quad \text{with} \quad h(\mu) = \mathrm{tr}(\mu\mathsf{X}) - \ell(\mathsf{X})\,.$$

Thus,

$$\dot{\mu} = -2\mathrm{sym}\Big(J\frac{\partial h}{\partial\mu}\mu\Big) = -J\frac{\partial h}{\partial\mu}\mu + \mu\frac{\partial h}{\partial\mu}J\,.$$

The Lie–Poisson bracket is obtained from

$$\begin{aligned}
\frac{d}{ds}f(\mu) &= \mathrm{tr}\left(\frac{\partial f}{\partial\mu}\frac{d\mu}{ds}\right)\\
&= -2\mathrm{tr}\left(\mu\,\mathrm{sym}\left(\frac{\partial f}{\partial\mu}J\frac{\partial h}{\partial\mu}\right)\right)\\
&= -\mathrm{tr}\left(\mu\left[\frac{\partial f}{\partial\mu}, \frac{\partial h}{\partial\mu}\right]_J\right)\\
&=: \Big\{f\,,h\Big\}_J\,.
\end{aligned}$$

The Jacobi identity for this Lie–Poisson bracket follows from that of the J-bracket discussed earlier.

The geodesic Bloch–Iserles Equation (C.4.2) is recovered when the Hamiltonian is chosen as $h = \frac{1}{2}\mathrm{tr}(\mu^2)$ and one sets $\mu \to \mathsf{X}$. ▲

Exercise. (A second Bloch–Iserles Poisson bracket) Show that the geodesic Bloch–Iserles Equation (C.4.2) may also be written in Hamiltonian form with Hamiltonian $h = \frac{1}{3}\mathrm{tr}(\mu^3)$. ★

Answer. Equation (C.4.2) may also be written as

$$\dot{\mu} = -2\mathrm{sym}\Big(J\frac{\partial h}{\partial \mu}\mu\Big) = -J\frac{\partial h}{\partial \mu} + \frac{\partial h}{\partial \mu}J$$

with Hamiltonian $h = \frac{1}{3}\mathrm{tr}(\mu^3)$. The corresponding Poisson bracket has constant coefficients,

$$\begin{aligned}
\frac{d}{ds}f(\mu) &= \mathrm{tr}\left(\frac{\partial f}{\partial \mu}\frac{d\mu}{ds}\right) \\
&= -2\mathrm{tr}\left(\mathrm{sym}\left(\frac{\partial f}{\partial \mu}J\frac{\partial h}{\partial \mu}\right)\right) \\
&= -\mathrm{tr}\left(\left[\frac{\partial f}{\partial \mu}, \frac{\partial h}{\partial \mu}\right]_J\right) \\
&=: \Big\{f\,,h\Big\}_{J2}.
\end{aligned}$$

▲

Exercise. (A parallel with the rigid body) The geodesic Bloch–Iserles Equation (C.4.2) may be written in a form reminiscent of the rigid body, as

$$\frac{d}{dt}X = [X, \Omega] \quad \text{with} \quad \Omega = JX + XJ = -\Omega^T.$$

This suggests the Manakov form

$$\frac{d}{dt}(X + \lambda J) = [X + \lambda J,\, JX + XJ + \lambda^2 J^2].$$

This seems dual to the Manakov form (2.4.28) for the rigid body, because the symmetric and antisymmetric matrices exchange roles.

Verify these equations and explain what the Manakov form means in determining the conservation laws for this problem.

★

Exercise. (The Bloch–Iserles G-strand) Refer to Chapter 10 and compute the G-strand equations for $G = Sp(2)$. ★

C.5 Two coupled rigid bodies

Formulation of the problem

In the centre of mass frame, the Lagrangian for the problem of two coupled rigid bodies may be written as depending only on the angular velocities of the two bodies $\Omega_1 = A_1^{-1}\dot{A}_1(t)$, $\Omega_2 = A_2^{-1}\dot{A}_2(t)$ and the relative angle $A = A_1^{-1}A_2$ between the bodies [GrKrMa1988],

$$l(\Omega_1, \Omega_2, A) : \mathfrak{so}(3) \times \mathfrak{so}(3) \times SO(3) \to \mathbb{R},$$

which we write as

$$l(\Omega_1, \Omega_2, A) = \frac{1}{2}\begin{pmatrix} \Omega_1 \\ \Omega_2 \end{pmatrix}^T \cdot M(A)\begin{pmatrix} \Omega_1 \\ \Omega_2 \end{pmatrix},$$

where $M(A)$ is a 6×6 block matrix containing both A and the two inertia tensors of the bodies.

Upon identifying $\mathbb{R}_3$ with $\mathfrak{so}(3) = T_eSO(3)$ by the hat map, this Lagrangian becomes

$$l = l(\widehat{\boldsymbol{\Omega}}_1, \widehat{\boldsymbol{\Omega}}_2, A)$$

and we may identify $SO(3)$ with its dual $SO^*(3)$ through the matrix pairing $SO(3) \times SO^*(3) \to \mathbb{R}$.

The Lagrangian is then a function

$$l : \mathfrak{so}(3) \times \mathfrak{so}(3) \times SO^*(3) \to \mathbb{R}$$

which may be written as

$$l(\Omega, A) = \frac{1}{2}\langle M(A)\,\Omega,\,\Omega\rangle =: \frac{1}{2}\langle \Pi, \Omega\rangle\,,$$

where a nondegenerate matrix trace pairing is defined in components by

$$\langle \Pi, \Omega\rangle := \mathrm{Tr}\left[\begin{pmatrix} \widehat{\mathbf{\Omega}}_1 \\ \widehat{\mathbf{\Omega}}_2 \end{pmatrix}^T \cdot \begin{pmatrix} \widetilde{\mathbf{\Pi}}_1 \\ \widetilde{\mathbf{\Pi}}_2 \end{pmatrix}\right]$$

for all $\Omega = (\widehat{\mathbf{\Omega}}_1, \widehat{\mathbf{\Omega}}_2) \in \mathfrak{so}(3) \times \mathfrak{so}(3)$, $\Pi = (\widetilde{\mathbf{\Pi}}_1, \widetilde{\mathbf{\Pi}}_2) \in \mathfrak{so}^*(3) \times \mathfrak{so}^*(3)$.

The Euler–Poincaré theory has been developed to treat Lagrangians of the form

$$l : \mathfrak{g} \times V^* \to \mathbb{R}$$

where V is a vector space on which the Lie algebra acts. This Appendix deals with the following.

Problem C.5.1 *Formulate the Euler–Poincaré equations for the problem of two coupled rigid bodies.*

Problem Solution

The direct-product Lie algebra

$$\mathfrak{g} = \mathfrak{so}(3) \times \mathfrak{so}(3)$$

is endowed with the product Lie bracket

$$\mathrm{ad}_\Omega \Xi = [\Omega, \Xi] = \left[\begin{pmatrix} \widehat{\mathbf{\Omega}}_1 \\ \widehat{\mathbf{\Omega}}_2 \end{pmatrix}, \begin{pmatrix} \widehat{\mathbf{\Xi}}_1 \\ \widehat{\mathbf{\Xi}}_2 \end{pmatrix}\right] = \begin{pmatrix} [\![\widehat{\mathbf{\Omega}}_1, \widehat{\mathbf{\Xi}}_1]\!] \\ [\![\widehat{\mathbf{\Omega}}_2, \widehat{\mathbf{\Xi}}_2]\!] \end{pmatrix} \quad \text{(C.5.1)}$$

where $[\![\,\cdot\,,\cdot\,]\!]$ indicates the standard $\mathfrak{so}(3)$ matrix commutator.

Formulating the Euler–Poincaré theorem for this problem will require a Lie algebra action of $\mathfrak{so}(3)\times\mathfrak{so}(3)$ on $SO^*(3)$, which fortunately is readily available. Indeed, from the definitions of the two body angular velocities and relative angle $A = A_1^{-1}A_2$, one finds

$$\frac{dA}{dt} = -\widehat{\Omega}_1 A + A\widehat{\Omega}_2\,, \tag{C.5.2}$$

which is the Lie algebra action we seek, abbreviated as

$$\frac{dA}{dt} = -\,\Omega(A). \tag{C.5.3}$$

The Euler–Poincaré variational principle is then $\delta S = 0$, for

$$\begin{aligned}
\delta\int_{t_0}^{t_1} l(\Omega, A)\,dt &= \int_{t_0}^{t_1}\left\langle \frac{\delta l}{\delta\Omega},\, \delta\Omega\right\rangle + \left\langle \frac{\delta l}{\delta A},\, \delta A\right\rangle dt\\
&= \int_{t_0}^{t_1}\left\langle \frac{\delta l}{\delta\Omega},\, \frac{d\Xi}{dt} + \mathrm{ad}_\Omega\Xi\right\rangle + \left\langle \frac{\delta l}{\delta A},\, -\Xi(A)\right\rangle dt\\
&= \int_{t_0}^{t_1}\left\langle -\frac{d}{dt}\frac{\delta l}{\delta\Omega} + \mathrm{ad}^*_\Omega\frac{\delta l}{\delta\Omega} + \frac{\delta l}{\delta A}\diamond A,\, \Xi\right\rangle dt
\end{aligned}$$

with $\delta\Omega = \dot{\Xi} + \mathrm{ad}_\Omega\Xi$ and $\delta A = -\,\Xi(A)$. As a result, the (left-invariant) Euler–Poincaré equations may be written as

$$\frac{d}{dt}\frac{\delta l}{\delta\Omega} = \mathrm{ad}^*_\Omega\frac{\delta l}{\delta\Omega} + \frac{\delta l}{\delta A}\diamond A\,. \tag{C.5.4}$$

This, of course, is the general form of the Euler–Poincaré equations with advected quantities.

The Euler–Poincaré equations for the present problem of coupled rigid bodies will take their final form, once we have computed the diamond operation ($\diamond$),

$$\diamond : SO(3)\times SO(3)^* \to \mathfrak{so}(3)^*\,. \tag{C.5.5}$$

The Lie algebra action (C.5.3) yields the following definition of diamond for our case,

$$\begin{aligned}\left\langle \frac{\delta l}{\delta A} \diamond A, \Xi \right\rangle &:= -\left\langle A, \Xi\left(\frac{\delta l}{\delta A}\right) \right\rangle \\ &= -\left\langle A, \left(\widehat{\Xi}_1 \frac{\delta l}{\delta A} - \frac{\delta l}{\delta A} \widehat{\Xi}_2 \right) \right\rangle \\ &= -\left\langle A \frac{\delta l}{\delta A}, \widehat{\Xi}_1 \right\rangle + \left\langle \frac{\delta l}{\delta A} A, \widehat{\Xi}_2 \right\rangle ,\end{aligned}$$

where the last step is justified by the cyclic property of the trace. Consequently, the components of the diamond operation are given by

$$\frac{\delta l}{\delta A} \diamond A = \left(-A \frac{\delta l}{\delta A}, \frac{\delta l}{\delta A} A \right)$$

and substituting them into the general form of the Euler–Poincaré equations in (C.5.4) gives the equations of motion of our problem:

$$\begin{aligned}\frac{d}{dt} \widetilde{\mathbf{\Pi}}_1 &= \mathrm{ad}^*_{\widehat{\mathbf{\Omega}}_1} \widetilde{\mathbf{\Pi}}_1 - A \frac{\delta l}{\delta A} , \\ \frac{d}{dt} \widetilde{\mathbf{\Pi}}_2 &= \mathrm{ad}^*_{\widehat{\mathbf{\Omega}}_2} \widetilde{\mathbf{\Pi}}_2 + \frac{\delta l}{\delta A} A .\end{aligned}$$

These along with the auxiliary Equation (C.5.2) comprise the Euler–Poincaré form of the Lie–Poisson equations that are derived for the motion of two coupled rigid bodies in [GrKrMa1988]. The corresponding Lie–Poisson equations in [GrKrMa1988] may be derived from the Euler–Poincaré equations here by applying a symmetry-reduced Legendre transform.

D

POINCARÉ'S 1901 PAPER

H. Poincaré (1901)

Sur une forme nouvelle des équations de la méchanique.[1]
C.R. Acad. Sci. **132,** 369-371.

Having had the opportunity to work on the rotational motion of hollow solid bodies filled with liquid, I have been led to cast the equations of mechanics into a new form that could be interesting to know. Assume there are n degrees of freedom and let $\{x^1, ..., x^n\}$ be the variables describing the state of the system. Let T and U be the kinetic and potential energy of the system.

Consider any continuous, transitive group (that is, its action covers the entire manifold). Let $X_i(f)$ be any infinitesimal transformation of this group such that[2]

$$X_i(f) = \sum_{\mu=1}^{n} X_i(x^\mu)\frac{\partial f}{\partial x^\mu} = X_i^1\frac{\partial f}{\partial x^1} + X_i^2\frac{\partial f}{\partial x^2} + \cdots + X_i^n\frac{\partial f}{\partial x^n}\,.$$

Since these transformations form a group, we must have

$$X_i X_k - X_k X_i = \sum_{s=1}^{r} c_{ik}{}^{s} X_s\,.$$

[1]Translation of [Po1901] into English by D. D. Holm and J. Kirsten

[2]For the finite dimensional case considered here, the $\{X_i\}$ may be regarded as a set of r constant $n \times n$ matrices that act linearly on the set of states $\{x\}$. Then, for example, $X_i(x^\mu) = \sum_{\nu=1}^{n}[X_i]^\mu_\nu x^\nu$. (Translator's note)

Since the group is transitive we can write

$$\dot{x}^\mu(t) = \frac{dx^\mu}{dt} = \sum_{i=1}^{r} \eta^i(t) X_i(x^\mu) = \eta^1(t)X_1^\mu + \eta^2(t)X_2^\mu + \cdots + \eta^r(t)X_r^\mu \,,$$

in such a way that we can go from the state $(x^1, \ldots, x^n)$ of the system to a state $(x^1 + \dot{x}^1 \mathrm{d}t, \ldots, x^n + \dot{x}^n \mathrm{d}t)$ by using the infinitesimal transformation of the group, $\sum_{i=1}^{r} \eta^i X_i(f)$.

T instead of being expressed as a function of the x and $\dot{x}$ can be written as a function of the η and x. If we increase the η and x by virtual displacements $\delta\eta$ and δx, respectively, there will be resulting increases in T and U

$$\delta T = \sum \frac{\delta T}{\delta \eta} \delta\eta + \sum \frac{\delta T}{\delta x} \delta x \quad \text{and} \quad \delta U = \sum \frac{\delta U}{\delta x} \delta x \,.$$

Since the group is transitive I will be able to write

$$\delta x^\mu = \omega^1 X_1^\mu + \omega^2 X_2^\mu + \cdots + \omega^r X_r^\mu \,,$$

in such a way that we can go from the state x^μ of the system to the state $x^\mu + \delta x^\mu$ by using the infinitesimal transformation of the group $\delta x^\mu = \sum_{i=1}^{r} \omega^i X_i(x^\mu)$. I will then write[3]

$$\delta T - \delta U = \sum_{i=1}^{r} \frac{\delta T}{\delta \eta^i} \delta\eta^i + \sum_{\mu=1}^{n} \left(\frac{\delta T}{\delta x^\mu} - \frac{\delta U}{\delta x^\mu} \right) \delta x^\mu = \sum_{i=1}^{r} \frac{\delta T}{\delta \eta^i} \delta\eta^i + \sum_{i=1}^{r} \Omega_i \omega^i .$$

Next, let the Hamilton integral be

$$J = \int (T - U) \, \mathrm{d}t \,,$$

so we will have

$$\delta J = \int \left(\sum \frac{\delta T}{\delta \eta^i} \delta\eta^i + \sum \Omega_i \omega^i \right) \mathrm{d}t \,,$$

[3]Here Poincaré's formula reveals that $\Omega_i = \sum_{\mu,\nu=1}^{n} \frac{\delta L}{\delta x^\mu} [X_i]^\mu_\nu x^\nu$ with $L = T - U$, or equivalently $\Omega = \frac{\partial L}{\partial x} \diamond x$ in the notation of Chapter 6. (Translator's note)

and can easily find

$$\delta\eta^i = \frac{d\omega^i}{dt} + \sum_{s,k=1}^{r} c_{sk}{}^i \eta^k \omega^s .$$

The principle of stationary action then gives

$$\frac{d}{dt}\frac{\delta T}{\delta\eta^s} = \sum c_{sk}{}^i \frac{\delta T}{\delta\eta^i}\eta^k + \Omega_s .$$

These equations encompass some particular cases:

1. The Lagrange equations, when the group is reduced to the transformations, all commuting amongst each other, which each shift one of the variables x by an infinitesimally small constant.

2. Also, Euler's equations for solid body rotations emerge, in which the role of the η_i is played by the components p, q, r of the rotations and the role of Ω by the coupled external forces.

These equations will be of special interest where the potential U is zero and the kinetic energy T only depends on the η in which case Ω vanishes.

Bibliography

[AbMa1978] Abraham, R. and Marsden, J. E. [1978] *Foundations of Mechanics, 2nd ed.* Reading, MA: Addison-Wesley.

[AbMaRa1988] Abraham, R., Marsden, J. E. and Ratiu, T. S. [1988] *Manifolds, Tensor Analysis, and Applications, 2nd Ed.* Applied Mathematical Sciences, Vol. 75. New York: Springer.

[AcHoKoTi1997] Aceves, A., Holm, D. D., Kovacic, G. and Timofeyev, I. [1997] Homoclinic orbits and chaos in a second-harmonic generating optical cavity. *Phys. Lett. A* **233**, 203–208.

[AlLuMaRo1998] Alber, M. S., Luther, G. G., Marsden, J. E. and Robbins, J. M. [1998] Geometric phases, reduction and Lie–Poisson structure for the resonant three-wave interaction. *Physica D* **123**, 271–290.

[AlLuMaRo1999] Alber, M. S., Luther, G. G., Marsden, J. E. and Robbins, J. M. [1999] Geometry and control of three-wave interactions. In *The Arnoldfest* (Toronto, ON, 1997), *Fields Inst. Commun.* **24**, 55–80.

[AlEb1975] Allen, L. and Eberly, J. H. [1975] *Optical Resonance and Two-Level Atoms.* New York: Wiley.

[AmCZ1996] Ambrosetti, A. and Coti Zelati, V. [1996] *Periodic Solutions of Singular Lagrangian Systems.* Boston: Birkhäuser.

[Ap1900] Appel, P. [1900] Sur l'intégration des équations du mouvement d'un corps pesant de révolution roulant par une arête circulaire sur un plan horizontal; cas parficulier du cerceau. *Rendiconti del circolo matematico di Palermo* **14**, 1–6.

[Ar350BC] Aristotle [350BC] *Physics*.

[Ar1966] Arnold V. I. [1966] Sur la géométrie différentielle des groupes de Lie de dimiension infinie et ses applications à l'hydrodynamique des fluides parfaits. *Ann. Inst. Fourier, Grenoble* **16**, 319–361.

[Ar1993] Arnold V. I. [1993] *Dynamical Systems III*. New York: Springer.

[Ar1979] Arnold V. I. [1979] *Mathematical Methods of Classical Mechanics*. Graduate Texts in Mathematics, Vol. 60. New York: Springer.

[ArKh1998] Arnold, V. I. and Khesin, B. A. [1998] *Topological Methods in Hydrodynamics*. New York: Springer.

[Ba2006] Batista, M. [2006] Integrability of the motion of a rolling disk of finite thickness on a rough plane. *Int. J. Mech.* **41**, 850–859.

[Be1986] Benci, V. [1986] Periodic solutions of Lagrangian systems on a compact manifold. *J. Differ. Equations* **63**, 135–161.

[BeShWy1999] Bendik, J. J. Jr., Shaw, L. J. and Wyles, R. H. [1999] Spinning/rolling disc. U. S. Patent No. 5863235.

[BeBuHo1994] Berman, G. P., Bulgakov, E. N. and Holm, D. D. [1994] *Crossover-Time in Quantum Boson and Spin Systems*. Springer Lecture Notes in Physics, Vol. 21. New York: Springer.

[BlBr1992] Blanchard, P. and Bruning, E. [1992] *Variational Methods in Mathematical Physics*. New York: Springer.

[Bl2004] Bloch, A. M. (with the collaboration of Baillieul, J., Crouch, P. E. and Marsden, J. E.) [2004] *Nonholonomic Mechanics and Control*. New York: Springer.

[BlBrCr1997] Bloch, A. M., Brockett, R. W. and Crouch, P. E. [1997] Double bracket equations and geodesic flows on symmetric spaces. *Comm. Math. Phys.* **187**, 357–373.

[BlCr1996] Bloch, A. M. and Crouch, P. E. [1996] Optimal control and geodesic flows. *Sys. Control Lett.* **28**, 65–72.

[BlCrHoMa2001] Bloch, A. M., Crouch, P. E., Holm, D. D. and Marsden, J. E. [2001] An optimal control formulation for inviscid incompressible ideal fluid flow. In *Proc. of the 39th IEEE Conference on Decision and Control*. New Jersey: IEEE, pp. 1273–1279.

[BlCrMaRa1998] Bloch, A. M., Crouch, P. E., Marsden, J. E. and Ratiu, T. S. [1998] Discrete rigid body dynamics and optimal control. In *Proc. 37th CDC IEEE*. New Jersey: IEEE. pp. 2249–2254.

[BlIs2006] Bloch, A. M. and Iserles, A. [2006] On an isospectral Lie-Poisson system and its Lie algebra. *Found. Comput. Math.* **6**, 121–144.

[BlKrMaMu1996] Bloch, A. M., Krishnaprasad, P. S., Marsden, J. E. and Murray, R. [1996] Nonholonomic mechanical systems with symmetry. *Arch. Rat. Mech. Anal.* **136**, 21–99.

[Bl1965] Bloembergen, N. [1965] *Nonlinear Optics*. New York: Benjamin.

[Bo1986] Bobenko, A. I. [1986] Euler equations in the algebras $e(3)$ and $so(4)$. Isomorphisms of integrable cases. *Funkts. Anal. Prilozh.* **20**, 1, 64–66. (English translation: *Funct. Anal. Appl.* **20**, 53–56.)

[Bo1985] Bogoyavlensky, O. I. [1985] Integrable Euler equations on Lie algebras arising in problems of mathematical physics. *Math. USSR Izvestiya* **25**, 207–257.

[BoMa2002] Borisov, A. V. and Mamaev, I. S. [2002] Rolling of a rigid body on a plane and sphere: Hierarchy of dynamics. *Regul. Chaotic Dyn.* **7**, 177–200.

[BoMaKi2002] Borisov, A. V., Mamaev, I. S. and Kilin, A. A. [2002] Dynamics of a rolling disk. *Regul. Chaotic Dyn.* **8**, 201–212.

[BoWo1965] Born, M. and Wolf, E. [1965] *Principles of Optics*. Oxford: Pergamon Press, p. 31.

[BoMa2009] Bou-Rabee, N. and Marsden, J. E. [2009] Hamilton-Pontryagin integrators on Lie groups part I: Introduction and structure-preserving properties. *Foundations of Computational Mathematics* **9**, 197–219.

[BoMaRo2004] Bou-Rabee, N., Marsden, J. E. and Romero, L. A. [2004] Tippe top inversion as a dissipation-induced instability. *SIAM J. Appl. Dyn. Syst.* **3**, 352–377.

[Bo1971] Bourbaki, N. [1971] *Variétés différentielles et analytiques. Fascicule de résultats*. Paris: Hermann.

[Bo1989] Bourbaki, N. [1989] *Lie Groups and Lie Algebras: Chapters 1–3*. New York: Springer.

[BrGr2011] Brody, D. C. and Graefe, E.-M. [2011] On complexified mechanics and coquaternions. *J. Phys. A: Math. Theor.* **44**, 072001.

[Ca1995] Calogero, F. [1995] An integrable Hamiltonian system. *Phys. Lett. A* **201**, 306–310.

[CaFr1996] Calogero, F. and Francoise, J.-P. [1996] An integrable Hamiltonian system. *J. Math. Phys.* **37**, 2863–2871.

[CaHo1993] Camassa, R. and Holm, D. D. [1993] An integrable shallow water equation with peaked solitons. *Phys. Rev. Lett.* **71**, 1661–1664.

[CeHoMaRa1998] Cendra, H., Holm, D. D., Marsden, J. E. and Ratiu T. S. [1998] Lagrangian reduction, the Euler–Poincaré equations, and semidirect products. *Arnol'd Festschrift Volume II*, **186**, Am. Math. Soc. Translations Series 2 (1999), 1–25, arXiv.chao-dyn/9906004.

[CeMa1987] Cendra, H., and Marsden, J. E. [1987] Lin constraints, Clebsch potentials and variational principles. *Physica D* **27**, 63–89.

[CeMaPeRa2003] Cendra, H., Marsden, J. E., Pekarsky, S., and Ratiu, T. S. [2003] Variational principles for Lie–Poisson and Hamilton–Poincaré equations. *Mos. Math. J.* **3**(3), 833–867.

[Ch1897] Chaplygin, S. A. [1897] On motion of heavy rigid body of revolution on horizontal plane. *Proc. of the Physical Sciences section of the Society of Amateurs of Natural Sciences* **9**, 10–16.

[Ch1903] Chaplygin, S. A. [1903] On the rolling of a sphere on a horizontal plane. *Mat. Sbornik* **XXIV**, 139–68 (in Russian).

[ChMa1974] Chernoff, P. R. and Marsden, J. E. [1974] *Properties of Infinite Dimensional Hamiltonian Systems.* Lecture Notes in Mathematics, Vol. 425. New York: Springer.

[CiLa2007] Ciocci, M. C. and Langerock. B. [2007] Dynamics of the tippe top via Routhian reduction. arXiv:0704.1221v1 [math.DS].

[Co1848] Cockle, J. [1848] A new imaginary in algebra. *Phil. Mag.* **33**, 345–349,

[CoLe1984] Coddington, E. A. and Levinson, N. [1984] *Theory of Ordinary Differential Equations.* New York: Krieger.

[Co2003] Cordani, B. [2003] *The Kepler Problem: Group Theoretical Aspects, Regularization and Quantization, with Application to the Study of Perturbations*. Basel: Birkhäuser.

[CoHo2006] Cotter, C. J. and Holm, D. D. [2006] Discrete momentum maps for lattice EPDiff. arXiv.math.NA/0602296.

[CoHo2007] Cotter, C. J. and Holm, D. D. [2007] Continuous and discrete Clebsch variational principles. arXiv.math/0703495. *Found. Comput. Math.* **9**, 221–242. DOI: 10.1007/s10208-007-9022-9.

[CoHo2009] Cotter, C. J. and Holm, D. D. [2009] Discrete momentum maps for lattice EPDiff. In *Computational Methods for the Atmosphere and the Ocean* (a special volume of the *Handbook of Numerical Analyis*), edited by R. Temam and J. Tribbia.

San Diego, CA: Academic Press, pp. 247–278. (Preprint at arxiv.org/abs/0901.2025).

[Cu1998] Cushman, R. [1998] Routh's sphere. *Rep. Math. Phys.* **42**, 47–70.

[CuBa1997] Cushman, R. H. and Bates, L. M. [1997] *Global Aspects of Integrable Systems*. Basel: Birkhäuser.

[CuHeKe1995] Cushman, R., Hermans, J. and Kemppainen, O. [1995] The rolling disc. In *Nonlinear Dynamical Systems and Chaos*, edited by H. W. Broer, S. A. van Gils, I. Haveijn and Fl. Takens. Basel: Birkhäuser, pp. 21–60.

[DaBaBi1992] Dandoloff, R., Balakrishnan, R. and Bishop, A. R. [1992] Two-level systems: Space curve formalism, Berry's phase and Gauss-Bonnet theorem. *J. Phys. A: Math. Gen.* **25**, L1105–L1110.

[Da1994] Darling, R. W. R. [1994] *Differential Forms and Connections*. Cambridge: Cambridge University Press.

[DeMe1993] Dellnitz, M. and Melbourne, I. [1993] The equivariant Darboux theorem. *Lectures in Appl. Math.* **29**, 163–169.

[doCa1976] do Carmo, M. P. [1976] *Differential Geometry of Curves and Surfaces*. Englewood Cliffs, NJ: Prentice-Hall.

[DuNoFo1995] Dubrovin, B., Novikov, S. P. and Fomenko, A. T. [1995] *Modern Geometry I, II, III*. Graduate Texts in Mathematics, Vols. 93, 104, 124. New York: Springer.

[El2007] Ellis, D. C. P. [2007] The Hopf fibration and the 1-1 resonant optical traveling wave pulse. Unpublished manuscript.

[EGHPR2010] Ellis, D. C. P., Gay-Balmaz, F., Holm, D. D., Putkaradze, V. and Ratiu, T. S. [2010] Dynamics of charged molecular strands. *Arch. Rat. Mech. Anal.* **197**(3), 811–902.

[Eu1758] Euler, L. [1758] Du mouvement de rotation des corps solides autour d'un axe variable. *Mem. de l'acad. xi. Berlin* **14**, 154–193.

[FaTa1987] Faddeev, L. D. and Takhtajan, L. A. [1987] *Hamiltonian Methods in the Theory of Solitons*. New York: Springer.

[Fl1963] Flanders, H. [1963] *Differential Forms with Applications to the Physical Sciences*. New York: Academic Press.

[Ga1632] Galilei, G. [1632] *Dialogue Concerning the Two Chief World Systems: the Ptolemaic and the Copernican.*

[GaMi1995] Galper, A. and Miloh, T. [1995] Dynamic equations of motion for a rigid or deformable body in an arbitrary non-uniform potential flow field. *J. Fluid Mech.* **295**, 91–120.

[GBRa2008] Gay-Balmaz, F. and Ratiu, T.S. [2008] The geometric structure of complex fluids. *Adv. in Appl. Math.* **42** (2), 176–275.

[GaVi2010] Gay-Balmaz F. and Vizman, C. [2011] Dual pairs in fluid dynamics. *Ann. Glob. Anal. Geom.* **41**, 1–24.

[GeDo1979] Gelfand, I. M. and Dorfman, I. Ya. R. [1979] Hamiltonian operators and algebraic structures related to them. *Funct. Anal. Appl.* **13**, 248 (in Russian).

[GiHoKeRo2006] Gibbons, J. D., Holm, D. D., Kerr, R. M. and Roulstone, I. [2006] Quaternions and particle dynamics in the Euler fluid equations. *Nonlinearity* **19**, 1969–1983.

[GiHoKu1982] Gibbons, J., Holm, D. D. and Kupershmidt, B. A. [1982] Gauge-invariant Poisson brackets for chromohydrodynamics. *Phys. Lett. A* **90**, 281–283.

[GlPeWo2007] Glad, S. T., Petersson, D. and Rauch-Wojciechowski, S. [2007] Phase space of rolling solutions of the tippe top. *SIGMA* **3**, 14 pages.

[Go1986] Gorringe, V. M. [1986] *Generalisations of the Laplace-Runge-Lenz Vector in Classical Mechanics*. Thesis, Faculty of Science, University of Witwatersrand, Johannesburg, Vol. I, II.

[GrNi2000] Gray, C. G. and Nickel, B. G. [2000] Constants of the motion for nonslipping tippe tops and other tops with round pegs. *Am. J. Phys.* **68**, 821–828.

[GrKrMa1988] Grossman, R., Krishnaprasad, P. S. and Marsden, J. E. [1988] The dynamics of two coupled rigid bodies. In *Dynamical Systems Approaches to Nonlinear Problems in Systems and Circuits*, edited by F. M. Abdel Salam and M. Levi, Engineering Foundation, United States Army Research Office, and U.S. National Science Foundation. Philadelphia, PA: SIAM, pp. 373–378.

[GuHo1983] Guckenheimer, J. and Holmes, P. [1983] *Nonlinear Oscillations, Dynamical Systems, and Bifurcations of Vector Fields.* New York: Springer.

[GuSt1984] Guillemin, V. and Sternberg, S. [1984] *Symplectic Techniques in Physics.* Cambridge: Cambridge University Press.

[Ha2006] Hanson, A. J. [2006] *Visualizing Quaternions.* San Francisco: MK-Elsevier.

[He1995] Hermans, J. [1995] A symmetric sphere rolling on a surface. *Nonlinearity* **8**, 493–515.

[Ho1983] Holm, D. D. [1983] Magnetic tornadoes: Three-dimensional affine motions in ideal magnetohydrodynamics. *Physica D* **8**, 170–182.

[Ho1991] Holm, D. D. [1991] Elliptical vortices and integrable Hamiltonian dynamics of the rotating shallow water equations. *J. Fluid Mech.* **227**, 393–406.

[Ho2002] Holm, D. D. [2002] Euler–Poincaré dynamics of perfect complex fluids. In *Geometry, Mechanics, and Dynamics: In Honor of the 60th Birthday of Jerrold E. Marsden*, edited by P. Newton, P. Holmes and A. Weinstein. New York: Springer, pp. 113–167.

[Ho2005] Holm, D. D. [2005] The Euler–Poincaré variational framework for modeling fluid dynamics. In *Geometric Mechanics and*

Symmetry: The Peyresq Lectures, edited by J. Montaldi and T. Ratiu. London Mathematical Society Lecture Notes Series 306, Cambridge: Cambridge University Press.

[Ho2008] Holm, D. D. [2008] *Geometric Mechanics I: Dynamics and Symmetry*. London: Imperial College Press.

[HoKu1988] Holm, D. D. and Kupershmidt, B. A. [1988] The analogy between spin glasses and Yang–Mills fluids. *J. Math Phys.* **29**, 21–30.

[HoMa1991] Holm, D. D. and Marsden, J. E. [1991] The rotor and the pendulum. In *Symplectic Geometry and Mathematical Physics,* edited by P. Donato, C. Duval, J. Elhadad, G. M. Tuynman. Prog. in Math., Vol. 99. Birkhäuser: Boston, pp. 189–203.

[HoMa2004] Holm, D. D. and Marsden, J. E. [2004] Momentum maps and measure valued solutions (peakons, filaments, and sheets) for the EPDiff equation. In *The Breadth of Symplectic and Poisson Geometry,* edited by J. Marsden, and T. Ratiu. Boston: Birkhäuser. arXiv.nlin.CD/0312048.

[HoMaRa1998] Holm, D. D., Marsden, J. E. and Ratiu, T. S. [1998] The Euler–Poincaré equations and semidirect products with applications to continuum theories. *Adv. Math.* **137**, 1–81.

[HoScSt2009] Holm, D. D., Schmah, T. and Stoica, C. [2009] *Geometric Mechanics and Symmetry: From Finite to Infinite Dimensions.* Oxford: Oxford University Press.

[HoVi2010] Holm, D. D. and Vizman, C. [2010] *The n:m resonance dual pair*. Preprint at http://arxiv.org/abs/1102.4377.

[HoJeLe1998] Holmes, P., Jenkins, J. and Leonard, N. E. [1998] Dynamics of the Kirchhoff equations I: Coincident centres of gravity and buoyancy. *Physica D* **118**, 311–342.

[Ho1931] Hopf, H. [1931] Über die Abbildungen der dreidimensionalen Sphäre auf die Kugelfläche. *Math. Ann.* **104**, 637–665.

[Is1999] Isham, C. J. [1999] *Differential Geometry for Physicists*. World Scientific Lecture Notes in Physics Vol. 61. Singapore: World Scientific.

[Iv2006] Ivanov, R. [2006] Hamiltonian formulation and integrability of a complex symmetric nonlinear system. *Phys. Lett. A* **350**, 232–235.

[Ja2011] Jachnik, J. [2011] Spinning and rolling of an unbalanced disk. Unpublished report.

[Je1872] Jellett, J. H. [1872] *A Treatise on the Theory of Friction*. Dublin: Hodges, Foster and Co.

[JoSa1998] José, J. V. and Saletan, E. J. [1998] *Classical Dynamics: A Contemporary Approach*. Cambridge: Cambridge University Press.

[Jo1998] Jost, J. [1998] *Riemannian Geometry and Geometric Analysis, 2nd ed.* New York: Springer.

[KaKoSt1978] Kazhdan, D., Kostant, B. and Sternberg, S. [1978] Hamiltonian group actions and dynamical systems of Calogero type. *Comm. Pure Appl. Math.* **31**, 481–508.

[KhMi2003] Khesin, B. and Misiolek, G. [2003] Euler equations on homogeneous spaces and Virasoro orbits. *Adv. in Math.* **176**, 116–144.

[Ki2001] Kilin, A. A. [2001] The dynamics of Chaplygin ball: The qualitative and computer analysis. *Regul. Chaotic Dyn.* **6**, 291–306.

[Ki2011] Kim, B. [2011] *Constrained Dynamics of Rolling Balls and Moving Atoms*. PhD Thesis, Colorado State University.

[KiPuHo2011] Kim, B., Putkaradze, V. and Holm, D. D. [2011] Conservation laws of imbalanced symmetric Chaplygin's ball. Unpublished notes.

[KnHj2000] Knudsen, J. M. and Hjorth, P. G. [2000] *Elements of Newtonian Mechanics: Including Nonlinear Dynamics*. New York: Springer.

[KoNo1996] Kobayashi, S. and Nomizu, K. [1996] *Foundations of Differential Geometry*, Vols. 1, 2. New York: Wiley-Interscience.

[Ko1984] Koiller, J. [1984] A mechanical system with a "wild" horseshoe. *J. Math. Phys.* **25**, 1599–1604.

[Ko1900] Korteweg, D. [1900] *Extrait d'une lettre à M. Appel: Rendiconti del circolo matematico di Palermo* **14**, 7–8.

[Ko1966] Kostant, B. [1966] Orbits, symplectic structures and representation theory. *Proc. US-Japan Seminar on Diff. Geom.*, Kyoto. Nippon Hyronsha, Tokyo, p. 71.

[Ko1985] Kozlov, V. V. [1985] On the integration theory of the equations in nonholonomic mechanics. *Adv. Mech.* **8**, 86–107.

[Ku1999] Kuipers, J. B. [1999] *Quaternions and Rotation Sequences: A Primer with Applications to Orbits, Aerospace and Virtual Reality*. New Jersey: Princeton University Press.

[Ku1978] Kummer, M. [1978] On resonant classical Hamiltonians with two equal frequencies. *Commun. Math. Phys.* **58**, 85–112.

[KuSt1965] Kustaanheimo, P. and Stiefel, E. [1965] Perturbation theory of Kepler motion based on spinor regularization. *J. Reine Angew. Math.* **218**, 204–219.

[La1999] Lang, S. [1999] *Fundamentals of Differential Geometry*. Graduate Texts in Mathematics, Vol. 191. New York: Springer.

[LeFl2003] Leach, P. G. and Flessa, G. P. [2003] Generalisations of the Laplace–Runge–Lenz vector. *J. Nonlinear Math. Phys.* **10**, 340–423.

[Le2003] Lee, J. [2003] *Introduction to Smooth Manifolds*. New York: Springer.

[LeLeGl2005] Le Saux, C., Leine, R. I. and Glocker, C. [2005] Dynamics of a rolling disk in the presence of dry friction. *J. Nonlinear Sci.* **15**, 27–61.

[LiMa1987] Libermann, P. and Marle, C.-M. [1987] *Symplectic Geometry and Analytical Mechanics*. Dordrecht: Reidel.

[Li1890] Lie, S. [1890] *Theorie der Transformationsgruppen. Zweiter Abschnitt*. Leipzig: Teubner.

[LuAlMaRo2000] Luther, G. G., Alber, M. S., Marsden, J. E. and Robbins, J. M. [2000] Geometric analysis of optical frequency conversion and its control in quadratic nonlinear media. *J. Opt. Soc. Am. B Opt. Phys.* **17**, 932–941.

[LyBu2009] Lynch, P. and Bustamante, M. [2009] Precession and recession of the rock'n'roller. *J. Phys. A: Math. Theor.* **42**, 425203.

[Man1976] Manakov, S. V. [1976] Note on the integration of Euler's equations of the dynamics of and n-dimensional rigid body. *Funct. Anal. Appl.* **10**, 328–329.

[Mar1976] Marle, C.-M. [1976] Symplectic manifolds, dynamical groups, and Hamiltonian mechanics. In *Differential Geometry and Relativity*, edited by Cahen, M. and Flato, M. Boston: Reidel, pp. 249–269.

[Ma1981] Marsden, J. E. [1981] *Lectures on Geometric Methods in Mathematical Physics*. CMBS, Vol. 37. Philadelphia: SIAM.

[Ma1992] Marsden, J. E. [1992] *Lectures on Mechanics*. London Mathematical Society Lecture Note Series, Vol. 174. Cambridge: Cambridge University Press.

[MaHu1983] Marsden, J. E. and Hughes, T. J. R. [1983] *Mathematical Foundations of Elasticity*. Englewood Cliffs, NJ: Prentice Hall. (Reprinted by Dover Publications, NY, 1994.)

[MaMiOrPeRa2007] Marsden, J. E., Misiolek, G., Ortega, J.-P., Perlmutter, M. and Ratiu, T. S. [2007] *Hamiltonian Reduction by*

Stages. Lecture Notes in Mathematics, Vol. 1913. New York: Springer.

[MaMoRa1990] Marsden, J. E., Montgomery, R. and Ratiu, T. S. [1990] Reduction, symmetry, and phases in mechanics. *Memoirs Amer. Math. Soc.* **88**(436), 1–110.

[MaRa1994] Marsden, J. E. and Ratiu, T. S. [1994] *Introduction to Mechanics and Symmetry*. Texts in Applied Mathematics, Vol. 75. New York: Springer.

[MaRaWe1984a] Marsden, J. E., Ratiu, T. S. and Weinstein, A. [1984a] Semidirect products and reduction in mechanics. *Trans. Amer. Math. Soc.* **281**(1), 147–177.

[MaRaWe1984b] Marsden, J. E., Ratiu, T. S. and Weinstein, A. [1984b] Reduction and Hamiltonian structures on duals of semidirect product Lie algebras. *Contemporary Math.* **28**, 55–100.

[MaWe1974] Marsden, J. E. and Weinstein, A. [1974] Reduction of symplectic manifolds with symmetry. *Rep. Math. Phys.* **5**, 121–130.

[MaWe1983] Marsden, J. E. and Weinstein, A. [1983] Coadjoint orbits, vortices and Clebsch variables for incompressible fluids. *Physica D* **7**, 305–323.

[MaWe2001] Marsden, J. E. and Weinstein, A. [2001] Comments on the history, theory, and applications of symplectic reduction. In *Quantization of Singular Symplectic Quotients*, edited by N. Landsman, M. Pflaum and M. Schlichenmaier. Boston: Birkhäuser, pp. 1–20.

[MaWi1989] Mawhin, J. and Willem, M. [1989] *Critical Point Theory and Hamiltonian Systems, 2nd ed.* Applied Mathematical Sciences, Vol. 74. New York: Springer.

[McDMcD2000] McDonald, A. J. and McDonald, K. T. [2000] The rolling motion of a disk on a horizontal plane. arXiv:physics/0008227v3.

[McSa1995] McDuff, D. and Salamon, D. [1995] *Introduction to Symplectic Topology*. Oxford: Clarendon Press.

[MeDe1993] Melbourne, I. and Dellnitz, M. [1993] Normal forms for linear Hamiltonian vector fields commuting with the action of a compact Lie group. *Math. Proc. Cambridge* **114**, 235–268.

[Mi1963] Milnor, J. [1963] *Morse Theory*. Princeton: Princeton University Press.

[MiFo1978] Mishchenko, A. S. and Fomenko, A. T. [1978] Euler equations on finite-dimensional Lie groups. *Izv. Akad. Nauk SSSR, Ser. Mat.* **42**, 396–415 (Russian); English translation: *Math. USSR-Izv.* **12**, 371–389.

[Mi2002] Misiolek, G. [2002] Classical solutions of the periodic Camassa–Holm equation, *Geom. Funct. Anal.* **12**, 1080–1104.

[Mo2002] Montgomery, R. [2002] *A Tour of Subriemannian Geometries, Their Geodesics and Applications*. Mathematical Surveys and Monographs, Vol. 91. Providence, RI: American Mathematical Society.

[Na1973] Nambu, Y. [1973] Generalized Hamiltonian mechanics. *Phys. Rev. D* **7**, 2405–2412.

[Ne1997] Needham, T. [1997] *Visual Complex Analysis*. Oxford: Clarendon Press.

[NeFu1972] Neimark, J. I. and Fufaev, N. A. [1972] *Dynamics of Nonholonomic Systems*. Translations of Mathematical Monographs, Vol. 33. Providence, RI: American Mathematical Society.

[No1918] Noether, E. [1918] 1861*Invariante Variationsprobleme. Nachrichten Gesell. Wissenschaft. Göttingen* **2**, 235–257. See also C. H. Kimberling [1972], Emmy Noether. *Am. Math. Monthly* **79**, 136–149.

[OcoRo1998] O'Connor, J. J. and Robertson, E. F. [1998] *Sir William Rowan Hamilton*. http://www-groups.dcs.st-and.ac.uk/~history/Mathematicians/Hamilton.html.

[Ol2000] Olver, P. J. [2000] *Applications of Lie Groups to Differential Equations, 2nd ed.* New York: Springer.

[Or1996] O'Reilly, O. M. [1996] The dynamics of rolling disks and sliding disks. *Nonlinear Dynamics* **10**, 287–305.

[OrRa2004] Ortega, J.-P. and Ratiu, T. S. [2004] *Momentum Maps and Hamiltonian Reduction*. Progress in Mathematics, Vol. 222. Boston: Birkhäuser.

[Ot1993] Ott, E. [1993] *Chaos in Dynamical Systems*. Cambridge: Cambridge University Press, p. 385.

[Pa1968] Palais, R. [1968] *Foundations of Global Non-Linear Analysis*. New York: W. A. Benjamin.

[PeHuGr2002] Petrie, D., Hunt, J. L. and Gray, C. G. [2002] Does the Euler Disk slip during its motion? *Am. J. Phys.* **70**, 1025–1028.

[Po1892] Poincaré, H. [1892] *Théorie Mathématique de la Lumière*, Paris: Georges Carré, p. 275.

[Po1901] Poincaré, H. [1901] Sur une forme nouvelle des équations de la mécanique. *C. R. Acad. Sci.* **132**, 369–371.

[Ra1980] Ratiu, T. [1980] The motion of the free n-dimensional rigid body. *Indiana Univ. Math. J.* **29**, 609–627.

[Ra1982] Ratiu, T. [1982] Euler–Poisson equations on Lie algebras and the N-dimensional heavy rigid body. *Am. J. Math.* **104**, 409–448.

[RaVM1982] Ratiu, T. and van Moerbeke, P. [1982] The Lagrange rigid body motion. *Annales de l'institut Fourier* **32**(1), 211–234.

[RaTuSbSoTe2005] Ratiu, T. S., Tudoran, R., Sbano, L., Sousa Dias, E. and Terra, G. [2005] A crash course in geometric mechanics. In *Geometric Mechanics and Symmetry: The Peyresq Lectures*, edited by J. Montaldi and T. Ratiu. London Mathematical Society Lecture Notes Series 306. Cambridge: Cambridge University Press.

[Ro1860] Routh, E. J. [1860] *A Treatise on the Dynamics of a System of Rigid Bodies*. London: MacMillan.

Ibid [1884] *Advanced Rigid Body Dynamics*. London: Macmillan. (Reprint: *A Treatise on the Dynamics of a System of Rigid Bodies*, parts I and II, Dover, New York, 1960.)

[Ru2002] Ruina, A. [2002] Rolling and sliding of spinning things: Euler's disk, Jellett's Egg and Moffatt's Nature. http://ruina.tam.cornell.edu/research/topics/miscellaneous/rolling_and_sliding/index.htm#Eulers_Disk.

[Ru2000] Rumyantsev, V. V. [2000] Forms of Hamilton's principle for nonholonomic systems. *Facta Univ. Ser. Mech. Automat. Control and Robot.* **2**(10), 1035–1048.

[SaCa1981] Sarlet, W. and Cantrijn, F. [1981] Generalizations of Noether's theorem in classical mechanics. *SIAM Rev.* **23**, 467–494.

[Sc2002] Schneider, D. [2002] Non-holonomic Euler–Poincaré equations and stability in Chaplygin's sphere. *Dynam. Syst.* **17**, 87–130.

[SeWh1968] Seliger, R. L. and Whitham, G. B. [1968] Variational principles in continuum mechanics. *Proc. R. Soc. A* **305**, 1–25.

[Se1992] Serre, J.-P. [1992] *Lie Algebras and Lie Groups*. Lecture Notes in Mathematics, Vol. 1500. New York: Springer.

[Se1959] Serrin, J. [1959] In *Mathematical Principles of Classical Fluid Mechanics*, Vol. VIII/1 of Encyclopedia of Physics, edited by S. Flügge. New York: Springer, sections 14–15, pp. 125–263.

[ShWi1989] Shapere, A. and Wilcek, F. [1989] *Geometric Phases in Physics*. Singapore: World Scientific.

[SiMaKr1988] Simo, J. C., Marsden, J. E. and Krishnaprasad, P. S. [1988] The Hamiltonian structure of nonlinear elasticity: The material and convective representations of solids, rods, and plates. *Arch. Rat. Mech. Anal.* **104**, 125–183.

[Sk1961] Skyrme, T. H. R. [1961] A nonlinear field theory. *Proc. R. Soc. A* **260**, 127–138.

[Sl1861] Slesser, G. M. [1861] Notes on rigid dynamics. *Q. J. Math.* **4**, 65–67.

[Sm1970] Smale, S. [1970] Topology and mechanics. *Inv. Math.* **10**, 305–331; **11**, 45–64.

[So1970] Souriau, J. M. [1970] *Structure des Systèmes Dynamiques*. Paris: Dunod.

[Sp1982] Sparrow, C. [1982] *The Lorenz Equations: Bifurcations, Chaos and Strange Attractors*. New York: Springer, p. 269.

[Sp1979] Spivak, M. [1979] *Differential Geometry, Volume I*. Houston, TX: Publish or Perish, Inc.

[St1852] Stokes, G. G. [1852] On the composition and resolution of streams of polarized light from different sources. *Trans. Camb. Phil. Soc.* **9**, 399–423.

[Sy1937] Synge, J. L. [1937] *Geometrical Optics: An Introduction to Hamilton's Method*. Cambridge: Cambridge University Press.
Ibid [1937] Hamilton's method in geometrical optics. *J. Opt. Soc. Am.* **27**, 75–82.
Ibid [1960] Classical dynamics. In *Handbuch der Physik*, edited by S. Flügge. Berlin: Springer-Verlag, Vol. III/1, p. 11.

[Va1996] Vaisman, I. [1996] *Lectures on the Geometry of Poisson Manifolds*. Progress in Mathematics, Vol. 118. Boston: Birkhäuser.

[Wa1685] Wallis, J. *Treatise of Algebra* (London, 1685), p. 272, quoted by William Rowan Hamilton in the Preface to his *Lectures on Quaternions* (Dublin, 1883), p. 34.

[Wa1983] Warner, F. W. [1983] *Foundation of Differentiable Manifolds and Lie Groups*. Graduate Texts in Mathematics, Vol. 94. New York: Springer.

[We1983a] Weinstein, A. [1983a] Sophus Lie and symplectic geometry. *Expo. Math.* **1**, 95–96.

[We1983b] Weinstein, A. [1983b] The local structure of Poisson manifolds. *J. Differ. Geom.* **18**, 523–557.

[We2002] Weinstein, A. [2002] Geometry of momentum (preprint). ArXiv:math/SG0208108 v1.

[Wh1974] Whitham, G. B. [1974] *Linear and Nonlinear Waves*. New York: John Wiley and Sons.

[Wi1984] Witten, E. [1984] Non-Abelian Bosonization in two dimensions. *Commun. Math. Phys.* **92**, 455–472.

[Wi1988] Wiggins, S. [1988] *Global Bifurcations and Chaos: Analytical Methods*. New York: Springer.

[YoMa2006] Yoshimura, H. and Marsden, J. E. [2006]. Dirac Structures and Lagrangian Mechanics, Part I: Implicit Lagrangian systems. *J. Geom. and Physics*, **57**, 133–156.

[ZaMi1980] Zakharov, V.E. and Mikhailov, A.V. [1980] On the integrability of classical spinor models in two-dimensional spacetime, *Commun. Math. Phys.* **74**, 21–40.

[Ze1995] Zenkov, D. V. [1995] The geometry of the Routh problem. *J. Nonlinear Sci.* **5**, 503–519.

Index

action integral, 37
AD operation
  for $SE(3)$, 144
Ad operation
  for $SE(2)$, 155
  for $SE(3)$, 144
Adjoint action, 119
  for $SE(2)$, 155
  for $SE(3)$, 145
  for $SO(3)$, 133
Adjoint motion
  equation, 120
adjoint operation
  Ad, 119
  ad, 119
alien life forms
  Bers, 334
  Bichrons, 328
  Tets, 338
alignment dynamics
  Newton's second law, 86
  quaternion, 88
angular momentum
  body, 184
  body vs space, 28, 138
  conservation, 28, 33
  definition, 27
  Poisson bracket table, 346
angular velocity
  body, 29
  hat map, 25
  space vs body, 30
  spatial, 23

Bloch–Iserles equation, 357
body angular momentum, 33
body angular velocity vector, 32
body frame, 21
bundle, 297
  cotangent, 299
  tangent, 295

Cartan
  structure equations, 329
Chaplygin's ball, 248
  Jellett's integral, 261
  Routh's integral, 263
Chaplygin's circular disk, 265
Chaplygin's concentric sphere, 265
chiral model, 217
Christoffel coefficients, 332
Christoffel symbols, 38
Clebsch action principle
  Euler–Poincaré, 199
  Hamiltonian formulation, 206
  heavy top, 187
  rigid body, 69
coAdjoint action, 122
  for $SE(2)$, 155
  for $SE(3)$, 145
  for $SO(3)$, 135
coadjoint motion
  equation, 129, 163
  Kirchhoff equations, 176
  on $se(3)^*$, 170, 174, 175
  relation, 122, 162
coadjoint operation, 121
commutation relations, 24
commutator
  group, 125
complex fluid, 215
conjugacy classes, 119
  Lie group action, 317
  quaternions, 117
conservation
  $se(3)^*$ momentum, 173

angular momentum, 28, 44, 90
energy, 35, 42, 283
Hamilton's vector, 91
Jellett's integral, 261
Lie symmetry, 39
linear momentum, 44
momentum map, 225
Noether's theorem, 39
Runge–Lenz vector, 92
constant of motion, 40
constrained reduced Lagrangian, 257, 278, 283
constrained variational principle
Clebsch, 69
Hamilton–Pontryagin, 50, 198
implicit, 70, 203
Kaluza–Klein, 188
nonholonomic, 273
coordinate charts, 288
coquaternion, 108
conjugate, 109
cotangent bundle, 299
cotangent lift, 207, 300, 318
momentum map, 227
covariant derivative, 332
curvature two-form, 330

diamond operation, 202
definition, 147
differentiable map, 301
derivative, 301
dual pairs, 240

equivariance, 228, 230, 299
Euler's disk, 268
Jellett's integral, 275
Euler's equations
rigid body, 35, 46
vorticity, 326
Euler–Poincaré equation, 194
continuum spin chain, 212
for $SE(3)$, 197, 218
for $Sp(2)$, 219
heavy top, 183, 259
on $se(3)^*$, 171
Euler–Poincaré theorem, 194
nonholonomic, 278
Euler–Rodrigues formula
rotation, 96, 112

fibre bundle, 105
fibre derivative, 53
flows, 296
force
noninertial, 29
frame
body, 21
spatial, 22
free rigid rotation, 21

G-strand, 210, 359
Galilean group, 4, 12, 17, 164
adjoint actions, 164
Lie group actions, 164
Lie algebra, 17
Lie group actions, 12
Galilean relativity, 3
Galilean transformations
boosts, 4
definition, 3
invariance, 3
Lie group property, 4
matrix representation, 14, 165
subgroups, 8
geodesic equations, 38
group
action, 8
automorphism, 126
commutator, 125
definition, 4, 306
Euclidean $E(3)$, 8
Galilean $G(3)$, 4
Heisenberg H, 124
matrix, 306
orthogonal $O(3)$, 8, 21
orthogonal $O(n)$, 312
semidirect $SL(2,\mathbb{R})Ⓢ\mathbb{R}^2$, 157
special Euclidean $SE(2)$, 153
special Euclidean $SE(3)$, 8, 15, 142, 197, 218, 313
special linear $SL(n,\mathbb{R})$, 312
special orthogonal $SO(3)$, 8, 21.

special orthogonal $SO(4)$, 338
special orthogonal $SO(n)$, 312
special unitary $SU(2)$, 101
special unitary $SU(n)$, 313
symplectic $Sp(l)$, 313
group action
homogeneous, 11
vector space, 11
group commutator, 125
group composition law
for $SE(3)$, 9
group orbit, 315

Hamilton's principle, 37
Hamilton–Pontryagin principle, 252
Euler–Poincaré, 198
heavy top, 259
nonholonomic, 273
rigid body, 50
Hamiltonian form, 62
continuum spin chain, 213
Lie–Poisson, 173, 206
hat map, 25, 132, 137, 323
isomorphism, 25
heavy top, 182
Casimirs, 171
Clebsch action principle, 187
Euler–Poincaré equation, 259
Hamilton–Pontryagin, 259
Kaluza–Klein, 188
Legendre transform, 184
Lie–Poisson bracket, 178, 186
variational principle, 183
Heisenberg group, 124
adjoint actions, 126
coadjoint actions, 127
coadjoint orbits, 130
Hermitian
Pauli matrices, 101, 229
Hopf fibration, 105, 228

immersed submanifold, 294
infinitesimal generator, 134
infinitesimal transformation, 40
injective immersion, 294
inner automorphism, 119
inverse Adjoint motion
identity, 121
isotropy subgroup, 307
iterated action
semidirect product, 167

Jacobi identity, 62, 133
Lie algebra, 311
matrix commutator, 311
Jacobi–Lie bracket, 320
Jellett's integral
Chaplygin's ball, 261
Euler's disk, 275

Kelvin–Noether theorem
for $SE(3)$, 176
for an arbitrary Lie group, 197
Kepler problem
Hamilton's vector, 91
monopole, 93
quaternionic variables, 90
Runge–Lenz vector, 91
Kirchhoff equations
underwater vehicle, 176
Kustaanheimo–Stiefel map, 101

Lagrange–d'Alembert, 275, 279
Lagrangian
constrained reduced, 257, 278, 282, 283
reduced, 256, 278
Legendre transform, 60
$se(3) \to se(3)^*$, 172
Clebsch variables, 206
constrained reduced, 283
heavy top, 184
Leibniz, 62
Lie algebra
Jacobi identity, 311
matrix, 310
Lie bracket
matrix commutator, 119
semidirect product, 187
Lie derivative, 323
fluid dynamics, 325
Lie group, 4, 306

action, 314
as manifold, 294
matrix, 118, 306
semidirect product, 12
Lie group action, 314
adjoint, 317
coadjoint, 317
conjugacy classes, 317
cotangent lift, 318
infinitesimal generator, 322
inner automorphism, 317
invariant vector fields, 319
left action, 314
left and right translations, 316
on vector fields, 318
properties, 315
tangent lift, 317
Lie symmetry, 39
Lie symmetry reduction, 242
Lie–Poisson
Hamiltonian form, 206
Lie–Poisson bracket, 64, 184
Casimirs, 173
continuum spin chain, 214
distinguished functions, 173
heavy top, 178, 186
momentum maps, 185
on $se(3)^*$, 172
spin chain, 210
lower triangular matrices, 128

Manakov
commutator form, 54, 170, 359
heavy top, 190
manifold
coordinate charts, 288
definition, 288
smooth, 287
submersion, 289
matrix commutator
Jacobi identity, 311
properties, 311
matrix Lie algebra, 310
matrix Lie group, 118, 306
$GL(n,\mathbb{R})$, 307
matrix representation
$G(3)$, 13, 14, 165
$SE(3)$, 9
$SE(3)$, 10, 142
$SL(2,\mathbb{R})$Ⓢ$\mathbb{R}^2$, 157
for quaternions, 80
hat map, 15
Heisenberg group, 124
mechanics on Lie groups, 79
moment of inertia tensor, 27, 32
principal axis frame, 33
momentum map, 71, 236
$GL(n,\mathbb{R})$ invariance, 348
$SU(2)$ action on $\mathbb{C}^2$, 227
$SU(3)$ action on $\mathbb{C}^3$, 348
$\mathbb{C}^2 \mapsto u(2)^*$, 236
$\mathbb{R}^6 \to sp(2)^*$, 354
1:1 resonance, 236
canonical symplectic form, 236
Clebsch variational principle, 208
cotangent lift, 208, 224, 227
equivariant, 230
examples, 231
Hopf fibration, 229
infinite dimensional, 240
Noether's theorem, 225
Poincaré sphere, 228, 229
standard, 222
motion equation
coadjoint, 129, 163

Nambu bracket, 65
Nambu form
geometric interpretation, 68
three-wave equations, 68
narrow-hat notation, 23
Noether's theorem, 39
angular momentum, 45
centre of mass, 46
energy and momentum, 42
Euler–Poincaré motion, 195
Galilean boosts, 46
Hamilton–Pontryagin, 199
nonholonomic, 282
rigid body, 49, 53
space and time translations, 42

nonholonomic
Chaplygin's ball, 248
Chaplygin's circular disk, 265
Chaplygin's concentric sphere, 265
Euler's disk, 268
Euler–Poincaré theorem, 278
Hamilton–Pontryagin, 252
Lagrange–d'Alembert, 279
Noether's theorem, 282
rolling constraint, 250
semidirect product, 276
notation
narrow hat, 23
wide hat, 23

orthogonal group, 21
orthogonal transformation, 4
oscillators
on $\mathbb{C}^2$, 228
on $\mathbb{C}^3$, 343
Poincaré sphere, 236

pairing
on SE(3), 145
Pauli matrices, 80, 101, 229
anticommutator relation, 81
commutator relation, 81
Hermitian, 229
quaternions, 80
Poincaré
1901 paper, 55, 65, 194, 363
Poincaré sphere, 104, 228
Hopf fibration, 105
momentum map, 229
oscillators, 236
Poisson bracket, 62
rigid body, 63
principal axis frame, 33
product rule, 62
prolongation, 41
pull-back, 318
push-forward, 318

quaternion, 78
action on a vector, 112
alignment dynamics, 88
basis, 79
conjugacy class, 95
conjugate, 83
conjugation, 93
dot product, 83
inner product, 83
inverse, 84
magnitude, 84
multiplication rule, 80
Pauli matrices, 80
pure vector, 85
tetrad, 79

reconstruction
formula, 50, 69, 152, 200
reduced Lagrangian, 256, 257, 278
reflection, 4
Riemann sphere, 289
rigid body, 21
Cayley–Klein parameters, 112
Clebsch approach, 69
coadjoint orbit of $SO(3)$, 135
coupled, 359
eigenvalue problem, 56
Euler's equations, 46
Hamiltonian form, 60
isospectral problem, 56
Lagrangian form, 46
Lie–Poisson bracket, 64
Manakov's formulation, 54
matrix Euler equations, 51
Newtonian form, 30
Noether's theorem, 53
Poincaré's treatment, 365
quaternionic form, 112, 116
spatial and body frames, 138
symmetric equations, 70
variations, 328
Rodrigues formula, 96
rolling constraint, 250
rotating motion
free rigid rotation, 21
in complex space, 334
in two times, 328
kinetic energy, 31

quaternions, 112
rotation, 4
Cayley–Klein parameters, 97
Euler parameters, 97
Euler–Rodrigues formula, 96, 113
in complex space, 334
in four dimensions, 338
Routh's integral
Chaplygin ball, 263

semidirect product
Adjoint action, 146
adjoint action, 148
coAdjoint action, 146
coadjoint action, 148
definition, 12
iterated, 167
Lie bracket, 149
vector notation, 149
Lie group, 143
semidirect-product group
$G(3)$, 14
$SE(3)$, 12
definition, 12, 143
semidirect-product structure
nonholonomic, 276
sigma model, 217
simple mechanical system, 37
Skyrme model, 217
sliding, 258
space-time translations, 3
spatial frame, 22, 31
special orthogonal group, 21
spin glass, 215
stereographic projection, 289
strand
$so(3)$-valued spins, 210
structure constants, 24
subgroup, 8
invariant, 11
isotropy, 307
normal, 11
submanifold, 288
submersion, 289
symplectic Lie group
$Sp(2)$, 354

tangent bundle, 295
tangent lift, 41, 300, 317
tangent space, 294
tangent vector, 295, 296
three-wave equations
Nambu form, 68
tilde map, 102
Hopf fibration, 106
translations
space-time, 3
transpose
of the Lie-derivative, 203
twist, 87

underwater vehicle
Kirchhoff equations, 176
variational principle, 176
uniform rectilinear motion, 3
upper triangular matrices, 124

variational derivative, 36, 47
left-invariant Lagrangian, 170
variational principle, 49
chiral model, 217
Clebsch, 199
Clebsch Euler–Poincaré, 202
constrained, 72, 279
continuum spin chain, 212
Euler–Poincaré, 194, 278
Hamilton–Pontryagin, 198, 252
heavy top, 183
implicit, 70, 200
Lagrange–d'Alembert, 279
Noether's theorem, 39
rigid body, 47
sigma model, 217
underwater vehicle, 176
vector field, 297
differential basis, 300
fluid dynamics, 325

wide-hat notation, 23

zero curvature relation, 330
continuum spin chain, 212